World Health Organization
Regional Office for Europe
Copenhagen

Nonionizing radiation protection

Second edition

Edited by

Michael J. Suess
World Health Organization
Regional Office for Europe
Copenhagen, Denmark

Deirdre A. Benwell-Morison
Environmental Health Directorate
Health and Welfare Canada
Ottawa, Canada

Published in cooperation with the WHO collaborating centre
at the Bureau of Radiation and Medical Devices
Environmental Health Directorate
Health and Welfare Canada, Ottawa

WHO Regional Publications, European Series, No. 25

ICP/CEH 005/m06
Text editing by Frank Theakston

ISBN 92 890 1116 5   2nd edition
(ISBN 92 890 1101 7   1st edition)
ISSN 0378-2255

PRINTED IN DENMARK

CONTENTS

	Page
Contributors	vii
Preface to the second edition	ix
Preface to the first edition	xi
Dedication	xiii
Note on terminology	xiv
Introduction to the second edition	1
Introduction to the first edition	7
1. Ultraviolet radiation	13
2. Optical radiation, with particular reference to lasers	49
3. Infrared radiation	85
4. Radiofrequency radiation	117
5. Electric and magnetic fields at extremely low frequencies	175
6. Ultrasound	245
7. Regulation and enforcement procedures	293
Glossary	311
Annex 1. Special acknowledgements: first edition	315
Annex 2. Special acknowledgements: second edition	327
Annex 3. Lists of working groups	333
Index	337

Contributors[a]

Dr L.E. Anderson, Program Manager, Bioelectromagnetics, Biology and Chemistry Department, Battelle Pacific Northwest Laboratory, Richland, WA, USA

Dr P.A. Czerski, Research Scientist, Molecular Biology Branch, Division of Life Sciences, Office of Science and Technology, Center for Devices and Radiological Health, Food and Drug Administration, Rockville, MD, USA

Dr J.A. Elder, Chief, Cellular Biophysics Branch, Experimental Biology Division, Health Effects Research Laboratory, US Environmental Protection Agency, Research Triangle Park, NC, USA

Mr R.J. Ellis, Physical Agents Effects Branch, National Institute of Occupational Safety and Health, US Public Health Service, Cincinnati, OH, USA

Dr M. Faber, Professor and Director, The Finsen Laboratory, The Finsen Institute, Copenhagen, Denmark

Dr L. Goldman, Director, Laser Laboratory, and Professor and Chairman, Department of Dermatology, College of Medicine, University of Cincinnati Medical Center, Cincinnati, OH, USA

Dr Gail ter Haar, Joint Department of Physics, Institute of Cancer Research, Royal Cancer Hospital, in association with Royal Marsden Hospital, Sutton, Surrey, United Kingdom

Dr C.R. Hill, Professor and Head, Joint Department of Physics, Institute of Cancer Research, Royal Cancer Hospital, in association with Royal Marsden Hospital, Sutton, Surrey, United Kingdom

Dr W.T. Kaune, Battelle Pacific Northwest Laboratory, Richland, WA, USA

Dr F. Kossel, Director and Professor, Division of Medical Radiation Technology and Radiation Protection, Institute of Radiation Hygiene, Federal Health Office, Neuherberg, Federal Republic of Germany

Dr J.C. van der Leun, Professor, Institute of Dermatology, State University of Utrecht, Netherlands

Dr K.H. Mild, First Research Engineer, National Board of Occupational Health and Safety, Umeå, Sweden

---

[a] The affiliation given for each contributor is that which obtained during the period of his or her most recent collaboration.

Mr C.E. Moss, Physical Agents Effects Branch, National Institute of Occupational Safety and Health, US Public Health Service, Cincinnati, OH, USA

Mr W.E. Murray, Physical Agents Effects Branch, National Institute of Occupational Safety and Health, US Public Health Service, Cincinnati, OH, USA

Dr W.H. Parr, Chief, Physical Agents Effects Branch, National Institute of Occupational Safety and Health, US Public Health Service, Cincinnati, OH, USA

Mr R.J. Rockwell, Associate Professor of Laser Sciences, Laser Laboratory, Department of Dermatology, College of Medicine, University of Cincinnati Medical Center, Cincinnati, OH, USA

Dr A.R. Sheppard, Research Physicist, Neurobiological Research, Research Service 151, Pettis Veterans' Administration Hospital, Loma Linda, CA, USA

Mr D.H. Sliney, Physicist and Chief, Laser Hazards Branch, Laser-Microwave Division, US Army Environmental Hygiene Agency, Aberdeen Proving Ground, MD, USA

Dr Maria A. Stuchly, Research Scientist, Non-ionizing Radiation Section, Research and Standards Division, Bureau of Radiation and Medical Devices, Environmental Health Directorate, Health and Welfare Canada, Ottawa, ON, Canada

Dr M.J. Suess, Regional Officer for Environmental Health Hazards, WHO Regional Office for Europe, Copenhagen, Denmark

Dr B.M. Tengroth, Professor and Chairman, Department of Ophthalmology, Karolinska Institute and Hospital, Stockholm, Sweden

Dr M.L. Wolbarsht, Professor of Ophthalmology and Biomedical Engineering, Department of Psychology, Duke University, Durham, NC, USA

Preface
to the second edition

This book represents a major milestone in the history of nonionizing radiation (NIR) protection both within and outside the WHO European Region. When the first edition was published in 1982 after several years of work, it was the first comprehensive publication in this field. The book has become a bestseller and has been sold worldwide. It has also served as a text for training at institutes of higher education and for special courses.

Since the completion of the first edition, much new research has been performed and experience gained in the field of NIR protection, particularly with respect to radiofrequency radiation and electric and magnetic fields. Consequently, it has become necessary and appropriate to update and/or revise various chapters; and Chapters 4 and 5 were rewritten anew. To accomplish this task in as short a time as possible, the WHO collaborating centre at the Bureau of Radiation and Medical Devices, Environmental Health Directorate, Health and Welfare Canada, Ottawa was requested to join the project. For the eagerness, cooperation and efforts of Deirdre A. Benwell-Morison, Head of Section, and the staff of the Non-ionizing Radiation Section at Health and Welfare Canada, I wish to extend my special thanks. While much work has been contributed by a number of people, the design, implementation and coordination of the work in putting this book together has represented a special effort by the unit on Control of Environmental Health Hazards of the Regional Office for several years. Major credit must be given to the unit chief, Michael J. Suess, without whose leadership this work would never have been possible.

The strengthening of this scientific field comes at a particularly opportune point in health development in Europe. With the adoption in 1984 by all the European Member States of WHO of the new European health for all policy and its 38 targets, very important new roles and tasks will fall on the shoulders of the health professions, including the radiation health personnel. The development at national level of methodologies and health criteria for the assessment of NIR data in relation to control and protection procedures; the advancement of adequate control measures, and their introduction and maintenance; and the training and utilization of sufficient numbers of competent personnel for all aspects of NIR protection will present many challenges in the years ahead. Consequently, I hope that the second edition of this book will be received as favourably as the first, and will continue to serve a useful purpose in furthering understanding of this still rather new field.

J.E. Asvall
WHO Regional Director for Europe

Preface to the first edition

As human populations multiply and industrialization increases and diversifies new hazards arise, some of which have become more and more critical. To limit and, as far as possible, reverse this trend, the WHO Regional Office for Europe mounted an intensive intercountry programme during the period 1969–1979, which has culminated in the successful completion of many projects in different sectors.

One of the newer environmental hazards is nonionizing radiation (NIR), which may lead to adverse effects on human health. Exposure to NIR extends from occupational health right into the field of public health. When considering exposure limits and setting up control programmes, the heterogeneity of the population to be protected has to be kept in mind. Possible genetic and carcinogenic effects, as well as effects on development, have all to be carefully considered and are of prime importance in protecting the public. Therefore, one of the major efforts within the radiation sector of the programme during these years was directed towards the development of this book, which I now introduce with pleasure to a broad professional audience. This publication represents the results of the collaborative effort of over 200 experts from 20 countries, to whom we are indebted not only for their professional competence but for their deep dedication. With equal satisfaction, I wish to commend the work of the WHO staff directly responsible for this activity. The financial assistance of the United Nations Environment Programme in the project on NIR protection, and in the publication of this book, is greatly appreciated.

Protection against exposure to NIR is a subject of increasing concern to European countries, but is also of growing importance to those in other parts of the world. I hope that this work will be of practical value to the many scientists, engineers, physicians and community leaders concerned with and responsible for protection against NIR and human health.

Leo A. Kaprio
WHO Regional Director for Europe

Dedication

This book is dedicated to the memory of Mr Frank Harlen, Principal Scientific Officer, National Radiological Protection Board, United Kingdom, who collapsed and died on 16 January 1987 aged only 60 years.

Frank was not only a good and helpful colleague, but also a friend. He became involved in the NIR activities of the WHO Regional Office for Europe (which eventually led to the first edition of this book) when participating in the first working group on nonionizing radiation (NIR) in October 1974, in Dublin, and three subsequent working groups in 1978. He also took part in a number of activities of WHO headquarters in Geneva, concerned with the preparation of Environmental Health Criteria publications on NIR subjects. Frank again demonstrated great ability in the latest working group in October 1985, in Ann Arbor, MI, USA, which reviewed the whole NIR subject and approved the changes that led to the present revised edition. In his humble and quiet way he succeeded in resolving many issues and gaining the support of his colleagues.

Frank's scientific knowledge, his unwavering objectivity and integrity, and the warmth and humour of his manner will be remembered by all those who knew him, and he will be missed by all of us.

Michael J. Suess

Note on Terminology

The policy of the World Health Organization in respect of terminology is to follow the official recommendations of authoritative international bodies such as the International Union of Pure and Applied Physics (IUPAP), the International Commission on Illumination (CIE) and the International Organization for Standardization (ISO). Every effort has been made in this publication to comply with such recommendations.

Nearly all international scientific bodies have now recommended the use of the SI units (*Système international d'unités*) developed by the Conférence générale des poids et mesures (CGPM)[a] and the use of these units was endorsed by the Thirtieth World Health Assembly in May 1977. Only SI units are used in this publication.

In the establishment of scientifically acceptable protection standards, both for workers and for the public, the first requirement is that internationally acceptable units for use in measurements must be available. The introduction of SI units means that this requirement has now largely been satisfied.

---

[a] An authoritative account of the SI system entitled *The SI for the health professions* has been prepared by the World Health Organization. Copies may be obtained through the sales agents listed at the back of this book, or direct from the Distribution and Sales Service, World Health Organization, 1211 Geneva 27, Switzerland.

Introduction
to the second edition

M.J. Suess

The first edition of this book, which was finally published in 1982 after a great deal of work, was received very enthusiastically if one is to judge from the demand for copies and the many letters received from both colleagues and strangers. A French translation, the preparation of which had begun even before the English version appeared, was published in 1985 by the Regional Office, and a Chinese translation followed in 1986. It took the People's Medical Publishing House in Beijing only one year from the time of receiving permission to translate until publication, for which they should be specially commended.

The Introduction to the first edition of this book refers to a survey of NIR-related national legislation and institutions and to the subsequent issuing of the results. Consequently, with very important support from the WHO collaborating centre at the Bureau of Radiation and Medical Devices, Environmental Health Directorate, Health and Welfare Canada, Ottawa, the first edition of a directory was issued in 1986.[a] It is intended that the directory will be updated and improved every few years. Since 1982, WHO headquarters has continued work on books in the Environmental Health Criteria series dealing with NIR. Following the four titles published up to 1982 (No. 14, 16, 22 and 23) another, on extremely low frequency fields (No. 35), was published in 1984 and a sixth, on magnetic fields (No. 69), in 1987.

Scientific and technical developments in the NIR field have continued apace, particularly with respect to lasers, microwave radiation, magnetic fields and ultrasound. While these developments have not necessarily had any new or unknown effect on the health of workers and the public, a new examination of the health-related issues was considered warranted and timely. Consequently, a quick survey among experts revealed that Chapters 4 and 5 in particular would require a new approach (and new titles) while all the other chapters except Chapter 7 would need some revision and updating.

---

[a] **Suess, M.J. & Benwell, D.A., ed.** *Institutions and legislation concerned with nonionizing radiation health-related research and protection — a directory.* Copenhagen, WHO Regional Office for Europe, 1986.

1

Table 1. Ranges of frequency, wavelength and energy for some types of electromagnetic radiation[a]

Type of radiation	Frequency range	Wavelength range	Energy range per photon
Ionizing	>3000 THz	<100 nm	>12.40 eV
Ultraviolet (UV) (nonionizing part)	3000— 750 THz	100 — 400 nm	12.40— 3.10 eV
extreme (vacuum)	3 × 10⁵ to 30 000	1 to 10	1240 to 124
far	1 580—1 000	190 — 300	6.53— 4.13
near	1 000— 750	300 — 400	4.13— 3.10
UV-C[b]	3 000—1 070	100 — 280	12.40— 4.43
UV-B[b]	1 070— 952	280 — 315	4.43— 3.94
UV-A[b] ("black light")	952— 750	315 — 400	3.94— 3.10
Visible radiation (light)[c]	750— 385 THz	400 — 780 nm	3.10— 1.59 eV
Infrared (IR)	385— 0.3 THz	0.78—1000 μm	1 590 — 1.24 meV
IR-A[b]	385— 214	0.78— 1.4	1 590 —886
IR-B[b]	214— 100	1.4 — 3	886 —413
IR-C[b]	100— 0.3	3 —1000	413 — 1.24
near	385— 100	0.78— 3	1 590 —413
middle	100— 10	3 — 30	413 — 41.3
far	10— 0.3	30 —1000	41.3 — 1.24
Lasers	1 500— 15	0.2 — 20	6 200 — 62

Class 1 — non-risk laser devices
Class 2 — low-risk, low-power laser devices
Class 3a — low-risk, medium-power laser devices
Class 3b — moderate-risk, medium-power laser devices
Class 4 — high-risk, high-power laser devices

	300 GHz—	0.1 MHz	1 mm—3 000	m	1 240 μeV—	0.41 neV
Radiofrequency (RF)						
extremely high frequency (EHF) ⎫	300—	30 GHz	1 — 10	mm	1 240 —124	μeV
super-high frequency (SHF) ⎬ "Microwaves" (MW)	30—	3	10 —100		124 —12.4	
ultra-high frequency (UHF) ⎭	3—	0.3	100 —1 000		12.4 —1.24	
very high frequency (VHF)	300—	30 MHz	1 —10	m	1 240 —124	neV
high frequency (HF)	30—	3	10 —100		124 —12.4	
medium frequency (MF)	3—	0.3	100 —1 000		12.4 —1.24	
Low frequency (LF)	300—	30 kHz	1 — 10	km	1 240 —124	peV
Very low frequency (VLF)	30—	3	10 —100		124 —12.4	
—	3—	0.3	100 —1 000		12.4 —1.24	
Extremely low frequency (ELF)	<0.3 kHz	>1 000 km			<1.24 peV	

[a] The limits shown are those of some of the more commonly used conventions. When conversion between columns is involved, the numbers have generally been rounded up or down to the third digit.

[b] Radiation bands of biological significance designated by the International Commission on Illumination (CIE) (Chapter 3, reference 11).

[c] The limits for the human eye vary among individuals between about 380–400 nm and 750–780 nm.

The revision process was begun in 1985 and was undertaken by the original and/or new authors. The newly drafted texts were circulated to reviewers for comments, and then submitted to a group of experts for discussion and finalization. Chapters 1–3 and Chapters 4 and 5 were discussed in plenary and by two subgroups, respectively, of the Working Group that met in October 1985 in Ann Arbor, MI, USA. Chapter 6, on ultrasound, had already been discussed by a third subgroup, which had met in September 1985 in Erice, Sicily, immediately following the sixth Course on Advances in Biological Effects and Dosimetry of Ultrasound of the International School of Radiation Damage and Protection at the Ettore Majorana Centre for Scientific Culture (see Annex 2). The chapters were then revised by the various contributors in accordance with the decisions of the Working Group, recirculated to the Working Group participants for final examination and comments, corrected, and then submitted for final thorough WHO editing.

The Working Group also reviewed new developments in the different NIR spectra ranges and discussed their present or potential impact on human health. The relevant conclusions and recommendations have been added at the end of each chapter. Table 1, which was first published in the Introduction to the first edition, was also re-examined during the meeting in Ann Arbor, and a revised version is presented on pp. 2 & 3.

For the benefit of new readers, the Glossary to the first edition has been reproduced in this edition with minimum change. It is, however, the intention to enlarge the Glossary significantly with the support of the WHO collaborating centre at Health and Welfare Canada, and then to publish it as a separate entity.

Special reference should be made to the essential contribution of the scientific, technical and secretarial staff of the NIR Section, Bureau of Radiation and Medical Devices, Environmental Health Directorate, Health and Welfare Canada, Ottawa in its capacity as a WHO collaborating centre, and its tireless efforts without which the preparation of this revised edition could not have been successfully undertaken. Deirdre Benwell-Morison, in her capacity as Head of the NIR Section, provided effective leadership and served as the major driving force in carrying the complex coordination of this project to a successful conclusion. She also took full command, both scientifically and administratively, of the organization and proceedings of the subgroup on ultrasound, meeting in Italy. Maria Stuchly contributed extensively from her vast scientific knowledge, and spared no effort in assembling all the necessary ingredients for the newly written Chapters 4 and 5. Yvon Deslauriers effectively served as the third member of the Canadian scientific team by providing the link in Chapters 1, 2 and 3.

The WHO collaborating centre at the Center for Devices and Radiological Health, Food and Drug Administration, Rockville, MD, USA played an important role in organizing the Working Group in the United States. Moris Shore of the Center, who strongly supported the preparation of both the first and the second editions, again demonstrated, as Chairman of the Working Group in Ann Arbor, his special talent in channelling the many views of the various experts in the direction of one successful finale.

All the participants in the meetings in Erice and Ann Arbor deserve great praise and appreciation for their efforts and contributions. But special mention should be made of the contributors who rewrote, or updated and revised, and corrected many times the chapters of this edition. Four invited participants whose contributions were expected to be of particular importance were, unfortunately, unable to attend at the last moment. Two of these, Sol Michaelson and Rudolf Hauf, the authors of Chapters 4 and 5 of the first edition respectively, could not attend because of illness. They did, however, extensively review the new draft Chapters 4 and 5 both before and after the meeting, and provided useful comments.

Professors Martino Grandolfo and A. Rindi deserve much appreciation for their courtesy, and for arranging the excellent free facilities which were provided for the subgroup meeting at the Ettore Majorana Centre in Erice. Similarly, credit goes to Nancy D. D'Angelo and Vivian H. Green of the University of Michigan for their excellent administrative support and arrangements. Their special attention to the comfort of the participants and to social events contributed greatly to the warm and pleasant atmosphere during the meeting on the University campus in Ann Arbor.

Finally, mention should be made of one staff member of the WHO Regional Office for Europe in Copenhagen who was very active and supportive in the production and distribution of the first edition of this book. In his latest capacity as Publications Promotion Officer, he was expected to be involved again in promoting the distribution of this second edition. Unfortunately this dear colleague — Martin Jones — passed away on 10 December 1986 at the age of 48, after a short but unsuccessful struggle against cancer. May this book be a tribute to his memory also.

Introduction
to the first edition

M.J. Suess

In recent years there has been an increase in the development and use of equipment that produces nonionizing radiant energy, and the question has been raised as to whether adequate measures are being taken to guard the user and the general public from its possible adverse effects. In contrast to ionizing radiation, radiation of longer wavelengths is intrinsically less energetic and usually interacts with human tissue primarily by generating heat. For want of a better collective term, "nonionizing radiation" (NIR) is used to encompass this group of electromagnetic radiations and also ultrasound. Nonionizing radiation pervades the entire environment but, except for the narrow spectrum of visible radiation, it is unperceived by any of the human senses unless its intensity becomes so great that it is felt as heat. Differences in wavelength, even within a single type of radiation, are particularly important when evaluating hazards from exposure to NIR. The ability of the radiation to penetrate into the human body and the sites of absorption will depend on this characteristic and will differ from one type of radiation to another.

In developed countries there has been a remarkable growth in the number of processes and devices that utilize or emit NIR. They find an ever increasing use in industry, engineering, telecommunications, medicine, research, education and the home. This gives rise to a number of questions. How serious are the problems linked with NIR, what are their dimensions, and what acute and chronic effects on the human body are involved? Is there sufficient knowledge of occupational risks and public health hazards? How can exposure be reduced? Are national protection standards adequate and, if not, how can better regulations be drafted and enforced to reduce exposure? Because of the rapidly developing technology and the associated health implications, there is a need to develop international cooperation in the use of NIR and measures to prevent overexposure. Moreover, governments will be expected to intensify the establishment of rules and regulations and means of enforcing them.

The NIR part of the electromagnetic spectrum is divided into five regions (Table 1).[a] No exact ranges for these regions can be defined, and those given

---

[a] Table 1 has been revised, and now appears in the Introduction to the second edition on pp. 2 & 3.

in Table 1 are only approximations. In some cases, and for various reasons, different international bodies have developed and agreed on slightly different ranges, depending on the purpose of the definition. This is similarly true for some of the working groups involved in the preparation of this book.

Ultraviolet (UV) radiation has been used extensively for sterilizing equipment and air, and in different types of medical apparatus. In recent years industrial use of UV sources has greatly increased, and a certain risk still exists for workers from open UV sources. Damage is confined to the eye and the skin, but there is a certain long-term risk of skin cancer. The largest exposed group comprises those who spend a great deal of time in the sun, and attention should be drawn to protective measures. As far as human cancer is concerned, little quantitative knowledge of dose–effect relationships and latency periods is available. Ultaviolet lamps for private use are widely distributed among the public and should be supplied with appropriate warnings.

Exposure to infrared (IR) radiation can occur in almost any industry from direct IR sources as well as from other heat sources, and the risks under certain working conditions are well known. Still unanswered is the question of whether IR radiation has produced lenticular cataract. In any case, the presence of well developed temperature sensors in the skin around the eye provides a good biological warning system.

Risks from the use of lasers should receive more attention because of the rapid development of these instruments and their increased use for both military and nonmilitary applications. The emitted light can damage the eye and the skin, and under certain conditions perhaps also internal tissues. The difficulty in evaluating the risks from lasers is due partly to the difficulty of extrapolating the results of animal experiments to man.

Microwave (MW) and radiofrequency (RF) radiation are recognized as the types of NIR that present the greatest perceived risk. The expansion in the use of MW in the communications field and of MW ovens, if appropriate safety devices are not present in such ovens, presents a possible public health hazard.

Some questions have been raised with respect to possible adverse effects of electric and magnetic fields, particularly those at low frequencies, in connection with high-voltage lines. However, no effects due to occupational exposure have been reported, nor are there any indications of injuries to human organs. At present, there is still a dearth of information, and there is a need for better experimentation and more objective and relevant observations.

The versatility of ultrasound has led to its widespread employment in industry, medicine and science for measurement, scanning and control applications, and for thermally modifying material. Ultrasound is relatively safe because of its inability to pass an air–water interface. Whether potential adverse effects exist from immersion of hands in ultrasonic cleaning baths is not known. It has been claimed that chromosome aberrations can be produced by ultrasound radiation, but the evidence tends to be negative. So far, no major adverse effects have been recognized following diagnostic exposure of children *in utero*, and there is general agreement that the risk is much less than from X-ray examinations.

The World Health Organization has always recognized that the protection of the environment is an integral part of protecting the health of the people who live in it. Conscious of the seriousness of the issue, the WHO Regional Committee for Europe decided in 1969 to adopt a comprehensive long-term programme on environmental health and pollution control, including NIR protection. The main aim of the programme was to develop management guides and decision aids for use by government administrations, executive agencies, scientific institutions and individual specialists concerned with the quality of the environment and the protection of public health.

Development of the NIR protection sector of the programme began with a working group which was held in The Hague in November 1971 to consider the health effects of ionizing and nonionizing radiation. This meeting could perhaps be considered the beginning of international activities on NIR protection. The working group reviewed the existing situation with regard to the use of NIR and concluded that although the existing codes and guidelines were sufficient to prevent injury, it was doubtful whether they would be adequate in the future in view of the expected growth in the use of all types of NIR. The group went on to make a number of specific recommendations with respect to surveys in the field of health protection, dissemination of information, and the preparation of codes of practice. The major project that followed this meeting consisted of a series of activities which have led to the publication of this book.

This book is the culmination of a decade of concerted effort by over 200 scientists from 20 countries in Europe, North America and Asia. Five scientific groups, each addressing the subject matter of one or two chapters, met at various times to review draft chapters and to recommend ways and means of finalizing them. Drafts went through several reviews before and after the group meetings, followed by revisions and final editing. The purpose of this book is to provide practical information on health aspects of exposure to NIR energies. Descriptions are given of the physical characteristics, biophysical principles and biological effects of exposure to these energies. Public and occupational health implications and protective measures against the hazards of exposure are also described. Existing standards for general public and occupational exposures are described to inform the reader about the maximum permissible exposure levels, or threshold limit values, at present accepted in various countries. The book is intended to provide information and recommendations that will assist in setting up NIR control programmes and establishing a unified system for the recording and evaluation of results. It is also designed to serve as a technical guide for scientists, engineers and medical personnel active in the field of NIR protection and control. Evidence is presented on possible adverse health effects of NIR, so that where appropriate standards and legislation may be enacted to protect health. Although the book was originally planned by the Regional Office for use in European countries, it should be of value to countries in other parts of the world and to the other international organizations concerned. Moreover, it should provide NIR specialists everywhere with helpful information based on up-to-date international practice.

9

Work began in 1973, when Professor Michaelson was invited to prepare a document on the potential hazards and safety considerations of human exposure to NIR energies. The first draft was ready in May 1973 and was sent to specialists for review. Moreover, advantage was taken of the International Symposium on Biologic Effects and Health Hazards of Microwave Radiation, held in Warsaw in 1973, to convene a small ad hoc group with the participation of Professors Czerski, Faber, Gordon, Michaelson and Schwan. It was decided to divide the document into two parts: that on MW and RF would be subjected to further revision and review, and a second on lasers would eventually be submitted to a working group on that subject. The final manuscript on MW and RF, following significant and elaborate changes, was ready for final editing in February 1976 and was issued by this Office in 1977. Some of the major contributions to the completeness and accuracy of this version were made by Professors Gordon and Schwan, and Dr Silverman.

However, this document had neither the advantage of being discussed by a working group, nor was it up to date by the time other chapters were completed. An ad hoc meeting in Freiburg in May 1978, attended by Czerski, Harlen, Kossel, Michaelson, Repacholi, Shore and Suess, decided to submit the MW/RF document for further review and revision. Consequently, Professor Michaelson convened a group of experts from Europe and North America for a review meeting, which was held in 1978 at the Bureau of Radiological Health in Washington, DC. This meeting led to a new, rather lengthy and detailed text, but one that was very comprehensive and acceptable to all the points of view represented. The new chapter was revised and corrected, and submitted for final editing in December 1979. Mr Harlen in particular was very helpful in checking the nonbiological sections of the manuscript for adequacy and accuracy.

The Working Group on Health Effects from Lasers, held in Dublin in October 1974, considered the revised working document provided by Professor Michaelson, and the authorship was then enlarged and the material redrafted. After further review and revision, the chapter was submitted for final editing in February 1976 and was issued by this Office in a provisional form in 1977. Mr Sliney and Dr Wolbarsht were particularly instrumental in giving this chapter its present form, while Professor Michaelson was the coordinator of the team and technical editor of the material.

The chapters on UV and IR radiation were both commissioned in 1976. The former was reviewed by correspondence and examined thereafter by a working group in Sofia, in February 1978. After further revision, the chapter was submitted for final editing in January 1979. The draft chapter on IR radiation went through a number of reviews and revisions before submission to the same working group. It was revised, reviewed and corrected thereafter, and submitted for final editing in October 1978.

The chapter on ultrasound was commissioned in 1975 and was passed through a similar process. It was discussed by a working group in London in October 1976, and after further review and corrections, was submitted for final editing in May 1977. A provisional edition of this chapter was issued by this Office that same year.

The two last chapters to be mentioned, which were discussed by the working group in Freiburg in May 1978, are concerned with electric and magnetic fields at low frequencies, and with regulations and enforcement procedures. They were commissioned in 1976, went through a similar review, revision, and correction process, and were submitted for final editing in December and October 1978, respectively.

Some chapters contain conclusions and/or recommendations of various types and purpose. These were agreed on, as were the contents of the chapters themselves, by the participants in the various working groups, and presented by the respective authors in their chapters. It will be obvious, considering the time spent in compiling this work, that some literature references are not as recent as one would wish. However, the reader will appreciate that the objective of this book is not to provide (even if it were possible) an up-to-date literature review. What is important is that the fundamental information contained in this publication is sound and will remain valid for some years. The updating of the material and the references will be dealt with in future revisions.

A great effort has been made to present a uniform text as regards style and terminology. Moreover, the general rule of WHO in its publications is to follow authoritative, internationally approved scientific terms and units, some of which are rather new. A short glossary provides definitions of some essential terms. However, terms in common use in related fields, such as physics, optics, radio-sciences, ionizing radiation and medicine, and their definitions in other glossaries, have been excluded.

This Office has also embarked on a survey of national legislation and regulations, and of institutions and specialists concerned with one or more sectors of the NIR field. The survey is based on replies to two questionnaires received from various governments, and on information from many experts involved in this project. The survey has revealed that only in a relatively small number of countries are there institutions dealing with NIR. In addition to the two countries that have long been involved in this area, namely the United States and the USSR, only about two dozen other countries in the world are known to have institutions that are concerned with the study of NIR, and of these about three quarters are European countries. Since the listing at present is incomplete, and the first attempt will inevitably contain errors and omissions, the results will be issued separately so that it can easily be revised and updated.

The World Health Organization is grateful to the Governments of Bulgaria, the Federal Republic of Germany, Ireland and the United Kingdom for support through their kind agreement to host working group meetings in their countries, and to the Government of the Netherlands for financial support. Also recognized with appreciation is the United Nations Environment Programme which, through its significant financial assistance from 1978 onwards, made the completion of this project and the publication of this book possible. The consistent support given to this project by Mr J.C. Villforth, Director of the Bureau of Radiological Health (BRH) of the US Public Health Service, and members of his staff, is greatly appreciated. In its capacity as the WHO collaborating centre for standardization of nonionizing radiation, the BRH kindly hosted a review meeting on health effects of exposure to microwave radiation. My thanks go also to Dr W.H. Parr, Chief of the Physical Agents Effects Branch, National Institute for Occupational Safety and Health (NIOSH) and his staff for their work on and contribution to the chapter on IR radiation.

Some of the figures and tables in this book have been reproduced from published works. In all cases this has been acknowledged by means of a suitable reference. The authors and WHO wish to thank the publishers for granting permission for such use to be made of copyright material.

I should like to record my personal indebtedness to all the colleagues, reviewers, and participants in the meetings who have contributed in many ways and at various stages. All of these are listed in Annex 1, and I offer my sincere apologies to anybody whose name may have been unintentionally omitted. The very close collaboration with the authors of the chapters and other particularly active contributors among the working group participants, some of whom have become personal friends in the course of the work, and their endless efforts to upgrade and update the material, are warmly acknowledged.

The friendly assistance and goodwill of various members of the Regional Office staff on different occasions is greatly appreciated. They include secretaries, draughtsmen, publications and reproduction staff, registry and mailing personnel, and the administration. Though too numerous to mention individually, all have contributed towards the successful implementation of this project. I am also grateful to Mr J. Kumpf and Mr J.I. Waddington, former Chief and present Director, respectively, of the environmental health team in the Regional Office, for their support during the implementation and completion of this work. The understanding of Dr Leo A. Kaprio, WHO Regional Director for Europe, and the late Dr F. Bauhofer, former Director of Health Services, of the potential value of this work in promoting NIR protection in many countries was a great encouragement.

Finally, acknowledgements and thanks should go to Dr Shore for his farsightedness and recognition of the significance and benefits of this work to international understanding and cooperation, and for his repeated encouragement and support; to Professor Michaelson for sparing no time and effort in providing special advice and essential assistance; to Dr R.B. Dean for his editorial assistance; and to Christiane Sørensen who, as my secretary from 1974 to 1978, worked tirelessly on behalf of this project and helped with the organization of all four working groups.

A revised and updated edition of this book may be prepared in the future and readers are invited to submit comments, corrections and observations, as well as suggestions for additional material, to the World Health Organization, Regional Office for Europe, Scherfigsvej 8, 2100 Copenhagen Ø, Denmark.

Ultraviolet radiation

M. Faber

Revised by J.C. van der Leun

CONTENTS

	Page
Introduction	13
Physical description	14
Production	15
Solar ultraviolet radiation	19
Transmission and absorption in biological tissue	23
Absorption and photochemical processes	23
Pathological effects in man	26
Non-stochastic effects	26
Chemical photosensitization	31
Immunological effects	31
Late effects	32
Hazards due to overexposure	34
Dosimetry	35
Safety standards	36
Protection	38
Solar ultraviolet radiation	38
Industrial sources	39
Conclusions and recommendations	39
Conclusions	39
Recommendations	40
References	41

INTRODUCTION

Of the various types of nonionizing radiation, ultraviolet (UV) is of special
interest because of its relatively high photon energy as compared to the other

types included in this group. This could lead to greater variation in biological response. On the other hand, the low penetration will restrict most of the direct biological responses to the superficial tissues.

Although UV radiation can arise from a large number of man-made sources, the sun is the main source and both the general public and people working out of doors will be exposed to it. This natural background radiation and the variations in its magnitude must be taken into account when exposure limits are discussed.

It is well known that UV can initiate photochemical reactions and that some of these take place in the skin. The best known is the production of vitamin D_3, which is necessary for the prevention of rickets in man. The full extent to which UV affects human wellbeing is difficult to quantify. Artificially produced UV has, however, been used in mines and cellars and in far northern latitudes as a supplement to combat functional impairment among people (1,2). Many of the observed effects, such as a decrease in the incidence of infectious diseases and in absenteeism, may be due to the bactericidal nature of the radiation (3). On the other hand, large doses of UV have an acute destructive effect on the skin and eye. Doses so low that they give rise only to normally acceptable or even desirable acute effects can, if repeated, induce changes resulting in late effects such as elastosis of the skin, keratosis and skin cancers. These effects will be of greater significance in people with lightly pigmented skin.

Our goal in protecting the population against the harmful effects of UV is to establish the most appropriate exposure limits based on a biological risk–benefit analysis of all these factors in as quantitative a fashion as possible (4).

PHYSICAL DESCRIPTION

Ultraviolet radiation is that part of the electromagnetic spectrum lying between the softest ionizing radiation on the one side and visible radiation on the other. For biological purposes, it is convenient to regard the range of wavelengths from 100 to 380–400 nm as constituting UV. The lower limit of 100 nm is equivalent to photon energies of 12.4 eV, which corresponds approximately to the limit for the production of ionization in biologically important materials. At the other end, the limit is the shortest visible wavelength; this varies slightly from individual to individual, and in adults lies between 380 and 400 nm.

Because of differences in physical properties and in biological effects, the UV region has been subdivided. Wavelengths shorter than approximately 180 nm are absorbed by air to such an extent that no biological effects would be expected, unless very powerful sources are used. The remainder can then be divided into the far-UV region between 180 and 300 nm and the near-UV region between 300 and 400 nm.

A somewhat different way of dividing the UV region takes some of the biological effects into account. In this arrangement the range 400–315 nm, the so-called "black light" region, is called UV-A. In this wavelength region,

14

fluorescence can be induced in many substances. UV-B covers the range 315–280 nm (the skin erythemal region). Most of the biologically active and potentially harmful UV from the sun reaching the surface of the earth falls within this spectral region. UV-C includes the radiation of wavelengths less than 280 nm (the germicidal region); it occurs in the radiation emitted by germicidal lamps and welding arcs, but not in sunlight reaching the earth's surface. These divisions are, however, arbitrary and usage varies from one worker to another.

PRODUCTION

Matter at a temperature of 2500 K or higher may emit a significant number of photons with energies inside the UV range. Such incandescent sources emit a smooth spectrum, a continuum, possibly with superimposed lines.

Most man-made sources of UV radiation can be grouped together in the categories shown below.

Incandescent sources
    tungsten halogen lamps

Gas discharges
    mercury lamps (low-, medium- and high-pressure)
    mercury lamps with metal halides
    xenon lamps
    hydrogen and deuterium lamps
    flash tubes

Electric discharges
    welding arcs
    carbon arcs

Fluorescent lamps
    fluorescent lighting tubes
    fluorescent sunlamps (UV-B emitters)
    fluorescent UV-A tubes

Lasers
    excimer lasers (several wavelengths)
    nitrogen lasers (337 nm)
    tunable UV lasers
    helium–cadmium lasers (325 nm)

The spectrum of the UV radiation emitted varies from one source to another. In the case of low-pressure mercury lamps, a line spectrum will be emitted. These lines are broadened into bands in high-pressure lamps (often called medium-pressure lamps by photobiologists) and there may also be emission of a continuum over a wide range of wavelengths. This continuum is most strongly marked in the highest pressure lamps. Addition of metal halides will increase both the continuum and the number of superimposed lines in the spectrum (Fig. 1 and 2). In high-pressure xenon lamps there is also a combination of distinct bands with a continuum (Fig. 3) *(5)*.

The emission spectrum of the arc produced during welding will depend not so much on the atmosphere in which the welding takes place but rather on the composition of the electrodes. An example of an emission spectrum produced by a welding process is shown in Fig. 4.

The spectrum of fluorescent lamps depends mainly on the properties of the fluorescent phosphors employed in the envelope. The amount of UV depends also on the absorption properties of the envelope glass used in the fluorescent tube (Fig. 5 and 6).

To permit the transmission of UV when the discharge does not take place in free air, gas discharge arcs and other UV sources must be contained within an envelope of quartz or UV-transmitting glass. On the other hand, sources that are designed primarily to emit visible radiation but which also produce significant but unwanted amounts of UV should be provided with an external filter that absorbs UV-B and UV-C radiation.

Fig. 1. Emission spectrum of a typical low-pressure
mercury vapour lamp

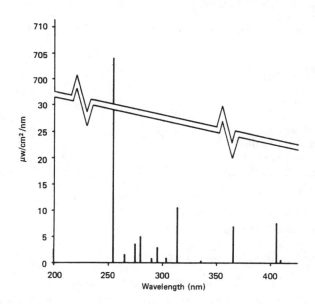

Note. 55% of output is at the 253.7-nm resonance line of mercury.

Source: Courtesy F. Urbach.

Fig 2. Emission spectrum of a typical mercury arc lamp operating at medium pressure

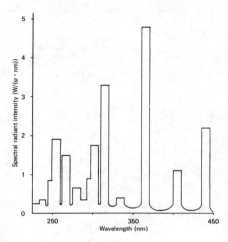

Note. The spectral lines of mercury are superimposed on a low continuum.

Source: Courtesy F. Urbach.

Fig 3. Emission spectrum of a compact xenon arc light source operating at high pressure

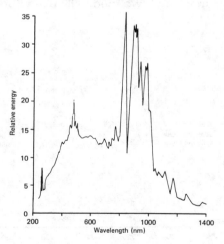

Note. Similarity to solar spectrum (above the atmosphere) and very high infrared output.

Source: Courtesy F. Urbach.

Fig. 4. Emission spectrum of gas tungsten arc welding

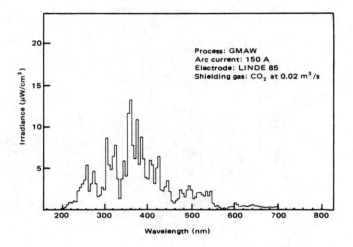

Note. The spectral emission from any kind of electric arc welding will depend on the composition of the electrodes, the plasma that is created and the shielding gas used.

Source: Adapted from Sliney & Wolbarsht *(6).*

Fig. 5. Emission spectra of two medically used fluorescent lamps

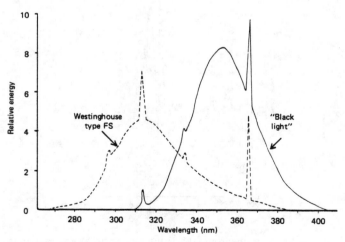

Note. Type FS emits primarily in the UV-B, type "Black light" emits primarily in the UV-A.

Source: Courtesy F. Urbach.

18

Fig. 6. Emission spectra of two fluorescent lamps emitting primarily in the UV-A.

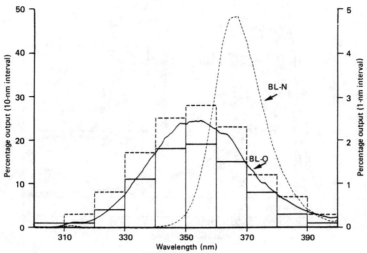

Note. Because of the use of two different phosphors, the spectral distribution between BL-O and BL-N differs markedly.

Source: Courtesy F. Urbach.

In the case of lasers, the emission lines will depend on the active medium and on the operating conditions.

SOLAR ULTRAVIOLET RADIATION

The sun is the main UV source. The broad spectrum and the intensity of the UV radiation from the sun are due to the high temperature at its surface and its size. The intensity is such that the UV radiation reaching the earth's atmosphere would probably be lethal to most living organisms on the surface. Fortunately, they are shielded by the atmosphere. The ozone layer in the upper atmosphere is particularly important in this connection. The spectrum, both before passage through the atmosphere and at sea level, is shown in Fig. 7. The path length traversed in the atmosphere by the UV radiation determines the irradiance at the surface of the earth; it is, therefore, affected by geographical latitude, altitude above sea level, and time of day and season. Scattering and absorption by dust, smoke and rain are also important. It should be noted that practically no UV radiation from the sun with wavelengths below 290 nm reaches the surface of the earth. The changes over the year in one location are shown in Fig. 8; it will be seen that large day-to-day changes occur.

19

Fig. 7. The sun's spectrum at the outer surface of the atmosphere and at sea level

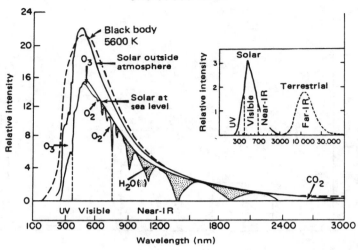

Source: Giese *(7)*.

Fig. 8. Daily total erythemally effective UV count for 1974, Minneapolis, USA

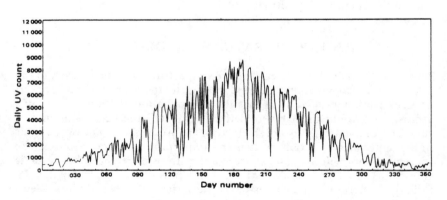

Note. The measurements were performed with a Robertson–Berger meter, which weights the radiation spectrally according to a sensitivity curve, approximating the action spectrum for UV erythema in human skin.

Source: Scotto *(8)*.

When the yearly changes for discrete wavelengths (Fig. 9) are given as monthly averages, it can be seen that such changes are not the same for all wavelengths. A comparison between the curves obtained in open country and in an adjacent city (Sofia) show the effect of polluted air in reducing the UV in the city. Again, there is a difference between the wavelengths, the shortest showing the greatest reduction. Dependence on latitude is shown in Fig. 10.

If the ozone layer were to decrease in thickness, the absorption in the critical UV bands would decrease, and an increase in the biological effects of UV-B would be expected (9).

Ozone is produced photochemically by UV from the sun at wavelengths largely below 242 nm. It is present in a concentration which is the outcome of a dynamic equilibrium between production and breakdown; there are daily, seasonal and geographical variations in this concentration. These variations add to the complexity of investigating long-term changes.

Volatile stable substances, released from the earth's surface as a result of human activities, can reach the ozone layer, and are decomposed there by photochemical reactions. Interest has concentrated on chlorofluorocarbons (CFCs), since photochemically produced chlorine can catalytically attack ozone and result in lower equilibrium concentrations (10). Nitrogen oxides have also been investigated in this connection, as well as many other compounds and their interactions in the atmosphere (11).

The computed long-term result, typically describing a development over several decades, depends strongly on the assumptions made for the future release of the various pollutants; projected changes of total ozone vary from an increase of 2–3% to a decrease of 10% or more. The latter possibility represents the case that the ozone chemistry would be dominated by chlorine; this is expected to occur only with a significant sustained growth in CFC emissions, of 3% or more per year. The production and release of CFCs have decreased since 1974, mainly as a result of restrictions on their use as propellants in spray cans. There is an increase, however, in their use for other purposes, such as refrigeration and foam blowing (9).

The possibility of a decrease in ozone with a concomitant rise in UV-B has important biological implications. The consequences for human health are easiest to evaluate, owing to the availability of quantitative data. The predominant effect expected is an increase in the incidence of skin cancer. For every 1% reduction of the average thickness of the ozone layer, the incidence of basal-cell carcinoma would increase by about 3% and the incidence of squamous-cell carcinoma by about 5% (12,13). It is uncertain whether there would be an influence on the incidence of malignant melanoma. UV-B radiation has been demonstrated to be harmful to several forms of plant and animal life; consequences for agriculture and fisheries are, therefore, likely. These are potentially at least as important as the direct effect on human health; because of lack of data it is as yet not possible to make quantitative predictions (9).

21

Fig. 9. Monthly average UV intensity in Sofia, Bulgaria and in adjacent open country

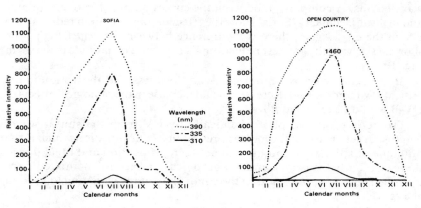

Fig. 10. Annual erythemal UV count by latitude

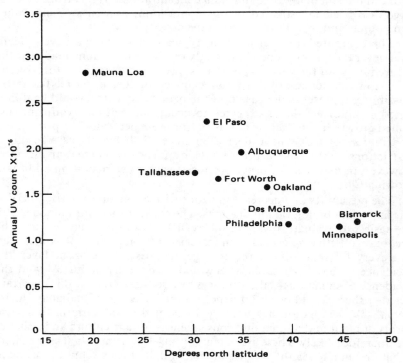

Source: Scotto *(8)*.

TRANSMISSION AND ABSORPTION
IN BIOLOGICAL TISSUE

Most of the UV radiation incident on the skin is absorbed in the epidermis. Absorption is generally greater for the shorter wavelengths (Fig. 11).

The same is true for the eye. Most of the UV radiation is absorbed in the tear film, the cornea and the lens. The lens and the tissues in the anterior part of the eye may, however, be exposed to UV at wavelengths above 295 nm and the retina is exposed to a certain extent to UV-A. The penetration of different wavelengths into the eye is given in Table 1. Some doubt exists, however, as to the high transmission given for the vitreous humour at the longer wavelengths.

ABSORPTION AND PHOTOCHEMICAL PROCESSES

The primary process in a photochemical reaction is the absorption of radiant energy by a chromophore. In most molecules the ground or relaxed state consists of two paired electrons, the singlet state (S_0). On absorption of radiant energy one of the electrons can make a transition to an excited state (S_1), provided that the photon energy corresponds to an existing energy level in the absorbing molecule. From the excited state the molecule can release its energy by *(a)* transition of the excited electron to the ground state, giving off the energy as fluorescence, or *(b)* transition to the generally long-lived triplet excited state (T_1) and then discharging energy as phosphorescence. In addition, molecules in the S_1 or T_1 states can relax either by forming photoproducts or by transferring the energy to an acceptor molecule. Formation of photoproducts, as well as transfer of energy to an acceptor molecule, requires compatible energy levels for the donor molecule and the energy receptor.

In photosensitized reactions the photon energy is absorbed by a photosensitizer and then transferred directly or via an intermediate, such as oxygen, to the biochemical compound.

The photochemical process can be impeded by "quenchers". The quencher either brings the excited state molecule directly back to the ground state or neutralizes the reactive intermediates in the photosensitized reactions.

The most important biological UV radiation absorbers are proteins and nucleic acids. Examples of biologically active photosensitizers are methoxsalen and chlorpromazine. Important quenchers in biological tissue are vitamin E, carotenes and ascorbic acid.

The absorption spectrum describes the absorption of radiant energy as a function of wavelength. Nucleic acids have their main absorption peak close to 265 nm, due to the pyrimidine structure. The aromatic amino acids are the absorbing sites in protein, with tyrosine at 275 nm and tryptophan at 280 nm.

The action spectrum gives the relative response of a system to irradiation at different wavelengths, and ideally will correspond to the absorption spectrum. In biological systems, however, the action spectrum function is

23

Fig. 11. Penetration of UV and visible radiation into human epidermis for lightly pigmented Caucasian skin, at several wavelengths and to several depths

Depth (μm)	254	270	280	290	297	313	365	436	546
0	100	100	100	100	100	100	100	100	100
10	42	30	28	30	50	67	80	85	89
20	18	9.1	7.9	14	25	44	64	72	80
30	5.0	2.4	2.5	6.5	15	33	50	62	72
40	1.4	0.65	0.78	3.0	9.6	24	39	54	65
50	0.39	0.17	0.25	1.3	6.0	18	31	46	59
60	0.11	0.047	0.077	0.60	3.8	13	24	40	53
70	0.030	0.012	0.024	0.27	2.4	9.5	19	34	48

Wavelength (nm)

Stratum corneum

Malpighian layer

Note. The data apply to collimated, perpendicular irradiation and are expressed as a percentage of incident irradiance.

Source: Bruls et al. *(15)*.

Table 1. Percentage of energy incident on the corneal epithelium that impinges on the anterior surface of the various ocular media

Wavelength (nm)	Corneal stroma	Aqueous humour	Lens	Vitreous humour	Retina
230	3	—	—	—	—
235	11	—	—	—	—
240	19	—	—	—	—
245	26	—	—	—	—
250	26	—	—	—	—
260	26	—	—	—	—
265	27	—	—	—	—
270	29	—	—	—	—
275	31	—	—	—	—
280	33	—	—	—	—
285	42	—	—	—	—
290	52	2	0.4	—	—
295	63	9	3	—	—
300	70	27	14	—	—
305	75	50	37	—	—
310	78	64	51	—	—
320	81	78	74	0.3	0.3
330	84	80	77	0.5	0.5
350	87	86	83	2	2
360	89	88	85	4	4
370	90	90	87	12	11
380	93	91	88	28	26
390	94	93	91	49	45
400	95	94	93	69	64
450	—	96	96	84	81
500	—	96	96	87	84

Source: Modified from Kinsey *(14)*.

modulated by the shielding (i.e. thickness) of overlying tissue, energy transfer, and the action of sensitizers and quenchers.

A number of biologically important photochemical reactions have been shown to occur in biological tissue. DNA strand breaks and pyrimidine dimers are produced during exposure to UV radiation. Hydrates of cytosine and uracil may be produced, but have not been shown to produce biological effects after direct absorption of UV energy. Of greatest interest, at least in the UV-B region, is the photoproduction of covalent dimers of thymidine and other pyrimidines *(16)*. The biological effects of these lesions are reviewed by Smith *(17)*.

Defects in DNA are repaired in living cells principally by excision repair (dark repair), a process in which it has been possible to isolate at least five participating enzymes. These act by excision of thymidine dimers and dimers of other pyrimidines and reconstitution of the intact DNA strand. The primary biochemical mechanism has been studied both in bacteria and in mammalian and human cells and is now reasonably well understood *(18)*. Defects in the repair system are present in a number of rare diseases, of which xeroderma pigmentosum is the most important in relation to UV damage. Evidence has been presented to show that the genetic constitution, as measured by the ability of cells to repair UV lesions, may be important in relation to the development of UV-induced diseases *(19)*. When repair is in progress, it is easy to recognize in autoradiograms as unscheduled DNA synthesis after uptake of tritiated thymidine. The unscheduled synthesis disappears gradually during the 24 hours following the irradiation. The repair system is limited in its efficiency. In HeLa cells, the initial rate of repair increases linearly after doses of up to $30 \, \text{J/m}^2$. At higher doses, no further increase in the rate of thymidine uptake is seen *(18)*. The incident dose needed for the induction of a given number of dimers is wavelength-dependent; at wavelengths of 254, 290 and 310 nm the doses are in the ratio $1:30:6000$. At longer wavelengths, other types of DNA lesion become of increasing and ultimately of dominant importance *(20)*.

In addition, post-replication repair is active in mammalian cells and may be a more error-prone step. This repair system can also be deficient in xeroderma pigmentosum.

Photoreactivation is another independent repair system where exposure to longer wavelengths, whether UV or visible, can induce enzymatic repair. Enzymatic photoreactivation in human cells has been reported by several investigators *(21,22)*. Full agreement has, however, not yet been reached as to the significance of these results. It has been claimed that the enzyme is absent in xeroderma pigmentosum cells *(23)*.

Protein-DNA cross-links can also be repaired. The mechanism is not yet understood *(24)*.

Photochemical reactions induced by UV that have biological consequences are, however, not restricted to cellular constituents alone. Photo-oxidation of atmospheric constituents (smog formation) may, for example, indirectly influence human health.

PATHOLOGICAL EFFECTS IN MAN

For protection purposes these effects can be divided into two main groups, namely non-stochastic and stochastic *(25)*. The non-stochastic effects are related directly to the radiant exposure, while the stochastic effects take the form of an increase in the risk of contracting certain diseases, of which the late appearance of cancer is the most important.

Non-stochastic effects
In this group, the severity of the effects in the exposed individual varies with dose and there may be a threshold below which no effects occur *(25)*. These

effects may be either acute or late and can occur in any cell or tissue that can be reached by radiation or by photochemical by-products.

Because of its limited penetration, the primary effects of UV in man will essentially be restricted to the skin and eyes; only under special circumstances may such effects also extend to the oral cavity.

Skin

Four types of immediate change occur in the skin: *(a)* immediate pigment darkening; *(b)* the production of erythema (sunburn); *(c)* the production and upward migration of melanin granules (suntanning); and *(d)* changes in epidermal cell growth. The structure of the skin is shown in Fig. 12.

Immediate pigment darkening takes place during and immediately after irradiation. It is most strongly marked in pigmented skin. The action spectrum displays a maximum at 360 nm and extends to wavelengths above 400 nm. The phenomenon has long been ascribed to oxidation of existing premelanin *(27,28)*, but recent investigations do not confirm an increased amount of melanin pigment *(29)*.

The vascular reaction of the skin to UV known as erythema consists of vasodilation and an increased blood content; it may be accompanied by augmented blood flow and increased vascular permeability leading to cellular exudation, e.g. neutrophil leukocytes, and in severe cases by blistering. The erythema appears after a latent period of 1–8 hours, and lasts for one or more days. Higher doses of UV radiation lead to a shorter latent period and a longer duration of the erythema. The intensity of the erythema increases with increasing UV dose; the steepness of the increase in redness with increasing dose is different for the erythemas caused by the various wavelengths in the UV *(30,31)*. It appears likely that there are at least two different erythemal mechanisms, one based on direct action of the radiation on the superficial blood vessels, whose dilation causes the reddening of the skin, and one based on an indirect action, where the radiation acts on the epidermis and the vasodilation is effected by an active substance diffusing to the blood vessels in the dermis *(32)*. The involvement of prostaglandins has been suggested *(33)*.

The minimal dose that will provoke an erythema (the minimal erythema dose, MED) is the best defined threshold dose for acute effects following UV exposure in man. The MED depends on the properties of the individual's skin, probably mainly the transmission of UV radiation through the epidermal layers, which in turn is determined by the thickness of the epidermal layers and also by the pigmentation. Both factors are influenced by previous UV exposures. For the same reasons, the MED also depends on the skin region. In lightly pigmented Caucasian skin, not recently exposed, the MED on the trunk is on the average about 200 J/m² for wavelengths between 250 and 300 nm; it rises sharply between 300 and 330 nm and is of the order of $2 \times 10^5 J/m^2$ between 330 nm and 400 nm *(34,35)*.

The third reaction of the skin to UV exposure is an increase in pigmentation or "suntan", with an action spectrum close to that of erythema *(35)*. It is initiated by a spreading of existing pigment granules into neighbouring

Fig. 12. Cross-section of human skin

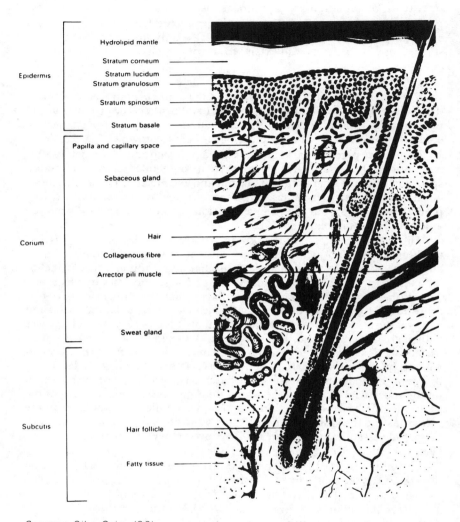

Epidermis
- Hydrolipid mantle
- Stratum corneum
- Stratum lucidum
- Stratum granulosum
- Stratum spinosum
- Stratum basale

Corium
- Papilla and capillary space
- Sebaceous gland
- Hair
- Collagenous fibre
- Arrector pili muscle
- Sweat gland

Subcutis
- Hair follicle
- Fatty tissue

Source: Ciba-Geigy *(26).*

cells throughout the exposed skin. Production of new pigment granules occurs later in the process.

UV radiation interferes with cell division in skin. Immediately after irradiation there is a cessation of cell division *(36,37)* for 24 hours or longer, depending on dose, followed by an increase in mitoses *(38)*. This increase reaches a maximum at 72 hours *(39)*. The increase in cell division after a single dose of only short duration gives rise to hyperplasia of the epidermis with a maximum at 5–6 days *(40)*, often accompanied by a shedding of superfluous cellular material (peeling). After single very large doses, the outcome of UV irradiation in animals such as mice is ulceration and scar formation, but this occurs in man only if there is a secondary bacterial infection.

Sunburn cells constitute a special histological feature; these are cells characterized by dense nuclei lying in isolation within the middle layers of the epidermis *(39)*. These severely damaged cells may be dead and will be removed, but little direct information on them is available. The wavelengths between 260 and 290 nm appear to be of equal importance and photo-sensitizers, such as 8-methoxypsoralen, will increase the yield *(41)*.

In recent years there has been a great increase in the use of UV-A sources to produce rapid skin tanning. An imperfect knowledge of the effect of high-dose UV-A on the skin and eye has resulted in widespread concern as to whether cosmetic UV exposure is advisable. There is certainly a risk to the eyes if these are not adequately protected. As far as the skin is concerned there is an acute risk especially to photosensitive individuals, and a potential long-term risk for all, especially in the case of immoderate use. In many countries, regulations on UV exposure are in preparation or already in effect. Such regulations should be based on the best available knowledge of the biological effects; however, better knowledge, particularly on the bio-logical effects of UV-A, is highly desirable *(42)*.

Mouth

As a consequence of dental practices, where plastic materials are hardened (or cured) by UV treatment, the mucous membrane of the mouth may be exposed to unwanted UV irradiation when defective apparatus is used *(43)*. Severe erythema of the skin around the mouth alone has been observed under such conditions. In recent years, the method has been largely replaced by visible light polymerization.

Eye

The structure of the eye is shown in Fig. 13. The main clinical effects of UV on the eye are photokeratitis and conjunctivitis, which appear 2–24 hours after irradiation. The symptoms are photophobia, foreign body sensation, and blepharospasm, which last from 1 to 5 days. In general, there is no residual lesion. Photokeratitis is caused preferentially by UV-C and UV-B but also by UV-A, though with less effectiveness. The peak sensitivity of the cornea is variously given as 270 nm *(44)* or 288 nm *(45)*. The effect is dose-related, which means that tissue damage depends on the total energy absorbed, not on the rate of absorption, at least if the exposure does not last for more than a few hours. Fairly accurate measurements have been made of the energy necessary for minimal photokeratitis. The threshold has a value of 50 J/m^2 at 270 nm, rising to 550 J/m^2 at 310 nm, followed by a steep increase to $22\,500 \text{ J/m}^2$ at 315 nm *(46)*. Photoconjunctivitis has a similar action spectrum. Anterior uveitis was not found with exposure

29

Fig. 13. Cross-section of the human eye

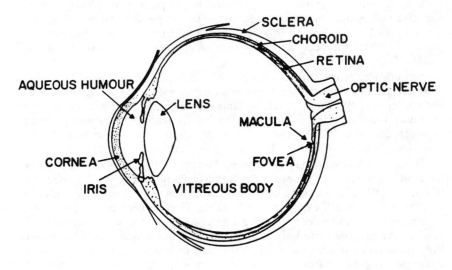

at 315–325 nm until the dose received was about four times that required for a minimal effect. Studies on primate eyes tend to indicate a slightly lower threshold *(46)*.

The problem of acute or lasting damage to deeper-lying ocular structures has become increasingly important with the availability of strong UV sources, including lasers. The effects will depend on the transmittance of UV through the ocular media, figures for which are given in Table 1, which shows that an effect on the lens is possible at wavelengths longer than 290 nm. The action spectrum for acute damage to the lens lies between 295 and 320 nm *(47)*. Animal experiments show that both a single, high radiant exposure and exposures of long duration but of low irradiance induce opacities in the lens. Some epidemiological studies on man suggest that sunlight, especially UV radiation, appears to be active in producing cataracts *(48,49)*. One particular type of cataract (nuclear cataract) may be due to photochemical processes in the lens involving tryptophan, resulting in the formation of a brown pigment (brunescent cataract). Discussions of the possible biochemical mechanism have been presented *(47,49)*. The most efficient wavelength for the production of transient lenticular opacities was 300 nm, but only when the exposure exceeded a threshold value of 1500 J/m² *(46)*. The damage became permanent at radiant exposure levels of approximately twice this figure. It has been suggested that a characteristic change in the conjunctiva, pterygium, is caused by exposure to UV radiation *(6)*.

Retinal damage by high-intensity UV-A has been demonstrated in experimental animals *(50–52)*. In the adult human eye such damage appears unlikely, due to the strong absorption of UV-A by the lens. Persons with the crystalline lens removed (aphakics) are, however, more vulnerable (6,53). The modern practice of implanting intraocular lenses in the eye of aphakic persons increases the risk of UV damage to the retina.

Chemical photosensitization

Photosensitization is both an area of great interest and one of importance in industrial hygiene. It manifests itself in a number of ways, involving somewhat different mechanisms.

The photodynamic effects of organic substances all appear to be based on the same principle and are associated with UV-A and extend into the visible region. The energy is first absorbed by the photosensitizer, a relatively low-molecular-weight substance present in the cells, e.g. a fluorescent dye (acridine orange, riboflavines). It can then be transferred to a target molecule, oxygen serving as an intermediate. The effect of this transfer is most serious when the target is DNA in the nucleus or in a virus, when chain breaks and cross-linking may result (54).

Other substances, such as the naturally occurring psoralens, will be bound to DNA after irradiation and will react with thymine and cytosine. After irradiation with UV-A, they form covalent photo-adducts sometimes leading to inter-strand linking. This results in a high degree of DNA damage.

From a clinical point of view, photosensitivity is a general term describing the combined action of UV and a chemical substance; this can lead to either phototoxic or photo-allergic reactions. Phototoxicity is the common response in all those whose skin is irradiated with sufficient energy and at the appropriate wavelength in the presence of a phototoxic substance. Although the result may only be the common sunburn reaction, it will be more marked with the photosensitizer. Photo-allergy is less common and is thought to depend on an acquired altered reactivity due to an antigen–antibody or cell-mediated hypersensitivity to the photosensitizer. Clinically immediate urticarial or delayed papular or eczematous reactions may appear. Photo-allergy may result in an increased sensitivity to light even without the sensitizer. The incident exposure may be relatively small. The UV of wavelengths 300–320 nm present in white fluorescent lamps (55) may suffice. With respect to the eye, photosensitizers such as psoralens do enhance cataract formation and chlorpromazine may have an effect on the retina.

Immunological effects

A great number of experiments performed during the past decade have shown that UV-B irradiation alters responses of the immunological system (56,57). UV-B depresses contact sensitivity reactions, reduces the number and functions of Langerhans' cells in the skin, and changes the distribution of subpopulations of circulating lymphocytes in mice and man. Repeated irradiation of mice with UV-B induces an immunological change that inhibits the animals' ability to reject transplanted tumours induced by UV-B radiation in syngeneic mice (58). Such an effect of UV-B radiation also appears to decrease the animal's defence against primary tumours induced by UV-B radiation in its own skin (59).

31

An action spectrum for UV-induced immune suppression was found to peak between 260 and 270 nm *(60)*. An entire field of new findings and insights is opening up. It is not yet clear what the implications will be for the ideas about benefit and risk of human exposures to UV radiation, and about protection against the risks.

Late effects

Late non-stochastic effects
These occur in the skin and in the eye.

Skin. After prolonged exposure to sunlight over a period of years, the dermis will begin to degenerate, with a decrease in elasticity due to degeneration of the collagen fibres combined with other histological changes *(61)*. The visible symptoms will be deep furrows in the skin giving an appearance of premature aging. No dose–effect relationship is known for this reaction in man.

The epidermis may also be involved, with the development of actinic keratosis. The importance of this lesion is difficult to evaluate, but the occurrence of an increased cellular proliferation rate *(62)* and a certain amount of cellular atypia *(63)* suggest that it may represent a precancerous stage in the development of squamous-cell carcinoma. Many of these carcinomas are surrounded by areas of actinic keratosis *(63)*. It has been suggested that actinic keratosis may be a universal disorder of epithelial growth *(64)*.

Eye. As mentioned earlier, some of the types of cataract seen in elderly people may be due to repeated exposure to UV radiation over many years. The quantitative dose–effect relationship is unknown.

Late stochastic effects
Late stochastic effects are those for which the probability of an effect occurring, rather than its severity, is a function of dose. A threshold cannot be expected in effects of this type *(25)*. The typical lesion following UV exposure is a malignant tumour. There is no reason to expect heritable damage since UV cannot reach the gonads but is absorbed in the overlying tissues.

Skin. Cancer of the skin is a well recognized effect of UV irradiation, both in experimental animals and in man.

Three types of cancer are concerned: basal-cell carcinoma, squamous-cell carcinoma and malignant melanoma. The evidence for the production of skin cancer by UV is good for squamous-cell carcinoma and reasonably good for basal-cell carcinoma *(65–68)*; it is still uncertain for malignant melanoma *(9)*. The evidence incriminating UV radiation comes basically in two steps. The first step is to establish the involvement of sunlight, from epidemiological or clinical observations. Such observations deal with influences of full-spectrum sunlight and therefore do not point to any particular wavelength region within the solar spectrum. The wavelength range involved has to be found by a second step, usually experimental investigations on animals.

Clinical observation shows that squamous-cell and basal-cell carcinomas appear preferentially on the most sun-exposed skin sites of lightly pigmented individuals. Malignant melanomas do not show such a convincing preference, though the incidence of all three types of skin cancer is higher in geographical areas with more intense sunlight. An inverse relationship between incidence and geographical latitude has been found for melanomas in white people in the United States. Similar, even stronger, correlations have been found for squamous-cell and basal-cell carcinomas. These and many other epidemiological observations suggest a causative role for sunlight in human skin cancer, including malignant melanoma *(69–76)*.

Animal experiments indicating a specific wavelength region have been performed mainly on mice. Such experiments yield mainly squamous-cell carcinomas and fibrosarcomas; the results apply directly only to these types of skin cancer, and not to basal-cell carcinomas or melanomas. That the wavelengths causing tumours are mainly below 320 nm was first shown by Roffo *(77,78)* and by Funding et al. *(79)*. The peak in the action spectrum appears to be between 280 and 320 nm *(80)* and the shorter wavelengths from the sun, i.e. 295–320 nm, are thus active *(67)*. Several observations suggest that the action spectrum for UV carcinogenesis is similar to that for UV erythema. Experiments with lamps emitting only UV-A showed that this can also induce skin tumours in mice *(81,82)*. An occasional clinical incident has been reported suggesting the induction of basal-cell carcinoma in man by UV-B radiation *(83)*.

Data incriminating a particular wavelength range for the induction of malignant melanoma are not available. There are no reports on the induction of melanoma in experimental animals by UV radiation alone.

In general, irradiation is effective only when the dose is spread over an extended period *(62,84,85)*, and the carcinogenic effect will depend on the number of doses and the duration of the irradiation. A single exposure may induce tumours in mice or rats, but only with doses so high that acute tissue damage is caused *(86,87)*. With regular daily exposure to UV-B, the tumours usually appear on nearly normal-looking skin *(88)*. Induction of tumours is mainly related to the doses of UV radiation administered, but dose delivery also plays a role. In general, doses are more effective if administered over longer periods of time *(89)*. With the dose delivery standardized, the relationship between the doses administered regularly and the time at which the tumours appear may be established quite accurately, and described by mathematical equations *(90)*. The same type of equations may be used, with different parameters, to describe skin cancer data for more or less homogeneous human populations *(91)*. The human data show much more spread than data from animal experiments and allow, therefore, different mathematical descriptions *(13,92)*. A dose–effect relationship in practical terms for human populations states that the incidence of skin cancer is approximately proportional to the square of the UV doses regularly received *(85)*. Such a relationship suggests that, although no dose of UV radiation is completely without risk, small doses have a minor effect; the important risk is caused by large doses received over long periods of time.

There are many factors complicating the straightforward relationship between UV radiation and skin cancer in man. Latitude correlations are significant only in reasonably homogeneous populations. In populations mixed with regard to pigmentation, the picture may be entirely different. Darkly pigmented skin is much less susceptible than white skin to UV carcinogenesis. The carcinogenic action of UV radiation also depends on dietary influences *(93–95)* and on skin temperature *(96,97)*. Diseases such as certain types of inherited albinism and xeroderma pigmentosum *(98)* lead to an increase in the incidence of skin tumours, the latter probably as a result of defects in DNA repair mechanisms *(99)*. Incomplete or erroneous repair of DNA damage appears to be important in the development of skin tumours *(100)*. Certain medical therapies also result in an increase in malignant skin tumours, such as immunosuppressants *(2)* and dermatological photochemotherapy with 8-methoxypsoralen and UV-A *(101)*.

Eye. So far no direct evidence has been presented that tumours in the anterior chamber of the eye, and especially melanomas at this site, can be produced by UV irradiation. It should be noted that melanoma of the eye is commonest in blue-eyed individuals, and that melanoma of the iris occurred only in blue eyes in a series of ocular melanomas *(102)*. Tumours of the cornea, both fibrosarcomas and haemangioendotheliomas, have been induced in experimental animals *(103)* and tumours, including melanomas, are known to occur in domestic animals and in man in tropical regions *(104)*.

HAZARDS DUE TO OVEREXPOSURE

The risk of damage following either acute or chronic exposure to UV is encountered in a number of situations.

Solar UV is the most important source of such exposures. All those who work out of doors are potentially at risk from overexposure, the consequences of which may be both acute and long-term effects. The fashion of exposing a large part of the body to sunlight has during recent years increased the exposure of the skin, resulting in quite high UV doses. This is true not only for outdoor work but is now also normal during leisure periods, as exemplified by the holiday exodus of a large part of the population of the northern European countries to the Mediterranean coast.

UV-emitting arcs are an integral part of the working conditions at a number of workplaces. In welding, such arcs constitute a serious risk. Not only UV, but also visible and infrared radiation as line spectra with a continuous component are emitted during welding. The shape of such spectra will be different for the different welding procedures *(105)*.

The use of UV in some graphic reproduction techniques also represents an exposure risk for the workers concerned.

UV, and especially UV-C, is used for the sterilization of food and air, and pathological effects due to accidental exposures, often as small doses but of long duration, may result.

A number of sources available to the general population are known to emit UV, either as a normal part of the emission or after accidental breakdown. The normal white-light fluorescent tubes usually emit small amounts of UV. This may be enough to induce a phototoxic reaction if a photosensitizer is present. Under certain conditions the UV output of lighting tubes is by itself large enough to contribute appreciably to a worker's annual UV dose *(106)*. In addition, UV sources have for many years been available for home use, partly for the semi-cosmetic purpose of tanning. This may result in overexposure if the instructions on the equipment are not followed. The spectrum of fluorescent sunlamps is not always restricted to the solar spectrum; it differs greatly from that of solar radiation in the UV-B range, and UV-C components may be present.

The use of black-light lamps to control the effectiveness of tooth brushing will, in general, not constitute a hazard as the sources used are weak and the exposure short. The same is true where fluorescent black-light tubes are used, for instance for crack detection, chromatography, philately, mineral identification or document inspection.

Accidental short-term exposure with resultant symptoms has, however, occurred after breakage of the protective shield around high-pressure mercury lamps used for lighting *(107)*.

Medical irradiation in the form of phototherapy is at present expanding. In the blue light phototherapy of infants with neonatal jaundice (hyperbilirubinaemia), incorrectly selected fluorescent tubes have given rise to erythema. The use of certain photodynamic dyes and light for the treatment of herpes, or of psoralens and UV-A for other skin diseases such as psoriasis (PUVA), will expose the patient to risk. Up to now the action spectrum that has been studied the most extensively is that for ultraviolet erythema *(108)*. The action spectrum of the effect on psoriasis appears to be similar *(109)* and the same applies to the carcinogenic action spectrum *(110)*. If adequate safety precautions are not taken, the treatment personnel may also be at risk.

Excimer lasers may pose new problems, as these produce high-intensity radiation of wavelengths as short as 193 nm, which have not been available to any great extent from conventional sources. At present little is known about the biological effects of these short wavelengths. UV lasers are considered in detail in Chapter 2.

DOSIMETRY

From the point of view of establishing standards, it is desirable not only to be able to make measurements of emitted radiation but also to be able to record the doses received by the persons to be protected. It must, however, be appreciated that the dose absorbed by the sensitive cells may be difficult to estimate. Most of the equipment available for dose measurements in relation to protection is cumbersome and delicate and not too well adapted for field use. In general, a phototube (photomultiplier) or photodiode detector is used. When it is necessary to determine the spectral distribution the different wavelengths can be separated, e.g. by a diffraction grating monochromator or by filters, and the transmitting optics must be made of quartz to allow all the UV to pass. The quality of broad-band measurements can be good, but considerable errors may be introduced when a full spectral description is required. One such error is the lack of precision in narrow wavelength bands when filters are used.

Instruments have been constructed that weight the radiation according to a sensitivity curve representative for certain biological effects, such as the action spectrum for UV erythema or a more generalized ultraviolet hazard curve *(8,111)*.

Personal dosemeters integrate the effective dose received by people over time, for instance during their work. The device consists, for instance, of a piece of polysulfone film in a badge carried by the person. The method is still under development, but has already produced several useful results *(111–115)*.

SAFETY STANDARDS

Any safety standards developed must take into consideration not only the harmful effects of UV radiation but also the need for a certain minimal irradiation, so as to ensure that sufficient vitamin D_3 is produced; this is of greatest importance during infancy and childhood. Vitamin D is supplied in two ways: through diet and by production in the skin. Yet deficiencies do occur, especially among children and elderly people. In children this leads to rickets and in the elderly to osteomalacia.

The synthesis of vitamin D_3 in the skin, its regulation and metabolism have been studied extensively in recent years *(116)*. A photoregulation process ensures that there is no danger of overproduction leading to vitamin D intoxication. The action spectrum for the production of vitamin D_3 shows some similarity to that for UV erythema *(117)*. It is therefore possible to estimate the UV doses needed for a sufficient production of vitamin D_3 in terms of erythemally effective radiation. From recent measurements *(118,119)* it may be calculated that for production in the skin of the daily vitamin D requirement of 400 IU, a UV dose on the head, neck and hands of about 60 MED per year is necessary.

Another reason that the skin needs at least some UV radiation, especially in long winters, is that it helps the skin to maintain some of its tolerance to UV. Many people have difficulty in adapting when UV irradiance increases again in the spring. This leads to many patients developing photodermatoses in areas with long winters. This difficulty may be prevented by maintaining some tolerance in winter *(120)*; doses required are of the same order of magnitude as those required for the production of vitamin D_3.

On the other hand, too much UV is not beneficial either. Up to now the standard established by the American Conference of Governmental Industrial Hygienists (ACGIH), which specifies a threshold limit value, has been used in preparing guidelines for other countries *(121)*. This is based on the action spectrum for photokeratitis and erythema in normal white-skinned individuals. This means that acute effects have alone been taken into consideration. For the UV-B region and at lower wavelengths, it states that the radiant exposure in an 8-hour period must not exceed the value given in Table 2. For the wavelength range 320–400 nm, the total irradiance on the unprotected skin or eye must not exceed 10 W/m² for periods exceeding 10³ seconds (about 17 minutes). For radiant exposures of shorter durations, it should not exceed 10 kJ/m².

A procedure has been suggested for characterizing the relative levels of UV from illumination sources and derived guideline numbers given for the maximum illumination level of the source that will not exceed the ACGIH standards *(122)*.

The values given in Table 2 apply directly only to sources emitting essentially monochromatic UV. The maximal permissible exposure for a broad-band source can be calculated by summing the relative contributions from all its spectral components, each contribution being weighted by means of the relative spectral effectiveness, as given in Table 2. In addition, the guidelines should not be used for determining exposure limits for photosensitive individuals.

Table 2. Threshold limit values (TLV)
and relative spectral effectiveness by wavelength
for any 8-hour period of exposure

Wavelength, λ (nm)	TLV (J/m²)	Relative spectral effectiveness, S_λ
200	1 000	0.03
210	400	0.075
220	250	0.12
230	160	0.19
240	100	0.30
250	70	0.43
254	60	0.50
260	46	0.65
270	30	1.00
280	34	0.88
290	47	0.64
300	100	0.30
305	500	0.06
310	2 000	0.015
315	10 000	0.003

The guidelines discussed so far do not take into account the long-term risk of skin cancer. As the action spectrum for UV carcinogenesis appears to be similar to that for UV erythema, the cancer risk may also be discussed in terms of erythemally effective doses. Thus the Health Council of the Netherlands *(123)* has tried to define "acceptable levels" for long-term unintended exposures to UV radiation. The reasoning was based on the clinical observation that skin cancer in the Netherlands occurs mainly in outdoor workers even though there are many more indoor workers. The difference in the UV doses received apparently brings the outdoor workers into the risk zone. The difference in erythemally effective UV doses received by outdoor workers and indoor workers was estimated with the help of data collected with Robertson–Berger meters *(8)* and personal dosimeters *(111)*. The acceptable level for long-term occupational exposures from man-made sources was defined as a small fraction of this difference. This led to an acceptable level corresponding to an average daily exposure of 3–9 minutes of full local summer sunlight.

This is a rather strict limit, but those given in Table 2 for short-term exposure are equally strict. Where these limits are observed there appears to be little need for additional limits for long-term exposure *(124)*.

PROTECTION

Solar ultraviolet radiation

The weak penetration of UV makes a simple form of protection possible, since it is excluded by most types of clothing. This, of course, may not provide protection for the face and hands during work out of doors. Furthermore, it should be remembered that not all clothing will adequately exclude UV. Relatively open-weave clothing or that made of UV-transparent material may result in sunburn being caused by sunlight penetrating the clothing. It is also the experience of dermatologists that synthetic material used for dresses and shirts permits sufficient UV to pass for a skin reaction to occur when phototoxic substances are being tested *(125)*. Apart from clothing, protection may also be afforded by the application of sun-blocking or sun-screening substances that act by absorbing or scattering the radiation. Of the former, *p*-aminobenzoic acid or some of its esters have proved to be the most successful. They can easily be applied as a lotion, a cream or preferably in an alcoholic solution *(97)*. The results are a decrease in UV-B-induced erythema and a slower rate of suntanning.

It has been suggested that β-carotene could act as a systemic protector *(126)*. Since even its possible effectiveness in erythropoietic protoporphyria, a photosensitive disease, has been questioned, the evidence in support of the general usefulness of this treatment is not very strong. The recent observation that the related retinoic acid decreases the dose necessary for the successful PUVA treatment of psoriasis *(127)* is interesting, in view of the inhibiting effect of this substance on skin tumour production *(128,129)*. Since animal experiments suggest that all transretinoic acids may increase cell growth *(128)*, some caution is necessary.

Industrial sources

Protection against UV in the working environment should preferably consist of containment of the radiation by appropriate design of the source or of the apparatus in which it is placed.

As mentioned previously, a large number of sources can be shielded by the use of an appropriate covering, which may be a filter that selects only those wavelengths corresponding to the purpose for which the lamp is to be used. In the case of high-intensity lamps containing UV sources, attempts are being made to introduce safety devices that will interrupt the emission if the covering glass is broken.

When containment is not possible and the irradiance is high, appropriate eye protection is mandatory together with protection of the skin. This may consist of appropriate clothing or the application of effective sunscreens. Standards for eye protection exist in most countries. Welding is an example of a type of work where sufficient protection can be obtained by suitably designed and fitted welding masks or hoods. When welding is started, however, the shield may have to be removed and this can give rise to photolesions of the eye.

It is fortunate that a certain degree of control of the hazard can be achieved by the welder himself, as the eye will reject filters of insufficient absorbing capacity in the visible light at wavelengths close to UV-A *(130)*. It is not sufficient to protect the welder himself; the surrounding area must also be monitored and screened so as to ensure that nobody is accidentally exposed. This requires fixed shielding between and around welders; such shielding work must be treated with non-reflecting paint in order to protect the neck of the welder from exposure by reflection.

CONCLUSIONS AND RECOMMENDATIONS[a]

Conclusions

All people are exposed to UV radiation from sunlight, and the risk to health varies with geographical, genetic and other factors. Similar risks are involved in the increasing exposure of people to UV radiation from artificial sources, such as those used for suntanning, in phototherapy and in industrial processes. The biological effects of a single exposure differ significantly from the effects of repeated and cumulative exposures. Both types of risk increase markedly with excessive exposure.

For UV radiation exposures the spectral composition of the source and action spectrum weighted irradiance (W/m^2) and radiant exposure (J/m^2) are the important parameters that determine the biological effect.

---

[a] These conclusions and recommendations are those pertaining to ultraviolet radiation made by the WHO Working Group on Health Implications of the Increased Use of NIR Technologies and Devices, Ann Arbor, USA, October 1985.

The essential measurements and determination of the hazard should include:

— careful estimation of the output spectrum of the source in narrow intervals (1–5 nm bands);
— knowledge of the action spectrum for the effect of concern;
— the irradiance in the individual wavelength bands;
— the exposure duration;
— the distribution of the radiation impinging over the exposed area;
— the characteristics of the reflecting surfaces;
— the frequency of the repetition of the exposure; and
— the effective irradiance (dose rate) in repeated exposures.

The interaction between biological tissues and UV radiation depends on:

— the spectral distribution of the source;
— the radiant exposure (dose) weighted against the action spectrum; and
— the number of exposures.

The interaction of UV radiation and biological tissues is mainly photochemical. However, there are some UV sources that emit sufficient energy to produce thermal effects in tissues.

Because of the relatively superficial absorption, the major biological effects are on the skin and the eye. Through its optical properties, the eye has an increased vulnerability to injury to UV radiation.

The major health hazards to the skin are both acute and chronic. Acute hazards are sunburn and photosensitized reactions, the chronic hazards accelerated aging and photocarcinogenesis. The major health hazards to the eye are photokeratitis, corneal burns, ocular inflammation, photochemical cataract, and photochemical injury to the retina.

Recommendations

Control measures should include education and training of all personnel working with UV sources. Engineering controls such as proper layout of working areas, enclosures, and ventilation of instruments should be selected as appropriate. Personal protection may consist of special eyewear, and appropriate clothing or protective sunscreens applied to the skin. Warning signs and other administrative controls may supplement these measures.

Various national and international bodies have dealt with the problem of maximum permissible exposures of the eye and skin. Despite the lack of data in some areas, there is a fair degree of consensus on the exposure limits, such as those recommended by the IRPA and ACGIH, and those in preparation by the IEC and CIE.

It is also recommended that additional specific studies be performed on the health effects of UV radiation. Epidemiological studies are recommended in the following areas:

— malignant melanoma in skin in relation to sunlight;

— cataract formation as produced by solar radiation, if possible separately for UV-A and UV-B radiation;

— risks to aphakic people and those with intraocular lens implants for retinal changes, mainly from UV-A and short wavelengths in the visible part of the spectrum;

— pterygium formation as the result of exposure to UV-A and UV-B radiation; and

— malignant melanoma in the uvea.

Experimental studies are recommended in the following areas:

— the action spectrum for photocarcinogenesis;

— the action spectrum for cataract formation, macular changes in aphakic animals, and pterygium formation;

— the chronic effects of UV-A radiation on human skin; and

— the effect of chronic exposures to UV-A and UV-B radiation on the lens, the retina and the uvea, as well as the cornea and conjunctiva.

There are widespread public misconceptions as to the benefits and risks to health of exposure to UV radiation. Information programmes are necessary to educate people about risks from excessive exposure.

REFERENCES

1. **Dantsig, N.M. et al.** *Ultraviolet installations of beneficial action.* Paris, Commission internationale de l'éclairage, 1968, p. 225 (CIE Publication No. 14A).
2. **Dantsig, N.M.** *Environmental health criteria for ultraviolet radiation.* Geneva, World Health Organization, 1977 (document HEE/EHC/WP/77.15).
3. **Ronge, H.E.** Ultraviolet irradiation with artificial illumination. *Acta physiologica scandinavica,* **15**(Suppl. 49) (1948).
4. *Ultraviolet radiation.* Geneva, World Health Organization, 1979 (Environmental Health Criteria, No. 14).
5. **Schafer, V. & Hinrich, G.** Erzeugung von UV-Strahlen. *In:* Kiefer, J., ed. *Ultraviolette Strahlen.* Berlin, de Gruyter, 1977, pp. 47–177.
6. **Sliney, D.H. & Wolbarsht, M.L., ed.** *Safety with lasers and other optical sources.* New York, Plenum Press, 1980.
7. **Giese, A.C.** *Living with our sun's ultraviolet rays.* New York, Plenum Press, 1976.
8. **Scotto, J. et al.** *Measurements of ultraviolet radiation in the United States and comparison with skin cancer data.* Bethesda, MD, National Institutes of Health, 1976 (DHEW Publication (NIH) 76-1029).
9. *Environmental assessment of ozone layer modification and its impacts.* Nairobi, United Nations Environment Programme, 1984 (Bulletin No. 8).

10. **Rowland, F.S.** The stratospheric photochemistry of chlorine compounds and its influence on the ozone layer. *In:* Castellani, A., ed. *Research in photobiology.* New York, Plenum Press, 1977, p. 579.

11. **National Academy of Sciences.** *Causes and effects of stratospheric ozone reduction: an update.* Washington, DC, National Academy Press, 1982.

12. **De Gruijl, F.R. & van der Leun, J.C.** A dose–response model for skin cancer induction by chronic UV exposure of a human population. *Journal of theoretical biology,* **83**: 487–504 (1980).

13. **Rundel, R.D. & Nachtwey, D.S.** Projections of increased non-melanoma skin cancer incidence due to ozone depletion. *Photochemistry and photobiology,* **38**: 577–591 (1983).

14. **Kinsey, V.E.** Spectral transmission of the eye to ultraviolet radiation. *Archives of ophthalmology,* **39**: 508 (1948).

15. **Bruls, W.A.G. et al.** Transmission of human epidermis and stratum corneum as a function of thickness in the ultraviolet and visible wavelengths. *Photochemistry and photobiology,* **40**: 485–494 (1984).

16. **Beukers, R. et al.** Isolation and identification of the irradiation product of thymine. *Recueil des travaux chimiques des Pays-Bas,* **78**: 883 (1959).

17. **Smith, K.C.** The radiation-induced addition of proteins and other molecules to nucleic acids. *In:* Wang, S.Y., ed. *Photochemistry and photobiology of nucleic acids.* New York, Academic Press, 1976, Vol. 2, pp. 187–215.

18. **Edenberg, H.J. & Hanawalt, P.C.** The time of DNA repair replication in ultraviolet-irradiated HeLa cells. *Biochimica et biophysica acta,* **324**: 206 (1973).

19. **Lambert, B. et al.** Ultraviolet-induced DNA repair synthesis in lymphocytes from patients with actinic keratosis. *Journal of investigative dermatology,* **67**: 594–598 (1976).

20. **Lohman, P.H.M.** Results of project No. 2. *In: Progress report 1977, programme radiation protection.* Brussels, European Economic Community, 1977, pp. 319–320.

21. **D'Ambrosio, S.M. et al.** Photorepair of pyrimidine dimers in human skin *in vivo. Photochemistry and photobiology,* **34**: 461–464 (1981).

22. **Sutherland, B.M.** Human photoreactivation enzymes. *In:* Castellani, A., ed. *Research in photobiology.* New York, Plenum Press, 1977, pp. 307–315.

23. **Sutherland, B.M.** Photoreactivation in normal and xeroderma cells. *In:* Magee, P.N. et al., ed. *Fundamentals in cancer prevention.* Tokyo, University of Tokyo Press, 1976, pp. 409–416.

24. **Kornhauser, A.** UV induced DNA-protein cross-links *in vitro* and *in vivo. Photochemistry and photobiology,* **23**: 457–460 (1976).

25. **International Commission on Radiological Protection.** *Recommendations of the International Commission on Radiological Protection.* Oxford, Pergamon Press, 1977 (ICRP Publication 26).

26. *Eczematous skin diseases. Part I. The skin — morphology and physiology.* Basle, Ciba–Geigy, 1967.

27. **Fitzpatrick, T.B.** Introductory lecture. *In:* Bowen, E.J., ed. *Recent progress in photobiology.* Oxford, Blackwell Scientific Publications, 1965, p. 365.

28. **Murphy, T.M.** Nucleic acids: interaction with solar UV radiation. *Current topics in radiation research quarterly,* **10**: 199 (1975).

29. **Hönigsmann, H.** Newer knowledge of immediate pigment darkening (IPD). *In:* Urbach, F. & Gange, R.W., ed. *The biological effects of UVA radiation.* New York, Praeger, 1986, pp. 221–224.

30. **Farr, P.M. & Diffey, B.L.** The erythemal response of human skin to ultraviolet radiation. *Journal of investigative dermatology,* **84**: 449–450 (1985).

31. **Hausser, K.W. & Vahle, W.** Sunburn and suntanning. *Wissenschaftliche Veröffentlichungen des Siemens Konzerns,* **6**: 101–120 (1927).

32. **van der Leun, J.C.** On the action spectrum of ultraviolet erythema. *In:* Gallo, U. & Santamaria, L., ed. *Research progress in organic, biological and medicinal chemistry.* Amsterdam, North-Holland, 1972.

33. **Snyder, D.S. & Eaglstein, W.H.** Intradermal anti-prostaglandin agents and sunburn. *Journal of investigative dermatology,* **62**: 47–50 (1974).

34. **Magnus, I.A.** *Dermatological photobiology.* Oxford, Blackwell Scientific Publications, 1976.

35. **Parrish, J.A. et al.** Erythema and melanogenesis action spectra of normal human skin. *Photochemistry and photobiology,* **36**: 187–191 (1982).

36. **Bowden, G.T. et al.** Excision of pyrimidine dimers from epidermal DNA and nonsemiconservative epidermal DNA synthesis following ultraviolet irradiation of mouse skin. *Cancer research,* **35**: 3599–3607 (1975).

37. **Epstein, W.L. et al.** Early effects of ultraviolet light on DNA synthesis in human skin *in vivo. Archives of dermatology,* **100**: 84–89 (1969).

38. **Blum, H.F.** On the mechanism of cancer induction by ultraviolet radiation. IV. The size of the replicated unit. *Journal of the National Cancer Institute,* **23**: 343–350 (1959).

39. **Daniels, F. et al.** Histochemical responses of human skin following ultraviolet irradiation. *Journal of investigative dermatology,* **37**: 351–357 (1961).

40. **Blum, H.F.** Hyperplasia induced by ultraviolet light: possible relationship to cancer induction. *In:* Urbach, F., ed. *The biologic effects of ultraviolet radiation.* Oxford, Pergamon Press, 1969, p. 83.

41. **Woodcock, A. & Magnus, I.A.** The sunburn cells in mouse skin: preliminary quantitative studies on its production. *British journal of dermatology,* **95**: 459–468 (1976).

42. **Urbach, F. & Gange, R.W., ed.** *The biological effects of UVA radiation.* New York, Praeger, 1986.

43. **Mills, L.F. et al.** *A review of biological effects and potential risks associated with ultraviolet radiation as used in dentistry.* Rockville, MD, Bureau of Radiological Health, 1975 (DHEW Publication (FDA) 76-8021).

44. **Pitts, D.G. & Gibbons, W.** Corneal light scattering measurement of UV radiant exposure. *American journal of optometry,* **50**: 187 (1973).

45. **Cogan, D.G. & Kinsey, V.E.** Action spectrum of keratitis produced by ultraviolet radiation. *Archives of ophthalmology,* **35**: 670–677 (1946).

46. **Pitts, D.G. et al.** *Ocular ultraviolet effects from 295 nm to 400 nm in the rabbit eye.* Cincinnati, OH, National Institute for Occupational Safety and Health, 1977 (DHEW Publication (NIOSH) 77-175).

47. **Kurzel, R.B. et al.** Ultraviolet radiation effects on the human eye. *Photochemical and photobiological reviews,* **2**: 133 (1977).

48. **Pirie, A.** Photooxidation of proteins and comparison of photooxidized proteins with those of the cataractous human lens. *Israel journal of medical science,* **8**: 1567 (1972).

49. **Zigman, S.** Near UV light and cataracts. *Photochemistry and photobiology,* **26**: 437 (1977).

50. **Zigman, S. & Vaughan, T.** Near-ultraviolet light effects on the lenses and retinas of mice. *Investigative ophthalmology,* **13**: 462–465 (1974).

51. **Zuclich, J.A. & Connolly, J.S.** Ocular hazards of near-UV laser radiation. *Journal of the Optical Society of America,* **66**: 79 (1976).
Zuclich, J.A. & Taboada, J. Ocular hazard from UV laser exhibiting self-mode-locking. *Applied optics,* **17**: 1482–1484 (1978).

53. **Kamel, I.D. & Parker, J.A.** Protection from ultraviolet exposure in aphakic erythropsia. *Canadian journal of ophthalmology,* **8**: 563–565 (1973).

54. **Smith, K.C. & Hanawalt, P.C.** *Molecular photobiology.* New York and London, Academic Press, 1969.

55. **Kobza, A. et al.** Photosensitivity due to the 'sunburn' ultraviolet content of white fluorescent lamps. *British journal of dermatology,* **89**: 351–359 (1973).

56. **Daynes, R.A. et al.** *Experimental and clinical photoimmunology.* Boca Raton, FL, CRC Press, 1983.

57. **Parrish, J.A., ed.** *The effect of ultraviolet radiation on the immune system.* Piscataway, NJ, Johnson and Johnson, 1983.

58. **Kripke, M.L. & Fisher, M.S.** Immunologic parameters of ultraviolet carcinogenesis. *Journal of the National Cancer Institute,* **57**: 211–215 (1976).

59. **De Gruijl, F.R. & van der Leun, J.C.** Systemic influence of pre-irradiation of a limited skin area on UV-tumorigenesis. *Photochemistry and photobiology,* **35**: 379–382 (1982).

60. **DeFabo, E.C. & Noonan, F.P.** Mechanism of immune suppression by ultraviolet irradiation *in vivo.* I. Evidence for the existence of a unique photoreceptor in skin and its role in photoimmunology. *Journal of experimental medicine,* **158**: 84–98 (1983).

61. **Daniels, F.** Ultraviolet carcinogenesis in man. *National Cancer Institute monographs,* **10**: 407–422 (1963).

62. **Winkelmann, R.K. et al.** Squamous cell tumor induced in hairless mice with ultraviolet light. *Journal of investigative dermatology,* **34**: 131–138 (1960).

63. **Hundeiker, M.** Entwicklung und Erkennung der Präkanzerosen. *Zeitschrift für Hautkrankheiten,* **52**: 1181–1199 (1977).

64. **Pearse, A.A. & Marks, R.** Actinic keratoses and the epidermis on which they arise. *British journal of dermatology,* **96**: 45–50 (1977).
65. **Blum, H.F.** *Carcinogenesis by ultraviolet light.* Princeton, NJ, Princeton University Press, 1959.
66. **Stenback, F.** Ultraviolet light irradiation as initiating agent in skin tumor formation by the two-stage method. *European journal of cancer,* **11**: 241–246 (1975).
67. **Urbach, F. et al.** Ultraviolet carcinogenesis: experimental, global, and genetic aspects. *In:* Pathak, M.A. et al., ed. *Sunlight and man.* Tokyo, University of Tokyo Press, 1974, pp. 259–283.
68. **Urbach, F., ed.** *The biologic effects of ultraviolet radiation.* Oxford, Pergamon Press, 1969, pp. 3–21.
69. **Elwood, J.M. & Lee, J.A.H.** Trends in mortality from primary tumours of skin in Canada. *CMA journal,* **110**: 913–915 (1974).
70. **Elwood, J.M. et al.** Relationship of melanoma and other skin cancer mortality to latitude and ultraviolet radiation in the United States and Canada. *International journal of epidemiology,* **3**: 325–332 (1974).
71. **Lancaster, H.O. & Nelson, J.** Sunlight as a cause of melanoma: a clinical survey. *Medical journal of Australia,* **1**: 452–456 (1957).
72. **Lancaster, H.O.** Some geographical aspects of the mortality from melanoma in Europeans. *Medical journal of Australia,* **1**: 1082–1087 (1976).
73. **Lee, J.A.H. & Merrill, J.M.** Sunlight and the aetiology of malignant melanoma: a synthesis. *Medical journal of Australia,* **2**: 846–851 (1970).
74. **Magnus, K.** Incidence of malignant melanoma of the skin in Norway, 1955–1970. *Cancer,* **32**: 1275–1286 (1973).
75. **Magnus, K.** Epidemiology of malignant melanoma of the skin in Norway with special reference to the effect of solar radiation. *In:* Castellani, A., ed. *Research in photobiology.* New York, Plenum Press, 1977, p. 609.
76. **Magnus, K.** Incidence of malignant melanoma of the skin in the five Nordic countries: significance of solar radiation. *International journal of cancer,* **20**: 477–485 (1977).
77. **Roffo, A.H.** Cancer et soleil. Carcinomes et sarcomes provoqués par l'action du soleil *in toto. Bulletin de l'Association française pour l'Etude du Cancer,* **53**: 59 (1934).
78. **Roffo, A.H.** Cancer y sol. *Boletín del Instituto de Medicina Experimental,* **10**: 417–444 (1933).
79. **Funding, G. et al.** Über Lichtkanzer. *In: Verhandlungen des 3. internationalen Kongresses für Lichtforschung, Wiesbaden, 1–7 September 1936,* p. 166.
80. **Freeman, R.G.** Data on the action spectrum for ultraviolet carcinogenesis. *Journal of the National Cancer Institute,* **55**: 1119–1121 (1975).
81. **Forbes, P.D.** UV-A and photocarcinogenesis. *Photochemistry and photobiology,* **41**: 7S (1985).
82. **van Weelden, H. et al.** Carcinogenesis by UVA, with an attempt to assess the carcinogenic risks of tanning with UVA and UVB. *In:* Urbach, F. & Gange, R.W., ed. *The biological effects of UVA radiation.* New York, Praeger, 1986, pp. 137–146.

45

83. **Ippen, H. & Goerz, G.** *Photodermatosen und Porphyrien, Tagung der Rheinisch-Westfalischen Dermatologen-Vereinigung, 26 August, 1972.* Düsseldorf, H. Ippen, 1974.

84. **Blum, H.F.** On the mechanism of cancer induction by ultraviolet radiation. II. A quantitative description and its consequences. *Journal of the National Cancer Institute,* **23**: 319–335 (1959).

85. **van der Leun, J.C.** Yearly review: UV-carcinogenosis. *Photochemistry and photobiology,* **39**: 861–868 (1984).

86. **Hsu, J. et al.** Induction of skin tumors in hairless mice by a single exposure to UV radiation. *Photochemistry and photobiology,* **21**: 185–188 (1975).

87. **Strickland, P.T. et al.** Induction of skin tumors in the rat by single exposure to ultraviolet radiation. *Photochemistry and photobiology,* **30**: 683–688 (1979).

88. **Blum, H.F. et al.** Relationships between dosage and rate of tumor induction by ultraviolet radiation. *Journal of the National Cancer Institute,* **3**: 91–97 (1942–1943).

89. **Forbes, P.D. & Davies, R.E.** Factors that influence photocarcinogenesis. *In:* Parrish, J.A. et al., ed. *Photoimmunology.* New York, Plenum Medical Book Company, 1983, pp. 131–153.

90. **De Gruijl, F.R. et al.** Dose–time dependency of tumor formation by chronic UV exposure. *Photochemistry and photobiology,* **37**: 53–62 (1983).

91. **Slaper, H. et al.** Risk evaluation of UVB therapy for psoriasis: comparison of calculated risk for UVB therapy and observed risk in PUVA-treated patients. *Photodermatology,* **3**: 271–283 (1986).

92. **Scotto, J. et al.** *Incidence of non-melanoma skin cancer in the United States.* Washington, DC, US Department of Health, 1981 (Publication (NIH) 82-2433).

93. **Black, H.S. et al.** Relation of dietary lipid level to photocarcinogenesis. *Photochemistry and photobiology,* **37**: 539 (1983).

94. **Mathews-Roth, M.M.** Carotenoid pigment administration and delay in the development of UV-B induced tumors. *Photochemistry and photobiology,* **37**: 509–512 (1983).

95. **Pauling, L. et al.** Incidence of squamous-cell carcinoma in hairless mice irradiated with UV light in relation to intake of ascorbic acid (vitamin C) and of D, L-α-tocopheryl acetate (vitamin E). *In: Third International Symposium on Vitamin C, São Paulo, Brazil, 1980.* Palo Alto, CA, Linus Pauling Institute of Science and Medicine, 1980 (Publication 150).

96. **Freeman, R.G. & Knox, J.M.** Influence of temperature on ultraviolet injury. *Archives of dermatology,* **89**: 858–864 (1964).

97. **Pathak, M.A. & Fitzpatrick, T.B.** The role of natural photoprotection agents in human skin. *In:* Pathak, M.A. et al., ed. *Sunlight and man.* Tokyo, University of Tokyo Press, 1974, p. 725.

98. **Keeler, C.E.** Albinism, xeroderma pigmentosum, and skin cancer. *National Cancer Institute monographs,* **10**: 349 (1963).

99. **Bootsma, D.** Defective DNA repair and cancer. *In:* Castellani, A., ed. *Research in photobiology.* New York, Plenum Press, 1977, pp. 455–468.

100. **Zajdela, F. & Latarjet, R.** Effet inhibiteur de la caféine sur l'induction de cancers cutanés par les rayons ultraviolets chez la souris. *Comptes rendus hebdomadaires des séances de l'Académie des Sciences. Série D: sciences naturelles,* **277**: 1073 (1973).

101. **Stern, R.S. et al.** Cutaneous squamous-cell carcinoma in patients treated with PUVA. *New England journal of medicine,* **310**: 1156–1161 (1984).

102. **Jensen, O.A.** *Malignant melanomas of UVA in Denmark 1943–1952.* Thesis, University of Copenhagen, 1963.

103. **Freeman, R. & Knox, J.M.** Ultraviolet-induced corneal tumors in different species and strains of animals. *Journal of investigative dermatology,* **43**: 431–436 (1964).

104. **Davies, J.N.P. et al.** Cancer of the integumentary tissues in Uganda Africans: the basis for prevention. *Journal of the National Cancer Institute,* **41**: 31–51 (1968).

105. **Tengroth, B. & Vulcan, J.** Welding light. *Strahlentherapie,* **153**: 267–272 (1977).

106. **Cole, C.A. et al.** Increased skin cancer risk from fluorescent lighting. *Photochemistry and photobiology,* **41**: 111S (1985).

107. *Notice of alert: possible hazards from high intensity discharge mercury vapor and metal halide lamps.* Washington, DC, US Department of Health, Education, and Welfare, 1977 (DHEW Publication (FDA) 77-8041).

108. **Johnson, B.E.** Reactions of normal skin to solar radiation. *In:* Jarrett, A., ed. *The physiology and pathophysiology of the skin.* London, Academic Press, 1984, Vol. 8, pp. 2413–2492.

109. **Parrish, J.A. & Jaenicke, K.F.** Action spectrum for therapy of psoriasis. *Journal of investigative dermatology,* **76**: 359–362 (1981).

110. **Cole, C.A. et al.** An action spectrum for photocarcinogenesis. *Photochemistry and photobiology,* **43**: 275–284 (1986).

111. **Davis, A. et al.** Possible dosimeter for ultraviolet radiation. *Nature,* **261**: 169–170 (1976).

112. **Corbett, M.F. et al.** Personnel radiation dosimetry in drug photosensitivity: field study of patients on phenothiazine therapy. *British journal of dermatology,* **98**: 39–46 (1978).

113. **Challoner, A.V.J. et al.** Personnel monitoring of exposure to ultraviolet radiation. *Clinical and experimental dermatology,* **1**: 175–179 (1976).

114. **Diffey, B.L. et al.** UV-B doses received during different outdoor activities and UV-B treatment of psoriasis. *British journal of dermatology,* **106**: 33–41 (1982).

115. **Leach, J.F. et al.** Measurement of the ultraviolet doses received by office workers. *Clinical and experimental dermatology,* **3**: 77–79 (1978).

116. **Holick, M.F.** The cutaneous photosynthesis of previtamin D: a unique photoendocrine system. *Journal of investigative dermatology,* **76**: 51–58 (1981).

117. **MacLaughlin, J.A. et al.** Spectral character of sunlight modulates photosynthesis of previtamin D3 and its photoisomers in human skin. *Science,* **216**: 1001–1003 (1982).

118. **Davie, M. et al.** Vitamin D from skin: contribution to vitamin D status compared with oral vitamin D in normal and anticonvulsant-treated subjects. *Clinical science,* **63**: 461–472 (1982).

119. **Toss, G. et al.** Oral vitamin D and ultraviolet radiation for the prevention of vitamin D deficiency in elderly. *Acta medica scandinavica,* **212**: 157–161 (1982).

120. **van Weelden, H. & van der Leun, J.C.** Lichtinduzierte Lichttoleranz bei Photodermatosen: ein Fortschrittsbericht. *Zeitschrift für Hautkrankheiten,* **58**: 57–59 (1983).

121. *Threshold limit values for chemical substances and physical agents in the workroom environment with intended changes for 1984–85.* Cincinnati, OH, American Conference of Government Industrial Hygienists, 1984.

122. **Bostrom, R.G. & Coakley, J.M.** Guide number for light sources that emit ultraviolet radiation. *Applied optics,* **15**: 574–575 (1976).

123. **Health Council of the Netherlands.** *Acceptable levels of micrometre radiation.* Leidschendam, Ministry of Health and Environmental Protection, 1979 (Report 65E).

124. *UV radiation: human exposure to ultraviolet radiation.* The Hague, Health Council of the Netherlands, 1986 (Report 86/9E).

125. **Pathak, M.A. et al.** Evaluation of topical agents that prevent sunburn: the superiority of PABA and its esters in ethyl alcohol. *New England journal of medicine,* **280**: 1459 (1969).

126. **Mathews-Roth, M.M.** Erythropoietic protoporphyria — the disease, and its treatment with beta-carotene. *In:* Castellani, A., ed. *Research in photobiology.* New York, Plenum Press, 1977, pp. 399–408.

127. **Fritsch, P.E. et al.** Augmentation of oral methoxsalen-photochemotherapy with an oral retinoic acid derivative. *Journal of investigative dermatology,* **70**: 178–182 (1978).

128. **Sporn, M.B. et al.** Prevention of chemical carcinogenesis by vitamin A and its synthetic analogs (retinoids). *Federation proceedings,* **35**: 1332–1338 (1976).

129. **Wilkinson, D.I.** Effect of vitamin A acid on the growth of keratinocytes in culture. *Archives of dermatological research,* **263**: 75–81 (1978).

130. **Sliney, D.H.** The ambient light environment and ocular hazards. *In:* Landers, M.B. et al., ed. *Retinitis pigmentosa.* New York, Plenum Press, 1976, pp. 211–221.

2

Optical radiation, with particular reference to lasers

L. Goldman, S.M. Michaelson, R.J. Rockwell,
D.H. Sliney, B.M. Tengroth, & M.L. Wolbarsht

Revised by D.H. Sliney

CONTENTS

	Page
Introduction	50
Lasers in medicine	52
Laser applications	52
Biological effects	53
Mechanisms	53
General pathophysiology	54
Adverse effects on the eye	55
Adverse effects on the skin	56
Exposure limits	57
Hazard evaluation	66
General procedures	66
Classification of laser device hazards	67
Laser output parameters required for hazard classification	68
Definitions of laser device hazard classes	68
Classification of multi-wavelength and multiple-source lasers	70
Detailed hazard analysis	70
Environment	71
Personnel	72
Control measures	73
Non-laser optical sources	75
Medical assessment	75
Recommendations for further investigations	76
Conclusions and recommendations	79
Conclusions	79
Recommendations	80
References	80

INTRODUCTION

The wavelengths covered by optical radiation range from 1 nm to 1 mm. This wavelength region includes not only the visible part of the electromagnetic spectrum, but also the ultraviolet (UV) down to the soft ionizing X-ray domain, and the infrared (IR) up to the radiofrequency domain. The region from 1 nm to 190 nm (vacuum UV) will not be discussed because it is completely absorbed in air and consequently has no direct biological effect.

There are quite a few reasons for treating this wavelength region as a separate entity, though any exact boundary is to a certain degree arbitrary. Optical radiation is produced by several radiation sources, such as conventional incandescent, fluorescent and phosphorescent lamps, electric arcs, and lasers. It is this last source that has engendered the greatest concern from the point of view of radiation protection.

As indicated, the boundaries of the optical radiation region cannot be precisely defined. Spectral designation schemes differ; the broad spectral regions defined by physicists such as "near" or "far" IR (1) are useful in discussing sources, whereas the CIE bands such as UV-A or IR-B relate to biological effects only (2). Both schemes are used in this chapter.

In contrast to X-rays, optical radiation is essentially nonionizing. Its action is either photochemical (as in the UV) or thermal (as in the IR); the visible region is a transition region characterized by both effects.

In contrast to radiofrequency radiation, optical radiation usually acts at the surface. Penetration of the skin is mostly restricted to a few millimetres or less. The eye is an exception in that it admits visible energy into the body. Even in this case the penetration rarely goes beyond the retinal pigment epithelium.

Until this century, the principal source of optical radiation was the sun, but solar radiation was not considered very dangerous, primarily because of protective avoidance reactions naturally built into the organism and the development of adaptive pigmentation. The development of artificial radiation sources, however, has made the protection problem more urgent. In the early 1920s the first health protection standards were laid down to protect against overexposure to the UV radiation and visible light produced by welding arcs, and against the near-IR radiation related to the so-called "glassblowers' cataract". When Meyer-Schwickerath in 1954 (3) produced retinal lesions with a carbon arc light source, it became impossible to avoid the conclusion that even commonly used equipment can be hazardous. Although hazards and protective procedures relating to lasers will be emphasized in subsequent parts of this chapter, most of the material is also applicable to non-laser optical sources.

The acronym "laser" (Light Amplification by Stimulated Emission of Radiation) is commonly applied to devices that emit an intense, coherent, directional beam of "light" as a result of a process whereby an electron or molecule undergoes a stimulated quantum jump from a higher to a lower energy state, causing a spatially and temporally coherent beam of optical radiation to be emitted. Some types of laser can deliver optical radiation in such short-duration pulses and/or be so restricted to non-visible radiation that the sensory systems are completely bypassed as a protective means. Although the biological effects to be described can be produced in principle by all types of optical radiation source, it should be particularly emphasized that laser sources often present new and more potent health hazards.

In general, a laser consists of an active medium in a resonant optical cavity and a source of excitation energy. To achieve laser action, the active medium is excited. A pattern of electromagnetic waves builds up to a very high intensity in the resonant cavity by reflections from the mirrored end-windows, one of which is deliberately made less reflective than the other. Most electromagnetic waves built up between these two windows radiate out through the less reflective window. The cavity length does not completely determine the output wavelength, which is small compared with the cavity dimensions, but for every system there is an opimum cavity dimension. The output wavelength will always be within the fluorescent line-width of the active material, but the cavity dimensions can favour a particular wavelength.

Various solids, liquids, gases and diode junctions have been found which, by proper choice, allow one to achieve stimulated emission at distinct wavelengths throughout the visible, UV and IR spectrum. Depending on the active medium and system design, the emission duration can vary from single pulses as short as 10^{-14} seconds to continuous wave. As a simplification, in addition to the continuous-wave mode, where the time of exposure may vary from a few milliseconds to seconds, minutes or longer depending on conditions, there are four types of pulsed laser depending on the mode of operation: ultrashort pulses (exposure duration 0.01–15 ps); mode-locked (an envelope of ultrashort pulses in which the envelope duration ranges from 10 to 100 ns); Q-switched (exposure duration between 1 and 100 ns); and normal multiple-spike mode (in which the envelope of random pulses ranges from $100 \mu s$ to 30 ms) (4).

Information on the potentially hazardous effects of optical radiation, and guidelines for practical protective measures against such hazards, are included. In these guidelines the acceptable exposure levels for irradiation are based on a consensus of opinion among experts after evaluation of the currently available biological threshold data. These form the basis for current decisions, permit investigators to judge where additional information is needed, and give responsible experts a starting point for assessing the value of new data. In practice, the use of optical radiation seldom permits personnel to measure readily the parameters relevant to exposure levels; hence more practical means of protection will be proposed, based on the establishment of classes of optical radiation sources defined in terms of the degree of hazard involved.

Secondary effects, which under some circumstances may be much more hazardous than the actual laser radiation, are not dealt with. These fall outside the scope of the present publication because many, if not all, are well recognized and are already the subject of properly codified safety procedures. These secondary effects include electric shock from insufficiently insulated high-voltage parts of lasers, injuries from the cryogenics, implosion or explosion accidents with arc lamps, respiratory hazards from vaporized materials, or falls at construction sites resulting from the dazzling effect of laser beams. All these examples and many more have been reported and are far from imaginary. Protection measures should be taken, of course, but relative to a particular apparatus and its specific method of operation.

LASERS IN MEDICINE

Lasers have been used in ophthalmology for a long time. For photocoagulation, argon and krypton lasers have replaced the ruby laser and the xenon-arc coagulators. With the growing practice of inserting intra-ocular lenses in patients after extra-capsular cataract surgery, the removal of secondary cataract (i.e. cells on the lens capsule that create opacities in the visual axis) with neodymium–YAG lasers has become common. By creating a plasma disruption of membranes, neodymium–YAG lasers have also been used for cutting vitrous strands and preretinal membranes.

The use of the excimer laser for corneal refractive surgery is now being developed.

In general surgery, as well as in urology, the use of CO_2 lasers to replace the scalpel in heavily vascularized tissues is now common. Development in these areas is rapid and other lasers, with or without the use of vital staining to enhance the effect, will soon be in use.

These lasers are also being used in plastic surgery, and a number of different lasers are already used in endoscopic treatment of various disorders. There are stringent regulations in force to protect the eyes of both patients and personnel.

LASER APPLICATIONS

The extremely collimated character and generally high degree of monochromaticity of the laser beam make this device of potential value in industrial, military and communications applications. It is beyond the scope of this chapter to include detailed discussions of all laser applications. Reference to the books by Charschan (5) or Goldman (6) would be useful. It can be generally stated that lasers are used in communications, precision measurement, symbol recognition, radar systems (lidar), guidance systems, range finding, firing simulation, metal working, photography, holography and medicine. The medical uses include treatment of the eye and skin, various diagnostic techniques, and surgery of the skin and internal organs; in dentistry, enamel scaling, bridge work, etc.; in industry, welding, drilling and cutting various materials; in communications, long-distance transmission; and in geodesy, accurate surveying. Some information on laser applications is given in Table 1.

Lasers do not at present constitute an environmental hazard to the uninformed public to the same degree as air pollution, noise or radioactive fallout, except under rather special circumstances such as laser illumination at entertainment spectacles and holographic public displays, range-finding at military installations or commercial airports, satellite tracking, air turbulence and pollution studies, and processing of materials such as metals and plastics. As the use of lasers increases, however, so will the possibility that a larger proportion of the population will be exposed.

52

Table 1. Typical laser applications

Area of use	Personnel involved
Business offices	Office workers
Communications	Communication engineers
Construction	Surveyors, equipment operators, sewer pipe installers, ceiling installers
Dentistry	Researchers, technicians
Geodesy	Aircraft pilots, surveyors
Holography	Researchers, photographers, artists
Materials processing	Processing engineers and machine technicians
Medicine	Physicians, surgeons, paramedical personnel, patients
Military	Researchers, troops during manoeuvres, aircraft pilots
Retail establishments	Sales clerks
Service and maintenance	Repair technicians

BIOLOGICAL EFFECTS

Mechanisms

Several mechanisms are involved in producing a laser lesion. The initial physical effects of laser irradiation are known to include thermal, thermo-acoustic, optical breakdown or photochemical effects. The initial physical trauma is followed by the biological reaction of the tissue itself. The types of physical trauma may differ, but only a few types of biological reaction occur, that is, different types of physical insult may call forth identical physiological reactions from the tissue. This tends to mask the differences in physical causation (7).

There may also be amplifying factors in the biological reactions to the physical trauma. These include reactions to thermally denatured protein or other parts of injured cells, and increased cellular activity resulting from increased tissue temperatures accompanied by diminished cell survival. In the case of the photoreceptors themselves, the stimulation by light itself may cause a similar increase in metabolic rate. This deleterious effect of light may be synergistic with a similar effect caused by a rise in temperature. Many models exist that include both physical and biological processes and each is supported by some data, but no such model explains all types of damage. Indeed, there is good reason to suppose that several types of damage mechanism operate simultaneously or sequentially, so that no single theory can be expected to predict all possible situations where laser damage will occur. Thus, accumulation of large amounts of experimental data has become a prime necessity for the largely empirical formulation of exposure limits.

One important consequence of interaction of a laser beam with tissue is denaturation of protein, the extent of which is related to the incident energy

53

per unit area or power per unit area and duration of exposure. The potential for injury to tissues also depends on the "accessibility" of the tissue to the radiation, which is a function of the depth of penetration of the radiant energy. When laser radiation impinges on tissue, the absorbed energy produces heat, and the resultant rise in temperature can easily denature tissue protein. The absorption of IR, UV or visible energy in tissue is not homogeneous, and the thermal stress is greatest around those portions of tissue that are the most efficient absorbers. Rapid and localized absorption may produce enough heat to boil the tissue water. The resultant steam production can disrupt cells or even produce dangerous pressure changes in an enclosed and completely filled space such as the eye or skull.

Another interaction mechanism is an elastic or thermo-acoustic transient pressure wave. As the laser pulse impinges on tissue the energy, through thermal expansion, produces waves (acoustic transients) that can rip and tear tissue and, if near the surface, can send out a plume of debris from the impact.

Non-linear optical breakdown with the creation of a plasma can occur even in the transparent media of the eye when local irradiances exceed 10^{16} W/m². This process has been employed surgically to cut membranes inside the eye.

Photochemical reactions result in activation of molecules by the capture of quanta of energy. Such capture constitutes the primary event in a photochemical reaction. Some of the photochemical reactions induced by laser exposure may be abnormal, or exaggerations of normal processes.

All of these mechanisms have been shown to operate in the retina, and indeed the safe exposure levels reflect their different effects.

General pathophysiology
An extensive bibiliography on the biological effects and hazards associated with lasers is available, including books by Gamaleja (8), Goldman (6), Sliney & Wolbarsht (9), and Wolbarsht (10–12). Sliney et al. (13) have prepared a bibliography consisting of 2795 references in the published literature organized into subject categories related to general biological effects, effects on the eye and the skin, laser safety, laser propagation in the atmosphere, and laser measurement.

So far, experimental work on animals has not indicated genetic changes, although malignant transformation has been suggested (14–16). Such effects have not as yet been seen in man.

It is generally considered that the biological effects resulting from exposure to lasers are primarily a manifestation of a thermal response or photochemical reaction. Of special interest is the information on acute exposures to the eye and skin which, for the most part, has been obtained from research on experimental animals. There is little documented evidence from accidental exposure of man, but the increasing numbers of incidents occurring during therapeutic use should provide additional information.

The pathophysiological effects of optical radiation may be divided into a small number of categories, depending on the section of the spectrum within the optical radiation domain and on the part of the body affected. Table 2 provides a brief summary of the effects on the eye and on skin.

54

Table 2. Pathophysiological effects of optical radiation

Photobiological spectral domain	Effects on:	
	eye	skin
UV-C (100–280 nm)	Photokeratitis	Erythema (sunburn), skin cancer
UV-B (280–315 nm)	Photokeratitis, photochemical cataract	Increased pigmentation, sunburn, skin cancer, accelerated skin aging
UV-A (315–400 nm)	Photokeratitis, photochemical cataract	Pigment darkening, sunburn, photosensitive reactions, skin cancer, accelerated skin aging
Visible (400–780 nm)	Photochemical and thermal retinal injury	Pigment darkening, photosensitive reactions, skin burn
IR-A (780–1400 nm)	Cataract, retinal burn	Skin burn
IR-B (1400–3000 nm)	Corneal burn, aqueous flare, cataract	Skin burn
IR-C (3000–10^6 nm)	Corneal burn	Skin burn

Adverse effects on the eye

Potentially the most serious hazard from optical radiation is exposure of the eye. This is particularly important in the visible and near-IR regions of the spectrum, but there are also serious problems in the other regions.

Excessive UV exposure produces photophthalmia (photophobia) accompanied by redness, tearing, conjunctival discharge, corneal surface exfoliation and stromal haze. This is the syndrome of photokeratitis, often called "snow blindness" or "welders' flash". In the UV-B and UV-C regions this photokeratitis is the primary result of excessive acute exposure *(17)*. The action of the UV-B and UV-C radiation is usually photochemical rather than thermal, since the resulting temperature rise from exposure appears to be negligible. UV-A can produce erythema *(18)* by both photochemical and thermal mechanisms.

In the visible region the cornea, lens and associated eye media are largely transparent as they neither absorb nor scatter light to any great degree *(19)*. Only a small part, perhaps 5%, of the radiation that passes through the eye media is used for vision; the greater part is absorbed in the pigment granules in the pigment epithelium and the choroid, which underlies the photoreceptors (the rods and cones). The absorbed energy is converted into heat. If this absorbed energy becomes too great, tissue damage (usually referred to as retinal burn) can develop. Until relatively recently, only the sun was bright enough to cause retinal injury, and then only as a result of prolonged viewing. However, the availability of compact arc sources and lasers has greatly increased the danger of retinal burns. There is little doubt that the temperature rise in the chorioretinal tissue is the major factor in causing threshold damage, at least for short exposures.

A retinal injury occurring in the macula must be considered a serious trauma, since the visual functions are most highly developed there. On the other hand, similar damage at the periphery will often have little if any functional significance; even a large blind spot at the periphery has only a trivial effect on vision unless a haemorrhage appears. The damage caused by optical radiation may be one of many degrees of severity, and the extent of subsequent recovery may vary from none to complete restoration of normal function.

A transitional zone between retinal effects and effects on the anterior portion of the eye begins at the far end of the visible region and extends into the IR-A region. In the IR-B region both lenticular and corneal damage is seen. The ocular media become opaque to radiation in the IR-C region as the absorption by water, a major constituent of all cells, is very high in this region. Thus, in this region the damage is primarily to the cornea. For the longer wavelengths the IR damage mechanism appears to be thermal. The CO_2 laser at $10.6\,\mu$m in its action on all materials containing water exemplifies the thermal nature of the damage. In the IR-C region, as in the UV-A and UV-B regions, the threshold for damage to the skin is comparable to that of the cornea. The damage to the cornea, however, is much more disabling and of greater concern (7).

Adverse effects on the skin
The large skin surface makes this structure readily accessible to both acute and chronic exposures to all forms of optical radiation, which can produce skin damage of varying degrees. The use of numerous different types of laser for the treatment of skin disorders in man has been explored rather extensively. Certainly, skin injury is of lesser consequence than eye damage; however, with the expanding use of higher-power laser systems, the unprotected skin of personnel using lasers may be exposed more frequently to hazardous levels of radiation.

The structural inhomogeneities of the skin cause internal scattering of optical radiation in tissues. As a result there will be multiple internal reflections in addition to absorption and transmission of the incident laser beam. For the common laser sources in the 300–1000 nm range, almost 99% of the radiation penetrating the skin will be absorbed in at least the first 3.6 mm of tissue (see Chapter 1).

For wavelengths greater than 400 nm, the reaction of the skin to absorbed optical radiation is essentially that of a thermal coagulation necrosis. This type of injury can be produced by any optical radiation source of similar parameters and is, therefore, not a reaction specific to laser radiation. It is similar in causation and clinical appearance to the tissue reaction of a deep electrical burn. For pulsed laser irradiations, including exposures in the picosecond domain, there may be other secondary reactions in the tissue. Studies (6) have shown that the plume of vaporized tissues or an optical plasma (20) produced by high-level irradiation with laser pulses can attenuate a significant portion of the incident energy. This effectively reduces the amount of absorbed radiation in the tissues.

The principal thermal effects of laser exposure depend on the following factors:

— absorption and scattering coefficients of the tissues at the laser wavelength;

— the irradiance or radiant exposure of the laser beam on the tissues;

— the duration of the exposure;

— the extent of the local vascular flow;

— the size of the area irradiated.

As shown in Table 2, the UV spectrum is divided into three specific regions related to the different biological responses of these regions. In the skin, UV-A can cause erythema and hyperpigmentation. In addition to thermal injury caused by UV energy, UV-A and UV-B, probably by acting directly on DNA, may give rise to photocarcinogenesis.

Few data are available on the reaction of skin exposed to UV radiation from highly monochromatic laser sources. It is known, however, that chronic exposure to narrow-band, non-laser UV wavelengths in this range can result in carcinogenic effects on the skin as well as in a severe erythematous and blistering response. On the basis of studies with non-coherent UV radiation, exposure to UV-B is most injurious to human skin; exposure to the shorter UV-C and the longer UV-A wavelengths seems less harmful. The shorter wavelengths are absorbed in the outer, dead layers of the epidermis (stratum corneum) and the longer wavelengths have an initial pigment-darkening effect followed by erythema and delayed pigmentation. It should be remembered that phototoxic and photosensitizing chemicals on the skin may potentiate the effects of lasers operating in the visible and UV regions.

Studies by Mešter et al. (14–16) and others on the stimulating effect of low doses of the ruby and helium–neon lasers on hair growth, phagocytosis index and wound healing are of interest in any consideration of chronic effects. Such effects have generally been labelled as biostimulation (21).

Studies (22–23) have provided detailed data on the exposure levels required to produce minimal reactions in human skin for six common laser types emitting in the visible and IR regions. The variations, or spread, in the data were found to be directly related to the degree of absorption in the tissues (Table 3).

EXPOSURE LIMITS

To establish a rationale for developing exposure limits (EL) from biological data, a careful analysis is required of the physical and biological variables that influence the spread of laboratory data, the factors related to the potential for injury in individuals exposed, the increase in severity of injury from greater-than-threshold exposures, and the reversibility of injury. In

57

Table 3. Minimal skin reaction levels[a]

Type of laser	Radiant exposure (kJ/m²)	Exposure duration, t
Ruby laser (694.3 nm)		
unpigmented skin	110–200	2.5 ms
pigmented skin	22–69	2.5 ms
Q-switched ruby laser (694.3 nm)	2.5–3.4	75 ns
Argon laser (458–515 nm)	40–82	1 s
Carbon dioxide laser (10.6 μm)	28	1 s
Neodymium glass laser (1060 nm) —		
Q-switch	25–57	75 ns
Neodymium–YAG laser (1064 nm)	460–780	1 s
Nitrogen (334 nm)	220	210 s

[a] All taken at 50% probability points.

Source: Parrish *(24)*.

addition, the accuracy and availability of measuring instruments and the desire for simplicity in expressing exposure levels influence exposure limits.

The development of adequate and operable EL requires comprehensive evaluation of information obtained from animal experiments and studies on man. Some data may be obtained from volunteers. Surveys of individuals engaged in laser work will give assurance that EL are safe, but may be of limited value in developing standards. The criteria to be used in evaluating experimental results of laser exposure and the interacting variables in such an evaluation require the exercise of informed judgement.

The availability of a broad spectrum of laser sources, although generally advantageous, is a disadvantage when attempting to provide a uniform "all-purpose" laser safety code, as more detailed and specific data are required to establish: *(a)* operational parameters of the lasers in use; *(b)* safe exposure criteria for each laser; and *(c)* protective devices necessary for these lasers. Similarly, those responsible for establishing levels of safe exposure as well as those who manufacture protection devices must also keep pace with advances in technology *(4)*.

Current protection guides for the eye are based on minimum exposure factors. These threshold values may vary depending on the criteria by which they are measured. In the retina, the criterion is the minimum level required to produce an ophthalmoscopically visible lesion. Other criteria are used for the cornea, lens, etc. The protection guides are based primarily on acute exposure effects, namely thermal injury to the retina, keratitis and dermatological effects. There is only a limited amount of information on long-term effects of chronic exposure of the eye/retina to monochromatic light. Results of recent research suggest a possible loss of small-angle visual acuity following repeated exposure to very low levels of diffusely scattered blue-spectrum laser radiation *(9,12,21)*.

Unfortunately, since the various experiments reported in the literature were not carried out as part of an integrated programme of investigation, there are numerous gaps and inconsistencies in reported results. Such parameters as pulse duration and irradiated spot area or diameter on the retina vary from one study to another.

As mentioned previously, there are at least three principal threshold mechanisms of retinal injury: thermo-acoustic, thermal and photochemical. It is now believed that each mode of injury has a particular temporal relationship or time domain in which it is the principal cause of threshold injuries. Threshold retinal lesions from very short exposures, such as those from mode-locked and Q-switched lasers, probably result in part from a thermo-acoustic transient which accompanies the localized heating in the vicinity of the highly absorbing pigment granules (25). Somewhere in the domain of pulse durations of the order of 1 μs, the acoustic transient no longer plays a significant role and the principal process is that of thermal denaturation of complex organic molecules, although the pigment granules are still localized hot spots (26,27). At ultrashort pulse durations characteristic of mode-locked lasers, optical breakdown can also occur.

Fig. 1 shows the most reliable information from several laboratories for minimum-image-size retinal injury. The two curves represent different

Fig. 1. Laser retinal injury threshold for minimal image condition

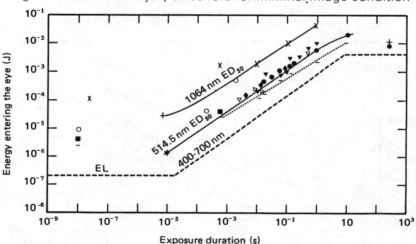

Note. Selected data from numerous experiments to determine the laser retinal injury threshold in the rhesus monkey for the minimal image condition. Plotted for comparison is the exposure limited (EL) applicable to intrabeam viewing (minimal image) condition. Data points represent ED$_{50}$ values of Ham et al. *(28)* for helium–neon (▼); Dunsky & Lappin *(29)* for krypton (●); Bresnick et al. *(30)* for argon (▲); Vassiliadis et al. *(31)* for neodymium 530 nm (■), ruby (○); Vassiliadis et al. *(32)* for argon 514.5 nm (◆); Lappin & Coogan *(33)* for helium–neon 632.8 nm (▽); Naidoff & Sliney *(34)* for welding-arc point source (+); Skeen et al. *(35)* for neodymium 1064 nm (✕); and Skeen et al. *(36)* for argon 514.5 nm (✳). Points marked (—) indicate lowest reported injury values.

laser wavelengths (514.5 nm and 1064 nm) with exposure durations from about 10 ns to 1000 s. The apparent break in the two curves in the microsecond region indicates the change from the thermo-acoustic to the purely thermal effect. The ocular EL for optical radiation at 514.5 nm is shown for comparison.

A similar display of data for extended sources is shown in Fig. 2. The ocular EL for extended sources and the smoothed threshold for a minimum image size are shown for comparison.

The differences in shape between the threshold curves for the two conditions necessitate different EL curves for each case. In Fig. 3 the basis for the spectral dependence of the EL is shown. The relative absorption by the retina, including the pigment epithelium, is closely approximated by the modification factor (M_A) line. At wavelengths below 450 nm and above 1100 nm the approximation diverges. This divergence is time-dependent and EL are adjusted to follow closely the data points on a wavelength and exposure–duration basis.

Exposures to visible radiation for durations greater than 1–100 seconds appear to result in injury if EL are much above those encountered in the natural environment. The most sensitive mechanism in this exposure domain appears to be some variety of photochemical reaction to the pigment epithelium. Several mechanisms have been proposed to account for damage

Fig. 2. Laser retinal injury threshold for extended sources

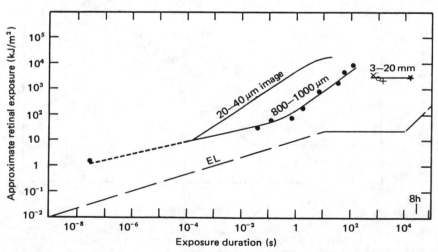

Note. Experimental retinal burn threshold determined by Ham et al. (37) in the rabbit for large image sizes, 0.8–1 mm diameter, are shown with long-term exposure data for larger image sizes obtained by Verhoeff & Bell (38) (+); Eccles & Flynn (39) (O); Kuwabara (40) (X); and Lawwill (41) (*). The exposure limits (EL) for extended sources and the threshold data curves for small image size at 514.5 nm from Fig. 1 are shown for comparison.

Fig. 3. Relative absorption by the retina and modification factor

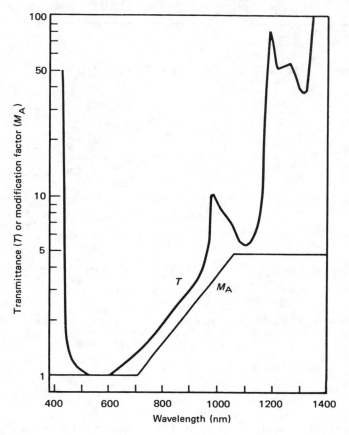

Note. A normalized plot of the reciprocal of the retinal absorption of optical radiation incident on the cornea, based on the data of Geeraets & Berry *(19)* and Boettner *(42)*. The M_A curve used to calculate the EL is shown for comparison. The plateau in the M_A curve from 1050 nm is based on a presumed effect on the ocular lens.

at this level. These are probably interrelated, but the most important appears to be absorption of light by the pigment epithelium or photo-receptor itself. Although ambient retinal temperature is important, it is contributory or synergistic rather than the principal factor *(43–45)*.

Fig. 4 shows the reduction in EL necessitated by the interaction of multiple pulses within a pulse train. The data from later work by Ham (personal communication) validate the continued high level of interaction at frequencies higher than 1000 Hz *(9)*.

Fig. 4. Modification factor for repetitively pulsed lasers having pulse durations shorter than 10^{-5} seconds

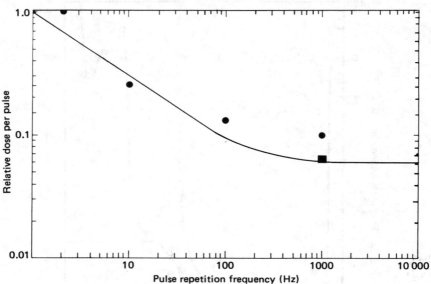

Note. The exposure level for a single pulse of the pulse train is multiplied by the above modification factor. The modification factor for a pulse repetition frequency greater than 276 Hz is 0.06. Experimental data: ● — argon; ■ — neodymium.

Source: Adapted from IEC *(46)* and Wilkening *(47)*.

The corneal damage levels in the UV are shown in Fig. 5; the EL is the envelope of the human data. It is interesting to note that the non-human primate (monkey) data correlate more closely with the human data than do the rabbit data.

Obviously more data were used as a basis for the EL than those cited above. This information is comprehensively reviewed by Wolbarsht & Sliney *(7)* and Clarke *(51)*. In certain regions, however, few data are available. The situation is discussed later in this chapter and some suggestions concerning the research needed in order to fill the gaps are included in this discussion.

As additional information becomes available the EL will be revised. The development of more or less permanent optical radiation protection standards is thus still not possible. In recent years, however, considerable progress has been made in many countries towards recommended levels. This progress has been greatest for the visible and IR regions, in which many lasers and the xenon arc lamps now operate and where the principal biological data have been collected. The members of the working group that discussed

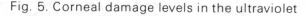
Fig. 5. Corneal damage levels in the ultraviolet

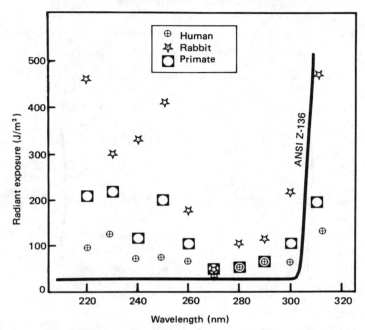

Note. The UV photokeratitis threshold obtained by Pitts & Gibbons *(48)* for a broadband (10 nm bandwidth) source is shown in comparison to the EL for UV laser radiation (solid line) *(49)*. The apparent discrepancy in the 303–315 nm range is due to the rapid increase of the biological threshold which could not be adequately tracked using a broad-band source *(50)*.

this chapter felt that, for the most part, the revised scheme of protection standards developed by the American National Standards Institute (ANSI) came closest to meeting the present need for EL. These values have also been adopted by IEC *(46)* and were recommended by the International Non-Ionizing Radiation Committee (INIRC) of the International Radiation Protection Association (IRPA) in 1985 *(52)*, although with some reservations where only a minimal amount of biological data exists and where the limits are based on extrapolation. These limits are listed in Tables 4–7. It is felt that these values are suitable for acute exposures, since the limits were based principally on considerations of immediate and easily assessed effects. However, where photochemical processes are involved (UV exposure of the eye and skin and long-term visible radiation exposure of the eye and skin) the EL should be considered only as guidelines to be used with caution if applied for long-term exposure conditions (i.e. frequently pulsed exposures each day for months or years, or chronic continuous-wave exposure for such periods.

Table 4. Exposure limits for direct ocular exposures (intrabeam viewing) from a laser beam[a]

Wavelength, λ (nm)	Exposure duration, t (s)	Exposure limit, EL
200– 302		0.03 kJ/m²
303		0.04 kJ/m²
304		0.06 kJ/m²
305		0.10 kJ/m²
306		0.16 kJ/m²
307		0.25 kJ/m²
308	10^{-9}– 3×10^{4}	0.40 kJ/m²
309		0.63 kJ/m²
310		1.00 kJ/m²
311		1.60 kJ/m²
312		2.50 kJ/m²
313		4.00 kJ/m²
314		6.30 kJ/m²

(For wavelengths 200–314 nm: Not to exceed $5.6\sqrt[4]{t}$ kJ/m²)

Wavelength, λ (nm)	Exposure duration, t (s)	Exposure limit, EL
315– 400	10^{-9} – 10	$5.6\sqrt[4]{t}$ kJ/m²
315– 400	10 – 10^{3}	10.0 kJ/m²
315– 400	10^{3} – 3×10^{4}	10 W/m²
400– 700	10^{-9}–1.8×10^{-5}	0.005 J/m²
400– 700	1.8×10^{-5}– 10	$18(t/\sqrt[4]{t})$ J/m²
400– 550	10 – 10^{4}	100 J/m²
550– 700	10 – T_{1}	$18(t/\sqrt[4]{t})$ J/m²
550– 700	T_{1} – 10^{4}	$100M_{B}$ J/m²
400– 700	10^{4} – 3×10^{4}	$0.01M_{B}$ W/m²
700–1050	10^{-9}–1.8×10^{-5}	$0.005M_{A}$ J/m²
700–1050	1.8×10^{-5}– 10^{3}	$18M_{A}(t/\sqrt[4]{t})$ J/m²
1050–1400	10^{-9}– 5×10^{-5}	0.05 J/m²
1050–1400	5×10^{-5}– 10^{3}	$90(t/\sqrt[4]{t})$ J/m²
700–1400	10^{3} – 3×10^{4}	$3.2M_{A}$ W/m²
1400– 10^{6}	10^{-9}– 10^{-7}	100 J/m²
1400– 10^{6}	10^{-7}– 10	$5.6\sqrt[4]{t}$ kJ/m²
1400– 10^{6}	10 – 3×10^{4}	1 kW/m²

[a] The limiting aperture for all EL for wavelengths in the range 100–1000 μm is 10 mm. For all other skin EL and for UV, and for IR-B and IR-C ocular EL, the limiting aperture is 1 mm. For ocular EL in the visible and IR-A region the limiting aperture is 7 mm. Modification factors are: $M_{A} = 1$ for $\lambda = 400$–700 nm, $M_{A} = 10^{(0.002(\lambda - 700))}$ for $\lambda = 700$–1050 nm, $M_{A} = 5$ for $\lambda = 1050$–1400 nm (and see Fig. 3); $M_{B} = 1$ for $\lambda = 400$–550 nm, $M_{B} = 10^{(0.015(\lambda - 550))}$ for $\lambda = 550$–700 nm; $T_{1} = 10$ seconds for $\lambda = 400$–550 nm, $T_{1} = 10 \times 10^{(0.02(\lambda - 550))}$ seconds for $\lambda = 550$–700 nm.

Source: ACGIH *(53)*.

Table 5. Exposure limits for viewing a diffuse reflection
of a laser beam or an extended source laser[a]

Wavelength, λ (nm)	Exposure duration, t (s)	Exposure limit, EL
400– 700	10^{-9} – 10	$100\sqrt[3]{t}$ kJ/(m²·sr)
400– 550	10 – 10^4	210 kJ/(m²·sr)
550– 700	10 – T_1	$38.3(t\sqrt[4]{t})$ kJ/(m²·sr)
550– 700	T_1 – 10^4	$210 M_B$ kJ/(m²·sr)
400– 700	10^4 –3 × 10^4	$0.021 t M_B$ kW/(m²·sr)
700–1400	10^{-9} – 10	$100 M_A \sqrt[3]{t}$ kJ/(m²·sr)
700–1400	10 – 10^3	$38.3 M_A (t/\sqrt{t})$ kJ/(m²·sr)
700–1400	10^3 –3 × 10^4	$6.4 M_A$ kW/(m²·sr)

[a] M_A, M_B and T_1 are the same as in the footnote to Table 4.

Source: ACGIH (53).

Table 6. Minimum limiting angle of extended
source viewing exposure limits[a]

Exposure duration, t (s)	Angle (mrad)
10^{-9}	8.0
10^{-8}	5.4
10^{-7}	3.7
10^{-6}	2.5
10^{-5}	1.7
10^{-4}	2.2
10^{-3}	3.6
10^{-2}	5.7
10^{-1}	9.2
1.0	15
10	24
10^2	24
10^3	24
10^4	24

[a] For all angles less than limiting angle, use intrabeam viewing exposure limits.

Source: ACGIH (53).

Table 7. Exposure limits for skin exposure from a laser beam[a]

Wavelength, λ (nm)	Exposure duration, t (s)	Exposure limit, EL
200– 400	10^{-9}–3×10^4	same as Table 4
400–1400	10^{-9}– 10^{-7}	$0.2M_A$ kJ/m²
400–1400	10^{-7}– 10	$11M_A\sqrt[4]{t}$ kJ/m²
400–1400	10 –3×10^4	$2M_A$ kW/m²
1400– 10^6	10^{-9}–3×10^4	same as Table 4

[a] The limiting aperture for all skin EL for wavelengths in the range 100–1000 μm is 10 mm. For all other skin EL and for UV, and for IR-B and IR-C ocular EL, the limiting aperture is 1 mm. M_A is the same as in the footnote to Table 4.

Source: ACGIH (53), IEC (46) and IRPA/INIRC (52).

It should be emphasized that where photochemical injury mechanisms are dominant, the presence of photosensitizers (e.g. drugs, foods, cosmetics) and diseased states or genetic abnormalities may render the EL inapplicable.

HAZARD EVALUATION[a]

General procedures
The applicaton of the EL in evaluating the potential hazard of high-intensity optical sources requires information regarding the use and frequency of exposure. The analysis is largely orientated towards laser sources, but is not restricted to them. The following four aspects of the use of the source influence the total hazard evaluation and thereby determine the application of control measures.

1. The intrinsic capability of the laser or other optical source to injure personnel.

2. The environment in which it will be used.

3. The personnel who operate the hazardous optical source and those who may be exposed.

4. The intended use of the laser or other optical source.

[a] Adapted from ACGIH (54).

66

The most practical general means for both evaluation and control of laser radiation hazards is to classify the laser systems according to their relative hazards and to draw up specifications for appropriate controls for each class. Reference to the classification scheme in most cases precludes any requirement for radiometric measurements and will generally reduce the need for calculations. In the standardized laser classification scheme, aspect 1 (the potential hazard of the laser or laser system) is defined. Aspects 2 and 3 vary with each laser usage and cannot be readily included in a general classification scheme. Although, in total hazard evaluation procedures, all four aspects must be considered, in most cases aspects 1 and 4 are sufficient to determine the control measures applicable.

Classification of laser device hazards

The hazard classifications specified below are defined by the output parameters and accessible levels of radiation. This classification (Table 8) is based largely on that of the IEC *(46)* the US Food and Drug Administration *(55)* and that used by ANSI *(49)*, with arabic numerals instead of roman. It should be noted that the laser device classification may appear on many commercial laser products manufactured subsequent to the adoption of these standards. It should be used (with conversion of roman numerals to arabic) unless the laser is modified so as to change its output power or energy significantly. The classes are:

1 — laser systems that are not hazardous (non-risk);

2 — laser systems (visible only) that are normally not hazardous by virtue of normal aversion responses (low-risk);

3 — laser systems where intrabeam viewing of the direct beam and specular reflections may be hazardous (moderate-risk), sometimes divided into two subcategories a and b, where class 3a represents a low risk (equivalent to class 2) and is hazardous only if the beam is re-collected by an optical instrument;

4 — laser systems where even diffuse reflections may be hazardous or where the beam produces a fire hazard or serious skin hazard.

The basis of the hazard classification scheme is the ability of the primary laser beam or reflected beam to cause biological damage to the eye or skin.

A class 2 laser or low-power system may be viewed directly under carefully controlled exposure conditions, but must have a cautionary label affixed to the external surface of the device. Similar controls are also required for a class 3a laser.

The moderate-risk class 3b category (or medium-power system) requires control measures to prevent viewing of the direct beam.

Class 4 high-risk (or high-power) systems require the use of controls that prevent exposure of the eye and skin to the direct and diffusely reflected beam. In addition to the possibility of eye damage, exposure to optical radiation from such devices could constitute a serious skin hazard.

Table 8. Laser device classification

Wavelength	Laser classes			
	Non-risk	Low-risk, low- or medium-power	Moderate-risk medium power	High-risk, high-power
Ultraviolet	1	3a	3b	4
Visible	1	2, 3a	3b	4
Near-infrared	1	3a	3b	4
Infrared	1	3a	3b	4

Laser output parameters required for hazard classification

Accessible emission limits (AELs) have been established for each class of laser. The following parameters are required for the classification of the different types of laser.

1. Essentially *all* lasers: wavelength(s) or wavelength range, and a determination of the exposure duration.

2. *Continuous-wave* or *repetitively pulsed* lasers: as above, but knowledge of average power output also required.

3. *Pulsed* lasers: as above, but knowledge of total energy/pulse (or peak power), pulse duration, pulse repetition frequency, and emergent beam radiant exposure also required.

4. *Extended-source* laser devices, such as injection laser diodes and those lasers having a permanent diffuser within the output optics: all of the above parameters, but knowledge of the laser source radiance or integrated radiance, and the maximum viewing angular subtense, α, also required.

Definitions of laser device hazard classes

Class 1 — non-risk laser devices

A non-risk laser device is defined as any laser, or laser system containing such a laser, that cannot emit laser radiation levels in excess of the AEL for class 1 (see below) for the classification duration. The exemption from hazard controls applies strictly to emitted laser radiation hazards and not to other potential hazards. The classification duration is the longest daily exposure duration expected.

The AEL for class 1 is defined by a "worst-case" analysis of a laser's potential for producing injury. In this "worst-case" analysis it is necessary

68

to consider not only the laser output irradiance or radiant exposure, but also whether a hazard would exist if the total laser output were concentrated within the defining aperture for the applicable exposure limit. For instance, the unfocused beam of a far-IR continuous-wave laser would normally not be hazardous if the beam irradiance were less than 0.1 W/cm^2; however, if the output power were 10 W and the beam were focused at some location to a spot 1 mm in diameter, a serious hazard could exist. The AELs for class 1 must be defined in two different ways, depending on whether the laser itself is considered an "extended source" (an unusual case).

For most lasers, the AEL for class 1 is the product of $a \times b$, where a is the intrabeam exposure limit for the eye (Table 4) for the exposure duration T_{max} and b is the circular area of the defining aperture for the exposure limit in cm^2.

For extended-source lasers (e.g. laser arrays, laser diodes and diffused-output lasers that emit in the spectral range 400–1400 nm) the AEL for class 1 is determined by a power or energy output such that the source radiance would not exceed the extended source exposure limit (Tables 5 and 6) if the source were viewed at the minimum viewing distance through a theoretically perfect optical viewing system with an entrance aperture of 8 cm which collected the entire laser beam output and which had a 7 mm exit pupil. This AEL is seldom necessary, and the point-source AELs can be applied to provide a conservative analysis.

Class 2 — low-risk, low-power visible laser devices
Low-risk, low-power visible laser devices are defined as follows:

(a) visible continuous-wave laser devices that can emit a power exceeding the AEL for class 1 for the classification duration (0.4μW for T_{max} greater than 0.25 seconds) but not exceeding 1 mW;

(b) visible scanning laser systems may be evaluated by specifying the AEL for class 1 at a point 10 cm from the exit port of the laser system; these and repetitively pulsed laser devices that can emit a power exceeding the appropriate AEL for class 1 for the classification duration, but not exceeding the AEL for class 1 for a 0.25-second exposure, are low-risk. This AEL can also be referred to as the AEL for class 2.

Any laser device in a low-risk classification by virtue of enclosure must have warning labels indicating "Higher-risk class when access panels are removed". These labels may be covered by a separate enclosure which must be removed before the main access panels can be removed.

Class 3a — low-risk, medium-power laser devices
This class of laser devices is defined as visible-frequency continuous-wave lasers, operating in a power range of 1–5 mW, which have an irradiance in the emergent beam of 25 W/m^2 or less. In some standards the class applies to non-visible frequency lasers within five times the AEL for class 1.

69

Class 3b — moderate-risk, medium-power laser devices
Moderate-risk, medium-power laser devices are defined as:

(a) UV and IR (1.4μm–1 mm) laser devices that can emit a radiant power in excess of the AEL for the classification duration but cannot emit:
— an average radiation power in excess of 0.5 W for T_{max} greater than 0.25 seconds; or
— a radiant exposure of 100 kJ/m² within an exposure duration of 0.25 seconds or less;

(b) visible continuous-wave or repetitively-pulsed laser devices that produce a radiant power in excess of the AEL for class 2 (1 mW for a continuous-wave laser) but cannot emit an average radiant power of 0.5 W for T_{max} greater than 0.25 seconds;

(c) visible and near-IR (400–1400 nm) pulsed laser devices that can emit a radiant energy in excess of the AEL for class 1 but cannot emit a radiant exposure that exceeds either 10 J/cm² or that required to produce a hazardous diffuse reflection as given in Table 5;

(d) near-IR (700–1400 nm) continuous-wave or repetitively-pulsed laser devices that can emit power in excess of the AEL for class 1 for the classification duration but cannot emit an average power of 0.5 W or greater for periods in excess of 0.25 seconds.

Class 4 — high-risk, high-power laser devices
High-risk, high-power laser devices are defined as:

(a) UV and IR (1.4μm–1 mm) laser devices that emit an average power in excess of 0.5 W for periods greater than 0.25 seconds, or a radiant exposure of 100 kJ/m² within an exposure duration of 0.25 seconds or less;

(b) visible and near-IR (700–1400 nm) laser devices that emit an average power of 0.5 W or more for periods greater than 0.25 seconds, or a radiant exposure in excess of either 100 kJ/m² or that radiant exposure output required to produce a hazardous diffuse reflection (3.14 times the radiance values) as given in Table 5.

Classification of multi-wavelength and multiple-source lasers
The classification of laser devices that can potentially emit at numerous wavelengths should be based on the most hazardous possible wavelength combination. Multiple sources are considered independent if separated by the appropriate limiting angle given in Table 6.

Detailed hazard analysis
Classification is the initial step in hazard analysis. However, it is not sufficient merely to classify the laser in terms of its power or energy output; the place and way that a laser is used (or abused) as well as the people who may operate it or be in the exposure zone must also be considered. The additional safety measures that such environmental and personnel factors may require must be taken into account, and are discussed below.

70

Environment

Environmental factors require careful consideration after the laser device has been classified, as their importance in the total hazard evaluation depends on the laser classification. The decision to employ additional hazard controls not ordinarily required for moderate-risk and high-risk laser devices may depend largely on environmental considerations. The probability of exposure of personnel to hazardous laser radiation must be considered separately since it is influenced by the laser's use: indoors, as in a machine shop, a classroom, a research laboratory, or a factory production line; or outdoors, as in a mine, at a highway construction site, on the open sea, on a military laser range, in the atmosphere above occupied areas, or in a pipeline construction trench. Other environmental hazards should also be considered. If exposure of unprotected personnel to the primary or specularly reflected beam is expected, calculations or measurements of either irradiance or radiant exposure of the primary or specularly reflected beam, or radiance of an extended source, at that specific exposure location are required.

Indoor laser operations

In general only the laser source itself is considered in evaluating an indoor laser operation if the beam is enclosed or is operated in a controlled area. The following step-by-step procedure is recommended for evaluation of moderate-risk laser devices indoors when this is necessary, since unprotected personnel may potentially be exposed with this particular class of laser devices.

Step 1. Determine the applicable AEL considering the maximum exposure duration from the intended use.

Step 2. Determine the hazardous beam path(s).

Step 3. Determine the extent of hazardous specular reflection as indicated in Fig. 6.

Step 4. Determine the extent of hazardous diffuse reflections (nominal hazard zone).

Step 5. Determine whether any non-laser hazards exist.

Outdoor laser operations over extended distances

The total hazard evaluation of a particular laser device depends on defining the extent of several potentially hazardous conditions. They may be done in a step-by-step manner as follows.

Step 1. Determine the applicable AEL considering the maximum exposure duration from the intended use.

Step 2. Estimate the nominal hazardous range of the laser.

71

Fig. 6. Specular reflection of laser beams

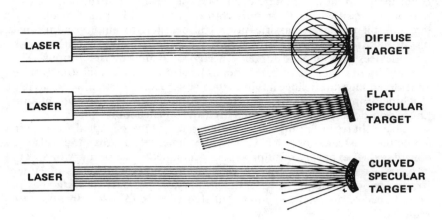

Source: ACGIH (53).

Step 3. Evaluate potential hazards from specular-surface reflections, such as those from windows and mirrors in vehicles, and hazards from retroflectors.

Step 4. Determine whether hazardous diffuse reflections exist, especially if the laser is operating in the 400–1400 nm band (nominal hazard zone).

Step 5. Evaluate the stability of the laser platform to determine both the extent of horizontal and vertical range control and which, if any, of the azimuth and elevation constraints need to be placed on the beam traverse.

Step 6. Determine the likelihood of people being present in the area of the laser beam.

Personnel

The individuals who may be in the vicinity of a laser and its emitted beam(s) can influence the decision to adopt additional control measures not specifically required for the class of laser being employed. This again depends on the classification of the laser device.

If children or others unable to read and/or understand warning labels are exposed to potentially hazardous laser radiation, the hazard evaluation is affected and control measures could require appropriate modification.

72

The type of personnel influences the total hazard evaluation, especially with the use of moderate-risk, medium-power lasers. The principal means of hazard control for certain lasers or laser systems, such as military laser range-finders and some moderate-risk lasers used in the construction industry, is for the operator to keep the laser beam away from personnel or flat, mirror-like surfaces.

The factors to be taken into account with regard to personnel who may be exposed are briefly as follows:

(a) the maturity and general level of training and experience of the laser users (e.g. students, master machinists, soldiers and scientists);

(b) the maturity of onlookers, their awareness that potentially hazardous laser radiation may be present, and their knowledge and ability to apply relevant safety precautions;

(c) the degree of training in laser safety of all individuals involved in laser operation;

(d) the extent to which individuals can be relied on to wear eye protection;

(e) the steps taken to ensure that intentional laser exposures are within the permissible range and that the fail-safe type of attenuator is used when required in the direct beam;

(f) the number and location of individuals relative to the primary beam or reflections, and the probability of accidental exposure.

CONTROL MEASURES

Approaches to laser safety vary greatly among individuals and groups who have an interest in the problem. Most programmes in industry, government and universities are still in course of development. Some organizations have written policies and practices outlining the responsibilities of management and of technical supervision, environmental health, safety and medical personnel. Such policies are usually broadly defined, with specific provisions for individual problems. All such policies and procedures should emphasize the need to rely primarily on appropriate education and training, both of the individual laser operator and of supervisory personnel, for the safe conduct of laser operations; when appropriate, engineering controls rather than personal protective equipment (goggles) should be stressed. Engineering measures should take into account the need for interlocks, proper layout of room areas, shielding materials and warning signs. The criteria for selecting protective eyewear involve many interrelated factors. It should be noted that commercially available protective eyewear is designed for protection against a specific wavelength or group of wavelengths *(47,56,57)*. Eye protection devices designed for protection against specific wavelengths and power from the laser system should be used when engineering and procedural controls are inadequate, so as to eliminate potential

73

exposure in excess of the applicable exposure limit. For cases in which long-term exposure to the eye by visible lasers (only) is not intended, the applicable exposure limit may be based on a 0.25-second duration.

The International Electrotechnical Commission (46), the American Conference of Governmental Industrial Hygienists (54) and the Laser Institute of America (58) have prepared guides for laser installations, and the American National Standards Institute (49) and other national bodies have developed a detailed personnel exposure standard for laser users. These documents give hazard controls for laser radiation that vary depending on the type of laser being used and the manner of its use. The control of laser operation should be entrusted to a knowledgeable laser operator under the supervision of personnel knowledgeable in laser hazards. A closed installation should be used when feasible.

In the above-mentioned guides and standards, only two general precautions are common to all laser installations:

— personnel should not look into the primary beam or at specular reflections of the beam, unless necessary, even if the exposure limit is not exceeded;

— the laser operator should be familiar with the type of laser used and act responsibly.

Fundamental responsibility for laser safety lies with the employing authority. In practice, the employing authority may delegate its responsibility for the establishment and surveillance of appropriate safety measures to a responsible individual. Provision should be made for the appropriate education and training of all individuals using laser devices.

Consideration should be given to the operation of laser devices in a controlled area according to laser classification. Special emphasis should be placed on control of the path of the laser beam. Only authorized personnel should operate laser systems. Spectators should not be allowed to enter a controlled area unless appropriate supervisory approval has been obtained and protective measures taken.

Laser optical systems (mirrors, lenses, beam deflectors, etc.) should be aligned in such a manner that the primary beam, or a specular reflection of the primary beam, cannot result in an ocular exposure above the exposure limit for direct irradiation of the eye.

Optical systems such as lenses, telescopes and microscopes may increase the hazard to the eye when viewing a laser beam, so that special care should be taken in their use. Microscopes and telescopes may be used as optical instruments for viewing, but should be provided with an interlock or filter, if necessary, to prevent ocular exposures above the appropriate exposure limit for irradiation of the eye.

With non-visible laser beams, extra vigilance is necessary to ensure that the beam path is properly positioned and that dangerous specular reflections do not occur. This may entail continuous environmental monitoring.

Laser medical instrumentation for surgery or for diagnostic purposes should have built-in safety devices, including special firing mechanisms, and warning notices as to the need for eye protection and protection of the patient, including the use of non-flammable gas anaesthesia. Safety precautions for any electrical equipment should be included. Laser surgical devices for training purposes should have dual controls. A suitable training programme should be provided for all potential users and operating room personnel. The control measures should not restrict or limit in any way the use of laser radiation of any type that may be intentionally administered to an individual for diagnostic, therapeutic or research purposes, by or under the direction of qualified professionals engaged in the healing arts. Precautions should be taken to ensure that any unnecessary exposure of organs or tissues is minimized.

With the increase in medical and industrial applications of high-power laser systems, there is an increased probability of accidental exposure of the skin to levels of laser radiation above the exposure limit for skin. It is recommended that, for personnel working with such high-power (class 4) laser systems, protection should be provided for the uninvolved skin wherever possible.

Non-laser optical sources
Although lasers pose the greatest potential hazard to the eye and skin among artificial light sources, some arc sources and other high intensity light sources may emit hazardous levels of visible radiation. UV radiation hazards are considered in Chapter 1 and IR radiation hazards in Chapter 3. The exposure limits and protective techniques developed for lasers can be adapted to the evaluation and control of non-laser sources. More sophisticated hazard criteria for evaluating a thermal and photochemical retinal injury hazard have been proposed, but have not been recommended as official standards (9,53).

MEDICAL ASSESSMENT

In the early days of laser use there was a general uncertainty about threshold concepts and associated safe exposure levels. This resulted in a conservative attitude towards possible health problems, and therefore in the widespread adoption of detailed and regular medical surveillance. In the past decade a large volume of empirical data has been collected concerning the possible risks involved in most common laser applications. In addition to threshold studies, the independent evaluation of medical examinations by the members of the WHO Working Group led to the following conclusions:

— it is unlikely that a near-threshold retinal lesion will be identified as such by ophthalmoscopic examination, even if carried out by an ophthalmologist experienced in laser problems;

— most near-threshold laser lesions will not be detected by the exposed individual when the macular region of the retina is unaffected; in most cases it is impossible to differentiate between laser-induced and other retinal lesions and pathologies if more than one week has elapsed since the possible exposure;

— if retinal change is identified, no therapy can be offered;

— if gross damage to the retina, or significant damage to other ocular components has occurred, the exposed individual will be aware of it.

In many countries medical examinations are performed regularly or are at least required for personnel handling laser equipment. In particular, an ophthalmological examination is performed, including tests of visual acuity and visual fields, together with funduscopy and sometimes even fundus photography. It must be realized that the expected ocular changes are often subtle, and that without any clear previous history of a laser hazard an ophthalmologist will have great difficulty in distinguishing an eclipse burn or an early macular degeneration from a laser-induced injury.

From a legal point of view, it will be difficult to relate any ocular or skin change to work with the laser if the hazardous situation cannot be reconstructed in a precise way.

An epidemiological analysis is a very important part of laser hazard evaluation, and an assessment of the individual's health status at the commencement of employment in the laser field is needed as the basis for all future investigations.

In view of the limited amount of information gained from surveillance examinations, and considering the amount of time that has to be devoted to them by highly qualified personnel, it is recommended that skin and eye examinations be carried out on laser workers only when a medical examination is a condition of employment; this requirement has, however, to be waived in the case of class 1 and class 2 lasers. It is also recommended that a medical examination by a qualified expert be carried out immediately after the alleged occurrence of a supra-threshold exposure. Such an examination should be supplemented by a full investigation of the circumstances under which the accident occurred. Results from both these studies should be referred to a central agency and the necessary steps taken to prevent recurrence of similar accidents.

RECOMMENDATIONS FOR FURTHER INVESTIGATIONS

Several laser radiation protection standards and guides have been published (21,45,49,54,55,58,59) and exposure limits have been formulated even though several unresolved problems remain with regard to such limits (7).

The present exposure limits are based on all available experimental data. However, detailed information is lacking for certain wavelengths or exposure conditions. The necessary extrapolations in these cases were based on current theories of the mechanism of injury. However, these theories have not been adequately tested.

The main questions remaining concern exposure to laser radiation delivered over extremely short periods (mode-locked lasers) or with exposures

76

delivered at low levels over long periods of time. It appears at present that retinal damage from short pulses is produced by a different mechanism from that producing minimal damage from long exposures. Also, no data are available to show whether interaction between these mechanisms exists, or whether the interaction, if present, is competitive or synergistic. In addition, exposure to UV and IR laser radiation has not been adequately studied, and there are still several large gaps in the available data regarding exposure to repetitive pulse trains of laser radiation. Also, some interesting questions remain with regard to non-circular images, especially when one dimension is small. The line images and elliptical images that may be produced by a laser diode or by a scanning laser are examples of this type of exposure.

All of the known injurious effects are strongly wavelength-dependent. However, little is known about the wavelength relationship in the ultra-short or extremely long time exposure domains. For example, although longer-term exposures to IR radiation result in injury at levels well above present protection standards, the rather sparse data seem to indicate a power-dependent thermal injury mechanism. This is in contrast to the dose-related photochemical model used to predict injury from visible radiation in this time domain. For most exposure durations, injury to the cornea and lens of the eye from UV radiation appears to be photochemical. However, there are indications that occasional thermally enhanced photochemical effects also occur.

Few data are available for long-term (chronic) exposures to laser radiation. Even exposure to non-laser sources, such as bright small-source lamps and high-luminance extended sources, has produced insufficient data to allow extrapolation to laser sources. Thus, the permissible exposure levels were based on the assumption that the total retinal dose from visible illumination levels normally encountered in the natural environment is not hazardous. Recent studies appear to substantiate the theory that injury from chronic low-level exposure is related to absorption by the visual pigment in the photoreceptors. This is particularly true of short-wavelength light (25). However, small temperature rises in the retina (of the order of 2 or 3 °C) appear to synergize with the photochemical process so that melanin pigment absorption will also play a role, albeit a secondary one (44).

The protection standard levels in the far-IR region were based on an understanding of the possible thermal effects of the cornea and a knowledge of exposures that have not resulted in adverse effects on the eye. Because of the lack of accurate data on exposures of the human eye to the IR laser, "worst-case" exposure conditions were assumed. Specifically, it was assumed that absorption takes place in a very thin layer at the anterior surface of the cornea. This condition is best represented by $10.6 \mu m$ CO_2 laser exposures. For that matter it will, as well, fit exposure of the eye to any wavelength beyond approximately $2.5 \mu m$. At wavelengths less than $3 \mu m$, the radiation penetrates into the cornea more deeply and significant absorption may take place in the aqueous humour and lens. For these wavelengths, short-term exposure to much greater irradiances can be permitted. However, the risk of IR cataracts is greatest for long-term exposure. The increased interest in certain near-IR lasers, such as hydrogen fluoride, deuterium fluoride, holmium, erbium and neodymium lasers, means that biological investigations to define more detailed permissible exposure conditions in the middle- and near-IR regions, and for long or repeated exposures, will be required.

Only scattered data points are available for damage thresholds to the cornea and lens of the eye from UV radiation. Extrapolations from studies using non-laser UV sources were made in order to arrive at reasonable values. The lack of availability of UV lasers has prevented extensive studies of laser injury thresholds and mechanisms of UV laser injury. Hopefully, the availability in the future of a tunable UV laser with a continuous-wave or nearly continuous output should permit such studies. Meanwhile, those using the present UV standards should consider them only as the best available guidelines. Exposures should be limited as much as possible.

It may legitimately be asked how it is possible to study delayed chronic effects on the retina and other portions of the eye. Will such studies have to be conducted for 20–30 years before we can know what may happen? Many patients who have had laser treatment for eye lesions have now been observed for over 20 years, and have not shown any significant chronic effects in the treated areas or in any adjacent areas as a result of the laser treatment. Fortunately, several powerful research tools are becoming increasingly available that permit study of the ultrastructural changes in tissues soon after exposure. The most valuable of such tools are electron microscopes, both scanning and fixed, and various spectroscopic probes. Studies incorporating such devices should yield a more fundamental understanding of the mechanisms of injury and permit an accurate prediction of chronic effects. As we develop a better understanding of retinal physiology and the fundamental photochemical mechanisms of vision, our predictions of chronic effects will become more reliable.

The interpretation of all such studies requires a sophisticated level of experience. It is important to remember in planning research studies that the availability of acute threshold data does not answer all questions. Clearly, the need exists for further biological studies of laser effects, especially where information is at present lacking. Laser manufacturers and others with lasers of unusual characteristics should be encouraged to make such equipment readily available to institutions conducting laser bioeffects research. In addition, and most importantly, an estimate of the gaps in present knowledge may facilitate a rational approach to future revisions of existing protection standards.

Much of the exposure data now available have been collected in an empirical fashion without any attempt to determine the underlying mechanism(s) of injury. The implications of repeated exposure at low levels below present protection standards cannot be evaluated without a far more complete understanding of the mechanisms of injury. Without such studies there can be no assurance that injurious effects will not appear long after exposure, perhaps many years following active use of lasers. In this regard, study of chronic effects is required since latent ocular effects could ensue from chronic exposure to IR radiation of the lens and anterior portion of the eye. Delayed skin effects might ensue from chronic exposures in the UV and

78

visible frequency ranges; this is of particular importance in the case of individuals who may be photosensitive as a result of drugs or other photo-sensitizing agents.

For the present, and with due regard to the lack of data in some areas, it is recommended that the maximum permissible exposures of the eye and skin recommended by ANSI Z-136.1 *(49)* be adopted as modified here. It is also recommended that additional studies of the effects of lasers on the skin be performed, especially for acute exposures in the UV spectral region and for chronic exposures over the entire spectral region.

It is felt that there is insufficient information on the effects of chronic exposure to laser radiation. Such information may be obtained through detailed and well controlled epidemiological studies. Such studies should be carried out on limited groups working in a laser environment in which the physical parameters are well defined. Members of these groups should undergo periodic examinations according to a standardized procedure. The studies must include control groups of personnel not exposed to intense optical radiation.

CONCLUSIONS AND RECOMMENDATIONS[a]

Conclusions

Although initially only research workers were exposed to laser radiation, today much of the public is now potentially exposed to lasers used in medicine, communications, entertainment and industry.

For short laser exposures, both peak irradiance (W/m^2) and radiant exposure (J/m^2), as well as the time pattern of exposure, are the important factors determining biological effects. For continuous-wave lasers irradiance is the important factor. In addition to irradiance (dose rate) and radiant exposure (dose), the quantity radiance ($W/(m^2 \cdot sr)$) are useful for calculations of the exposure dose to the retina.

The essential measurements for hazard evaluation should include:

— output level

— divergence

— wavelength

— exposure duration

— pulse repetition frequency

— pulse duration

— beam geometry

— characteristics of any reflecting surfaces.

---

[a] These conclusions and recommendations are those pertaining to lasers made by the WHO Working Group on Health Implications of the Increased Use of NIR Technologies and Devices, Ann Arbor, USA, October 1985.

Laser emissions cover wavelengths from 10 nm to 1 mm and are used in a wide variety of applications. The interactions between biological tissues and laser radiation depend on irradiance and radiant exposure, exposure duration and wavelength. The coherence properties of laser radiation *per se* are not considered important.

These interactions can take four main forms: photochemical, thermal, thermo-acoustic, and multiphoton-dependent and non-linear processes such as optical breakdown.

Because of the relatively superficial absorption, the major biological effects are on the skin and the eye. However, because of its unique optical properties, the eye is more vulnerable to injury: visible and IR-A radiation are focused sharply on the retina, and the corneal irradiance may be amplified by as much as 100 000 times.

The health hazards to the skin are both acute (thermal burns, sunburn, photosensitized reactions) and chronic (accelerated aging and photocarcinogenesis). The main health hazards to the eye are photokeratitis, corneal burns, photochemical and thermal cataract, ocular inflammation, and photochemical, thermal and thermo-acoustic retinal injury and optical breakdown within the eye. Secondary haemorrhage in the vitreous humour with complete loss of vision could be produced by severe damage to the retina.

Recommendations

Control measures should include education and training of all personnel working with lasers. Engineering controls, such as proper layout of working areas and enclosure of instruments, should be selected as appropriate. Personal protection may consist of special eyewear and appropriate clothing. Warning signs and other administrative control measures can supplement these measures. Public exposure to hazardous laser radiation should be precluded by the use of laser beam stops and limits on beam paths during laser light shows and outdoor laser use. Applicable control measures are very different depending on the classification of laser equipment: class 1 lasers are totally safe, whereas class 4 lasers are very dangerous and require extreme control measures and precautions.

Several national and international bodies have dealt with the problem of permissible exposure limits. Although there is a lack of data in some areas, there is general agreement on exposure limits such as those recommended by IEC and IRPA. It is also recommended that additional studies of the effects of lasers on the eye and skin be performed, especially for acute exposures in the UV spectral region and for chronic exposures over the entire spectral region.

REFERENCES

1. **Suess, M.J.** The development of a long-term programme of non-ionizing radiation protection. *Health physics,* **27**: 514 (1974).
2. *International lighting vocabulary.* Paris, International Commission on Illumination, 1970 (CIE No. 17).

3. **Meyer-Schwickerath, G.** Lichtkoagulation [Coagulation effects of light]. *Von Graefes Archiv für klinische und experimentelle Ophthalmologie,* **156**: 2 (1954).

4. **Rockwell, R.J.** Developments in laser instrumentation and calibration. *Archives of environmental health,* **20**: 149 (1970).

5. **Charschan, S.S.** *Lasers in industry.* New York, Reinhold, 1972.

6. **Goldman, L.G., ed.** *The biomedical laser.* New York, Springer-Verlag, 1981.

7. **Wolbarsht, M.L. & Sliney, D.H.** The formulation of protection standards for lasers. *In:* Wolbarsht, M.L., ed. *Laser applications in medicine and biology.* New York, Plenum Press, 1974, Vol. 2.

8. **Gamaleja, N.F.** *Lasers in experiments and clinical practice.* Moscow, Medicina, 1972.

9. **Sliney, D.J. & Wolbarsht, M.L.** *Safety with lasers and other optical sources.* New York, Plenum Press, 1980.

10. **Wolbarsht, M.L., ed.** *Laser applications in medicine and biology.* New York, Plenum Press, 1971, Vol. 1.

11. **Wolbarsht, M.L., ed.** *Laser applications in medicine and biology.* New York, Plenum Press, 1974, Vol. 2.

12. **Wolbarsht, M.L., ed.** *Laser applications in medicine and biology.* New York, Plenum Press, 1977, Vol. 3.

13. **Sliney, D.H. et al.** *Laser hazards bibliography — October 1984.* Aberdeen Proving Grounds, MD, US Army Environmental Hygiene Agency, 1984.

14. **Mešter, E. et al.** Laserstrahlenwirkung auf das Wachstum des Ehrlichschen Ascitestumors [Effect of laser radiation on the development of Ehrlich ascites tumour]. *Archiv für Geschwulstforschung,* **32**: 201 (1968).

15. **Mešter, E. et al.** Die Wirkung über längere Zeit wiederholt verabreichter Laserstrahlung geringer Intensität auf die Haut und inneren Organe von Mäusen [The long-term effects of repeated laser irradiation at low intensity on the skin and internal organs of mice]. *Radiobiology and radiotherapy,* **10**: 371 (1969).

16. **Mešter, E. et al.** Stimulation of wound healing by laser rays. *Acta chirurgiae academiae scientiarum hungaricae,* **13**: 315 (1972).

17. **Sliney, D.H. & Freasier, B.C.** Evaluation of optical radiation hazards. *Applied optics,* **12**: 1 (1973).

18. **Eckoldt, K.** Die optischen Eigenschaften der menschlichen Haut im IR-Bereich [The optical properties of human skin in the IR region]. *In:* Christensen, B.C. & Buchmann, B., ed. *Proceedings of the 3rd Congress of Photobiology.* Amsterdam, Elsevier, 1961.

19. **Geeraets, W.J. & Berry, E.R.** Ocular spectral characteristics related to hazards from lasers and other light sources. *American journal of ophthalmology,* **66**: 15 (1968).

20. **Mainster, M.A. et al.** Laser photodisruptors, damage mechanisms, instrument design and safety. *Ophthalmology,* **90**: 973–991 (1983).

21. **Karu, T.** Molecular mechanism of the therapeutic effect of low-intensity laser radiation. *Lasers in the life sciences,* **2**(1): 53–74 (1988).

22. **Rockwell, R.J. & Goldman, L.** *Research on human skin laser damage.* School of Aerospace Medicine, Brooks Air Force Base, Texas, 1974 (Final report, University of Cincinnati, Contract F41609-72-C-0007).

23. **Rockwell, R.J. & Moss, E.C.** Optical radiation hazards of laser welding processes, Part I: neodymium-YAG laser. *American Industrial Hygiene Association journal,* **44**: 572–579 (1983).

24. **Parrish, J.A. et al.** Cutaneous effects of pulsed nitrogen gas laser irradiation. *Journal of investigative dermatology,* **67**: 603–608 (1976).

25. **Ham, W.T. Jr. et al.** Ocular hazard from picosecond pulses of Nd:YAG laser radiation. *Science,* **185**: 362 (1974).

26. **Hayes, J.R. & Wolbarsht, M.L.** Thermal model for retinal damage induced by pulsed lasers. *Aerospace medicine,* **40**: 474 (1968).

27. **Hayes, J.R. & Wolbarsht, M.L.** Models in pathology — mechanisms of action of laser energy with biological tissues. *In:* Wolbarsht, M.L., ed. *Laser applications in medicine and biology.* New York, Plenum Press, 1971, Vol. 1.

28. **Ham, W.T. Jr. et al.** Retinal burn thresholds for the helium–neon laser in the rhesus monkey. *Archives of ophthalmology,* **84**: 797 (1970).

29. **Dunsky, I.L. & Lappin, P.W.** Evaluation of retinal thresholds for CW laser radiation. *Vision research,* **11**: 733 (1971).

30. **Bresnick, G.H. et al.** Ocular effects of argon laser radiation. *Investigative ophthalmology,* **9**: 901 (1970).

31. **Vassiliadis, A. et al.** *Ocular laser threshold investigations.* Menlo Park, CA, Stanford Research Institute, 1971 (SRI Report No. 8209).

32. **Vassiliadis, A. et al.** *Research on ocular laser thresholds.* Menlo Park, CA, Stanford Research Institute, 1969 (SRI Report No. 7191).

33. **Lappin, P.W. & Coogan, P.S.** Relative sensitivity of various areas of the retina to laser radiation. *Archives of ophthalmology,* **84**: 350 (1970).

34. **Naidoff, M.A. & Sliney, D.H.** Retinal injury from welding arc. *American journal of ophthalmology,* **77**: 663 (1974).

35. **Skeen, C.H. et al.** *Ocular effects of near infrared laser radiation for safety criteria.* San Antonio, TX, Life Sciences Division, Technology Inc., 1972 (US Air Force Contract No. F41609-71-C-0016).

36. **Skeen, C.H. et al.** *Ocular effects of repetitive laser pulses.* San Antonio, TX, Life Sciences Division, Technology Inc., 1972 (US Air Force Contract No. F41609-71-C-0018).

37. **Ham, W.T. Jr. et al.** Effects of laser radiation of the mammalian eye. *Transactions of the New York Academy of Sciences,* **28**: 517 (1966).

38. **Verhoeff, F.H. & Bell, L.** The pathological effects of radiant energy on the eye — an experimental investigation. *Proceedings of the American Academy of Arts and Sciences,* **51**: 629, 819 (1916).

39. **Eccles, J.C. & Flynn, A.J.** Experimental photoretinitis. *Medical journal of Australia,* **1**: 399 (1944).

40. **Kuwabara, T.** Retinal recovery from exposure to light. *American journal of ophthalmology,* **70**: 187 (1970).

41. **Lawwill, T.** Effects of prolonged exposure of rabbit retina to low intensity light. *Investigative ophthalmology,* **12**: 45 (1973).

42. **Boettner, E.A.** *Spectral transmission of the eye.* School of Aerospace Medicine, Brooks Air Force Base, Texas, 1967 (University of Michigan Contract AF41(609)-2966).

43. **Ham, W.T. Jr. et al.** Action spectrum for retinal injury from near-ultraviolet radiation in the aphakic monkey. *American journal of ophthalmology,* **93**: 299–306 (1982).

44. **Noell, W.K. & Albrecht, R.** Irreversible effects of visible light on the retina, role of vitamin A. *Science,* **172**: 72 (1971).

45. **Ward, B. & Bruce, W.R.** Chorioretinal burn: body temperature dependence. *Annals of ophthalmology,* **3**: 898 (1971).

46. *Radiation safety of laser products: equipment classification and users' guide.* Geneva, International Electrotechnical Commission, 1984 (Publication WS 825).

47. **Wilkening, G.M.** Laser hazard control procedures. *In: Electronic product radiation and the health physicist.* Washington, DC, Bureau of Radiological Health, 1970 (Publication No. BRH/DEP 70-26).

48. **Pitts, D.G. & Gibbons, W.D.** *The human, primate, and rabbit ultraviolet action spectra.* Houston, TX, College of Optometry, University of Houston, 1972.

49. *American national standard for the safe use of lasers.* New York, American National Standards Institute, 1980 (ANSI Z-136.1).

50. **Sliney, D.H.** Nonionizing radiation. *In: Industrial environmental health — the worker and the community.* New York, Academic Press, 1972.

51. **Clarke, A.M.** Ocular hazards from lasers and other optical sources. *Critical reviews in environmental control,* **1**: 307 (1970).

52. **International Radiation Protection Association.** Guidelines for human exposure to laser radiation (180 nm–1 mm). *Health physics,* **49**: 341–359 (1985).

53. *Threshold limit values for chemical substances and physical agents in the workroom environment.* Cincinatti, OH, American Conference of Governmental Industrial Hygienists, 1985.

54. *A guide for control of laser hazards.* Cincinnati, OH, American Conference of Governmental Industrial Hygienists, 1981.

55. *Laser products performance standard 21CRF1910.* Rockville, MD, US Food and Drug Administration, 1974.

56. **Sliney, D.H.** Laser protective eyewear. *In:* Wolbarsht, M.L., ed. *Laser applications in medicine and biology.* New York, Plenum Press, 1974, Vol. 2.

57. **Wilkening, G.M.** A commentary on laser-induced biological effects and protective measures. *Annals of the New York Academy of Sciences,* **168**: 621 (1970).

58. *Laser safety guide.* Cincinnati, OH, Laser Institute of America, 1974.

59. *Guide on protection of personnel against hazards from laser radiation.* London, British Standards Institution, 1972 (BS 4803).

3

Infrared radiation

C.E. Moss, R.J. Ellis, W.E. Murray & W.H. Parr

Revised by B.M. Tengroth & M.L. Wolbarsht

CONTENTS

	Page
Introduction	85
Production and characteristics	86
Sources	86
Occupational exposure	87
Instrumentation	88
Biological effects	89
Ocular hazards	89
Skin hazards	98
Other hazards	104
Existing standards	105
Control measures	107
Problems and recommendations	109
Conclusions and recommendations	111
Conclusions	111
Recommendations	111
References	111

INTRODUCTION

Infrared (IR) radiation is that part of the electromagnetic spectrum associated with energy levels such that thermal effects are produced when it is absorbed by matter. The IR region encompasses wavelengths from 0.78 to 1000μm, and most sources that emit ultraviolet or visible radiation will probably emit IR. This is important in considering the potential occupational hazards from the multitude of artificial radiation sources.

Over the past 50 years, human exposure to IR, formerly associated only with the glassmaking, metal smelting and foundry industries, has become more widespread and now includes exposure to welding arcs and to many specialized industrial heat sources. Consequently, increasing numbers of workers throughout the world are being exposed to broad bands of IR for long periods of time under unique conditions. The rapid growth in the development and application of IR devices has led to the awareness that meaningful biological information is lacking. While the obvious action sites or target organs (skin and eye) have been identified, questions persist concerning the mechanisms of damage in man, threshold levels for acute and chronic effects, effects on tissues and organs other than skin and eye, potential synergistic effects, and the role of heat stress.

PRODUCTION AND CHARACTERISTICS

The IR region has been subdivided by the International Commission on Illumination (CIE) into three biologically significant bands, IR-A (0.78–1.4μm), IR-B (1.4–3μm) and IR-C (3–1000μm) (1).

Infrared radiation is generated by the vibration and rotation of atoms and molecules within a material whose temperature is above absolute zero. All objects emit IR as a function of their temperature. Physical matter emits IR in accordance with the black body radiation laws, but with the inclusion of a factor called the emissivity. This is a fraction of unity, except for a black body, for which it equals unity. Many IR sources emit a continuum of wavelengths (Fig. 1) and the wavelength of maximum spectral power is determined by:

$$\lambda_m = \frac{2898}{T}$$

where T is the temperature (K), and λ_m is the wavelength in μm. Fig. 1 shows that, as the source temperature increases, the peak of the radiation curve moves towards shorter wavelengths while the intensity of the emitted radiation increases.

Infrared, like all types of electromagnetic radiation, can undergo a number of interactions, including reflection, absorption, transmission, refraction and diffraction. An understanding of all these interactions is required for measurement and control purposes, while the biological aspects are affected only by refraction, absorption and transmission.

SOURCES

All IR sources can be classified as either artificial or natural sources. Both types are encountered in the working environment. Artificial sources include various types of commercially manufactured incandescent, fluorescent, and high-intensity discharge lamps, flames, heaters and artificial black body sources. These are all broad-band sources and require the use of filters to limit the output to a particular wavelength band.

86

Fig. 1. Emission of optical radiation from black bodies
at various temperatures

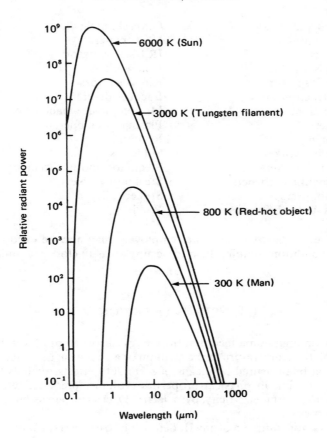

Numerous sources also occur in nature and cannot be directly controlled by man. The most significant natural source is the sun. The average total radiant power of the sun at the edge of the earth's atmosphere is approximately 1.35 kW/m², of which half is in the IR region. The sun resembles a black body source at a temperature of about 6000 K, reaching a peak near 500 nm, although the radiation extends from the near-UV through the visible radiation range well into the IR region.

Occupational exposure
A wide range of exposures under conditions involving large temperature differences occur in industry from both artificial and natural IR sources.

The occupations associated with potential IR exposures include the following:

Bakers and cooks	Glass skimmers
Blacksmiths	Heat treaters
Braziers	IR laser operators
Chemists	Iron workers
Cloth inspectors	Kiln operators
Construction workers	Lacquer dryers
Electricians	Motion picture machine operators
Farmers	Plasma torch operators
Firemen	Roofers
Foundry workers	Solderers
Furnace workers	Steam locomotive firemen
Gas mantle hardeners	Steel mill workers
Glassblowers	Stokers
Glass furnace workers	Welders

It must be emphasized that other employees may also be exposed when working conditions require them to occupy areas in close proximity to IR sources.

INSTRUMENTATION

When IR impinges on a medium, the energy absorbed produces either an increase in the temperature of the medium or a change in its structure. This change can be measured by means of secondary effects, such as the consequent variation in physical properties (volume, pressure, refractivity, conductivity, thermoelectricity, pyroelectricity, electron emission) or chemical properties.

The essential function of any IR detector is to convert the radiant energy into another form of energy that can be processed more readily. There are two types of IR detector: thermal and photonic. The first relies on an increase in temperature, which can be measured as a corresponding change in electrical resistance or other physical characteristics that provide a signal proportional to the radiant power absorbed. This type of detector responds to a broad spectrum, generally requires no cooling, has low sensitivity, and has a response time of the order of milliseconds to seconds. Thermal detectors are relatively inexpensive. The spectral response of such thermistor, bolometer, thermopile and pyroelectric detectors depends on the absorption properties of the detecting medium.

Most IR photon detectors are semiconductors in which photons interact to produce free charge carriers (e.g. photoelectric effect). These detectors respond to a relatively narrow range of wavelengths. In general, this type requires cooling and has a fast response time. The essential difference between the two types is that the photon detector determines the number of quanta per second absorbed whereas the thermal detector depends on the total power absorbed.

A special filter is often inserted in the incident beam so that only specific wavelengths are transmitted to the detector. Filters can be made by the vacuum deposition of thin films on to suitable transparent substrates. Almost any desired band width and peak wavelength can be obtained.

Various models of spectroradiometer are commercially available that will measure the spectral energy distribution of IR sources. However, most of these devices become expensive when spectral information is required beyond 1.2μm. Radiometers can then be used with special filters. Additional information on IR detectors can be found in Keyes[2] and Wolfe & Zissis [3].

When laboratory measurements and hazard evaluations are made, consideration of the effect of mechanical, electrical and thermal changes on the detector, changes in source emission levels, spectral response, aging of components, calibration, atmospheric conditions and contaminants, and reflections are of the utmost importance. In order to make accurate and reliable measurements, the investigator must be aware of the characteristics and limitations of the instrument used.

BIOLOGICAL EFFECTS

It is generally assumed that IR photons, because of their low energy levels, do not react photochemically in biological tissue. A review of the literature reveals that most research on the bioeffects of IR is on ocular effects. Far fewer studies have been concerned with skin and other effects.

Infrared radiation from all sources, including the sun and industrial IR sources, constitutes an important component of the microclimate, together with temperature, water vapour, pressure and air velocity. Traditionally, far-IR has been synonymous with radiant heat, and is usually associated with the older IR sources encountered in glassblowing, foundries, furnaces, etc. Occupational exposure to high levels of radiant heat from these sources may precipitate a thermal stress condition[4]. Quite apart from any specific effects of IR, therefore, there is an important effect of industrial IR on thermal stress in the workplace. It should also be appreciated that IR can be a major contributor to and cause of thermal stress; the normal precautions against such stress should therefore be borne in mind.

Ocular hazards
In general, the eye is as effective as the skin in protecting itself against IR. It has certain protective mechanisms, and these are adequate for the natural environment, since IR is usually accompanied by intense visible radiation. This evokes the blink and pupil reflexes, which limit the radiant exposure (dose) penetrating into the eye. However, with some industrial sources, IR can be present without intense light and those reflex mechanisms may not be activated.

One important feature of the eye is its focusing ability, which is, of course, not possessed by the skin. The optic media focus the incident radiation so that a significant concentration of energy is produced.

In 1962 Boettner & Wolter *(5)* carried out a comprehensive study of the transmission characteristics of the separate ocular components, using nine enucleated human eyes. Subsequent investigations into the transmission properties of the eye have been limited. In 1968 Geeraets & Berry *(6)* examined the transmittance of 28 intact human eyes in a similar manner. The results of these two studies are compared in Fig. 2, from which it can be seen that the difference between the upper and lower curves is substantial. This difference remains a problem. Sliney *(7)* has suggested that it is probably due to the measuring techniques used to obtain the total spectral transmission of the eye as opposed to the transmittance of each of the ocular media separately, and to the correction factors used for scatter.

Previous work by Fischer et al. *(8)* and Franke *(9)* led Ruth *(10)* to propose that the aqueous humour, lens and vitreous humour should be collectively termed the "inner eye" and that their absorption should be viewed compositely, as illustrated by Fig. 3. This figure, in turn, is extremely useful in that it can be used to relate the absorption of the eye to the spectral distribution of the various IR sources, as shown by Fig. 4 and 5. When Fig. 3 is used, it must be remembered that it does not take into account the absorption by the cornea; this must be considered whenever IR is evaluated relative to ocular absorption. The part of the IR spectrum that might be hazardous to the eye is confined to the near-IR region.

Effects on the eyelid

The presence of the eyelid and its associated blink reflex protects the eye from excessive radiation exposure and replenishes the liquid on the anterior surface of the eye. This dual action aids in cooling the organ by shielding it from radiant energy. Investigations into the degree of transmission of IR by the eyelid have not been reported in the literature. Since the anatomical structure of the eyelid is more or less similar to that of body skin, it is possible to estimate the transmission of IR radiation. This estimate may be important, since it is often assumed by safety personnel that protection against near-IR is provided by closure of the eyelids.

Effects on the cornea

Since exposure to high intensities of far-IR can produce corneal pain, the eyes are reflexively closed and the head averted. Sliney *(7)* has stated that the sensory nerve endings in the cornea are quite sensitive to small temperature elevations and that a temperature of 45 °C (corresponding to approximately 100 kW/m^2 absorbed in the cornea) elicits a pain response in humans within a small fraction of a second. Hence, he suggests that a thermally mediated response is initiated before the actual pain stimulus. For this reason, burn lesions are not commonly seen in the usual industrial exposures *(7)*. If a burn does occur from an intense radiation source that is limited to the corneal epithelium, the normal repair process will generally preclude any permanent

Fig. 2. Spectral transmittance of the ocular media of the human eye

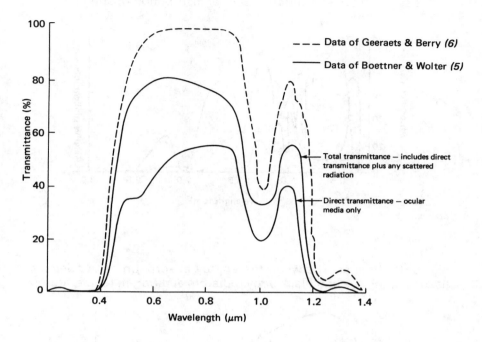

Fig. 3. Composite absorption of optical radiation by the deeper tissues of the eye (aqueous humour, lens, vitreous humour) after passage through the cornea

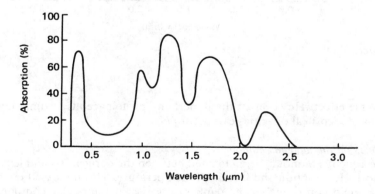

Source: Ruth *(10)*.

Fig. 4. Comparison between the energy absorption of the deeper tissues of the eye and the spectral distribution of sunlight

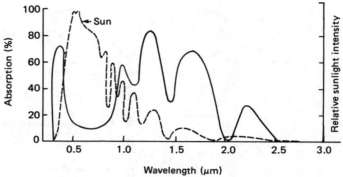

Source: Ruth *(10).*

Fig. 5. Comparison between the energy absorption of the deeper tissues of the eye and black body radiators at different temperatures

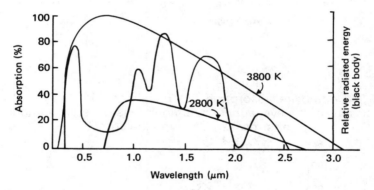

Source: Ruth *(10).*

adverse effect. However, if the underlying proteinaceous stromal layer is damaged, corneal opacities can occur *(11).*

Effects on the aqueous humour
The aqueous humour, being located between the cornea, iris and lens, will absorb IR radiation and increase in temperature. This increased temperature could contribute to the temperature rise of the ocular components, most notably the lens.

Effects on the iris

The effects of IR on the iris have been summarized by Duke-Elder *(12)*. In his view, the iris is very susceptible to IR because of heavy absorption by its pigment. Moderate doses result in constriction of the pupil (hyperaemia miosis) and the formation of an aqueous flare.

Effects on the lens

The lens of the eye is optically clear up to early adulthood. It is a unique body tissue, being both avascular and lacking innervation. Furthermore, the lens is a growing and actively metabolizing tissue throughout life. Probably as a result of disturbances in this continuing metabolic activity, the optical clarity can become progressively reduced as a result of a number of factors, including metabolic disorders, ocular inflammation, blunt trauma and different types of electromagnetic radiation. This reduction in optical clarity is caused by different types of opacity, generally referred to as cataract, which do not, however, necessarily result in lowered visual acuity. Such opacities should be defined as dark spots visible against the light reflected from the fundus of the eye, e.g. with an ophthalmoscope but not with a slit lamp. Cataracts interfering with vision to such an extent that surgical intervention is necessary occur mostly in those over 65 years of age. The frequency in a European population of that age group is up to 5 per 1000 per year.

In the literature prior to 1920, a number of reports suggested that glassblowers and furnace workers had a higher incidence of cataracts than the non-exposed population *(8,13,14)*. Modern epidemiological studies have confirmed these results among glassblowers *(15)*. The beginning of the twentieth century brought intense interest and debate as to the etiology of this kind of cataract *(16–20)*. These investigations do not show any specific type of cataract, true capsule exfoliation being the only specific sign. On the basis of animal experimentation and theoretical analyses, two theories have been advanced to account for the formation of what are now recognized as IR cataracts *(16,17,20,21)*. These theories suggest that cataract formation is due to direct absorption of IR in the lens, or is secondary to heating of the aqueous humour and iris by absorption of IR. Data from Pitts & Cullen *(22)* and Wolbarsht *(23)* indicate a photochemical type of lens damage with a constant dose relationship based on reciprocity between time and power. This is for acute *in vivo* exposures only, spanning hours, days or even weeks. How this relates to low-level chronic exposures over a period of years in industrial situations remains to be seen. Wolbarsht *(23)* has also shown that acute high-level exposures direct to the iris only can produce cataracts in the areas of the lens behind the exposed iris, presumably through heat transfer. It should be pointed out, however, that many other factors such as heredity, race, drug use, disease, and immunological and nutritional factors can predispose towards cataractogenesis, or be synergistic among themselves or with other factors. These synergistic effects are supported somewhat by the data, as all the epidemiological studies show an increased number of cataracts in older age groups.

93

Effects on the retina
Difficulties arise concerning the effects of IR on the retina. Since absorption by the retina of the shortest IR wavelengths differs only slightly from that of visible radiation, it is difficult to see that there are any specific IR effects apart from the thermal effects usually attributed to visible radiation.

Experimental threshold exposures
Another area of uncertainty has been the determination of the minimum energy level (threshold) at which tissue damage occurs. Past investigations have sought to establish the lowest dose level (acute exposure) at which the stated damage criterion occurs. Such studies have been made on the cornea, iris, lens and retina. The more significant experimental broad-band threshold studies for the above ocular structures are indicated in the work of Jacobson et al. *(24)*.

It must be remembered, in addition, that laboratory threshold experimentation has usually involved acute exposure. This poses a problem for the determination of the radiation energy level that causes cataracts in industry, since their development involves long latent periods and chronic exposure. Furthermore, it is debatable whether experimentally determined damage thresholds can be correlated with those occurring in industry. This is especially difficult because of the lack of extensive research in industrial situations to establish the energy level at which cataracts are produced. Fewer problems occur in determining the damage thresholds for the cornea and retina, since experimental conditions coincide fairly closely with those of the occupational acute exposures required to cause damage.

Industrial exposures
Typical data from the literature for worker exposure levels are given in Table 1. While damage thresholds determined experimentally are based on exposure to the lowest single dose causing the damage, determinations of industrial damage thresholds should be based on many exposures to the lowest dose over many years that will produce such damage. In the past, there has been a failure in industrial research to take into account irradiance and exposure duration. Nevertheless, some investigators have attempted to make some sort of threshold estimation under industrial exposure conditions for the cornea, lens and retina. The occurrence of lens damage in the form of "glassblowers' cataract" following exposure to IR has long been recognized. Unfortunately, the earlier studies do not include any data on the absolute temperature of the source, the wavelength region, or the irradiance level.

A more recent study *(25)* of glassblowers exposed for a number of years to an irradiance level of approximately $1.4 \, kW/m^2$ has shown no evidence of cataract formation. In addition, Keatings et al. *(26)* were unable to find any posterior cortical changes in the lens of iron-rolling millworkers exposed to irradiance levels of $0.8–4.2 \, kW/m^2$. However, they did report a higher incidence of posterior capsular opacities originating in the capsular plane and extending to the cortex. This differs from what has been defined as cataract.

94

In a recent large-scale epidemiological investigation, Wallace et al. *(27)* studied 1000 workers in a large steel mill, 900 of whom were exposed to IR; exposure was classified as high, intermediate, low and no-risk (scored as 3, 2, 1 and 0, respectively). Different job classifications were scored according to relative exposure. The exposure index was multiplied by the number of years of risk to arrive at a total exposure estimate in "exposure-years". Wallace et al. defined cataract type III (of which no cases were found) as that producing gross disturbance of vision and requiring surgery. Type II cataracts were defined as posterior polar subcapsular saucer-shaped cataracts capable of producing some interference with visual acuity. Type I cataracts were not true cataracts but small inhomogeneities which did not interfere with visual acuity. Fig. 6, which is based on the data of Wallace et al., shows the percentage of people with bilatral cortical cataract of type I as a function of age in this population.

When Wallace et al. *(27)* compared the percentage of type I cataracts with the number of "exposure-years" for the whole population, they found that there was a slight increase in such type I cataracts with increasing exposure, as shown in Fig. 7. While it appears as though the incidence of IR cataracts caused by the older industrial sources, e.g. in the glass industry, has decreased over the years, recent investigations of other industrial sources (both old and new) give rise to some concern. In 1971, Hager et al. *(29)* found several cases of "fire" cataract among locomotive firemen at temperatures of 1300–1500 K and wavelengths in the range 0.8–1.4 μm. Irradiance levels were found to be 0.5–1.8 kW/m². New sources of worker exposure to IR include welding arcs and industrial "heating" lamps. A comprehensive study by Hubner et al. *(28)* of different welding processes revealed irradiance levels as high as 34 W/m² in the IR. Despite these high radiance levels, however, there have been very few reported cataracts from welding operations. Another source of worker exposure is provided by the IR lamps used in paint and enamel drying operations. In 1975, Ruth *(10)* reported on what were designated as eye risk levels of 9–500 W/m² for various types of heating lamp for which most of the spectral distribution was in the near-IR region. To date, the literature indicates little or no incidence of cataracts from exposure to these newer industrial sources. It may be that the lack of information on the effects of welding and heating lamps is due to the comparative paucity of investigations into these questions and also to the possibility that IR cataractogenesis may have a long latent period.

As a result of worker exposure to certain industrial sources (welding arcs, arc lamps, xenon arcs, etc.) retinal injury in the form of burns and/or other lesions may occur because of the focusing effect of the cornea and the lens on the retina. These injuries are most probably due to radiant energy in the visible region of the optical radiation spectrum. Moreover, the size of the image on the retina and the absorbed irradiance are the predominant factors, as discussed by Sliney *(7)*.

Sensintaffar et al. *(30)* have reported a case of conjunctivitis and decreased lachrymation associated with exposure to near-IR from an IR heating device. The effective wavelength of the radiation from the heating device was 980 nm. The total irradiance from the device at the eye position was

Table 1. Data for ocular infrared radiation hazards

Study and source	Subject and exposure time	Effect	Wavelength (nm) (black body radiator temperature)	Radiant exposure (kJ/m²)	Irradiance level (kW/m²)
Goldmann, 1933 (21): electric furnace	Man:	Increase in temperature of aqueous humour (above 36 °C) of:	760–72 500 (1733 K with peak at 1 500)		
	30 seconds	3 °C		42	1.4
	231 seconds	9 °C		322	1.4
	Rabbit:				
	90 seconds	11 °C temperature increase behind pupil		408	4.53
	30 seconds	1.5 °C temperature increase behind lens		272	9.06
Dunn, 1950 (25): glass furnace	Glass workers, 20 years	None reported	1 500 (peak) (2000 K)	—	1.4
Keatings et al., 1955 (26)	Iron-rolling mill workers, 17 years	Lens posterior capsular opacities	—	—	0.84–4.18
Hubner et al., 1970 (28): welding arc	Welders	Not investigated	400–2 000	—	to 0.034

Reference	Subjects	Effect	Wavelength (nm)			
Hager et al., 1971 (29): glass and locomotive furnaces	Glass workers (G) and locomotive firemen (F). 10 years:	Cataract	800–1 400			
	G			(1300 K)	3 050	0.50
	F			(1300 K)	3 220	0.22
	G			(1400 K)	5 980	0.98
	F			(1400 K)	6 170	0.43
	G			(1500 K)	10 740	1.75
	F			(1500 K)	11 160	0.76
Wallace et al., 1971 (27)	Steel workers	Increase in opacities	—		—	—
Ruth, 1973 (10): molten brass	Man, 15 minutes	Increase in temperature of aqueous humour and lens of:	760–2 500 (1250 K with peak at 1 800)			
		3.9 °C			900	1.00
		6.4 °C			1 490	1.65
		8.5 °C			1 980	2.20
		All cases had lens opacities				
Lydahl, 1984 (15): IR heating lamps, automobile paint and enamel dryers	Not investigated		400–3 200		—	0.009–0.5

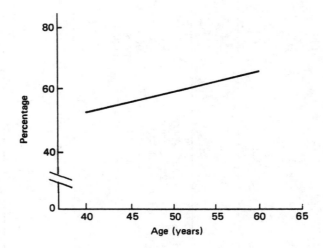

Fig. 6. Percentage of people with bilateral cortical cataract (type I)
as a function of age

Source: Wallace et al. *(27)*.

650 W/m² and the radiance between 400 and 1400 nm was 30 kW/(m²·sr).
Other reports *(31)* mention conjunctivitis and decreased lachrymation. It is
clear that exposure to IR will increase evaporation of the tear film and
therefore aggravate pre-existing deficiencies in accessory lachrymation
("dry-eye").

In recent years, a great deal of experimental research has been carried
out in an attempt to establish the threshold for retinal damage; this has been
summarized by Clark *(32)* and Sliney *(7)*. Unfortunately, laser sources were
used in most of these studies and the results are thus appropriate only to
laser hazards (see Chapter 2).

Skin hazards
To understand the effects of IR on skin, it is necessary to be familiar not only
with the optical and thermal properties of skin but also with other related
characteristics. Because of its high water content (60–70%) skin may be
regarded as having absorption properties similar to those of water.

The skin is one of the largest organs of the body. In an adult "standard"
man it comprises 4% of the body weight with a surface area of 1.6–2.0 m²
(33). It is generally 1–2 mm thick, although some areas may be as thick as
6 mm *(34)*. The skin is composed of an outer and thinner layer (epidermis)
and an inner and thicker layer (dermis). The epidermis is an epithelium and
varies from 0.07 to 0.12 mm in thickness over most of the body, except for

Fig. 7. Percentage of people with bilateral cortical cataract (type I) as a function of exposure assessment

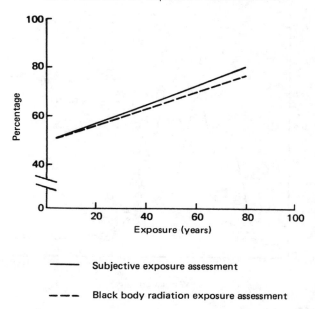

Source: Wallace et al. (27).

the palms and soles where it is thicker. The dermis may be collectively viewed as consisting of loose and dense connective tissue (collagenous bundles) containing hair and follicles, sebaceous and sweat glands, diffuse blood vessels, nerve endings and muscle. The thickness of the dermis is approximately 1–2 mm; it is thinner on the eyelids and much thicker on the palms and soles. (See Chapter 1, Fig. 9.)

Physiologically, skin functions are complex and diverse, and include functions as different as protection, excretion and sensation. Skin also plays an important role in maintaining fluid and electrolyte balance and body temperature. It is apparent that skin cannot be considered merely as "water", but rather is an extremely inhomogeneous tissue. This is an important factor in the determination of its optical and thermal properties, especially in the near-IR region.

The reflection and absorption characteristics must be considered in evaluating skin bioeffects. The reflection curves for human skin, as determined by Jacques et al. (35), are shown in Fig. 8. Beyond 2μm the reflectivity is variable and depends largely on skin pigmentation and blood flow. The maximum reflectivity occurs between 0.7 and 1.2μm, which is comparable to the wavelength of maximum intensity for some IR heating devices (Fig. 9). The spectral reflectance curve of skin is close in shape to the spectral

Fig. 8. Human skin reflectance as a function of wavelength for different pigmentation

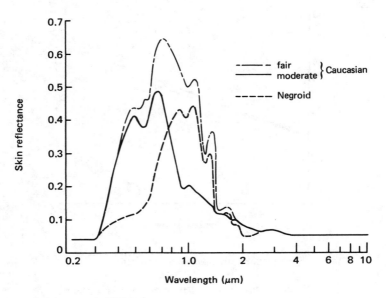

Source: Jacques et al. *(35)*.

Fig. 9. Emission curve for high-temperature infrared heater

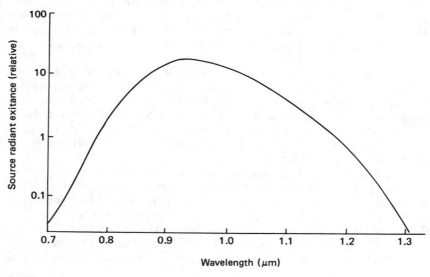

irradiance curve for the sun (Chapter 1, Fig. 4). This explains why the skin can reflect solar radiation effectively and yet have substantial thermal emissivity in the far-IR so that body heat can be gained or lost by thermal radiation. Furthermore, Fig. 8 shows that exposed skin surfaces in the darkly pigmented individual may be heated more by solar radiation than those of the fair-skinned individual. This difference can, however, be minimized by clothing. Beyond 1.5 μm skin pigmentation does not affect reflectivity.

The second important optical factor is the depth of penetration of IR into skin. As previously mentioned, the skin is a dynamic and non-homogeneous tissue and will scatter transmitted radiation. This effect, shown by Hardy et al. (36), is illustrated in Fig. 10. The curves given

Fig. 10. Scattering of transmitted energy under normal incident radiation with skin samples 0.43 mm thick (top) and 2.1 mm thick

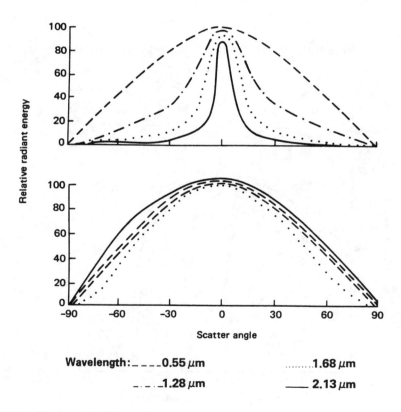

Wavelength: _ _ _ _0.55 μm   ....1.68 μm

_ . _ .1.28 μm   ____ 2.13 μm

Source: Hardy, et al. (36).

in Fig. 10 show the transmission for the indicated wavelengths through two samples of excised skin. As Hardy points out, if the skin were perfectly diffuse, the transmitted radiation would follow Lambert's cosine law for all wavelengths. However, it is apparent from Fig. 10, firstly that short wavelengths are scattered more than long wavelengths, and secondly that the differences due to wavelength are minimized as skin thickness increases. It is clear, therefore, that IR absorption by the skin depends not only on skin pigmentation, blood pigments and other substances that absorb specific spectral bands, but also on the degree of scattering due to the microstructures of the skin. The transmission spectrum of the skin, together with the absorption bands due to water, which is the principal IR absorber in biological tissues, are shown in Fig. 11. Although the skin is essentially opaque to wavelengths beyond 2μm, shorter wavelengths can penetrate a considerable distance below the skin surface. The maximum penetration is at a wavelength of approximately 1.2μm. Fig. 12 shows the average penetration of IR below the skin surface for both Negroid and Caucasian skin *(36)* and indicates that at least 50% of the radiation penetrates to a depth of about 0.8 mm, thus interacting with nerve endings and capillaries.

Fig. 11. Spectral transmittance of excised white human breast skin

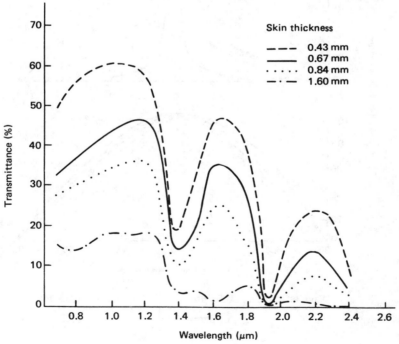

Source: Hardy et al. *(36)*.

102

Fig. 12. Average percentage penetration into Caucasian and Negroid skin of near-infrared radiation (1.23 µm)

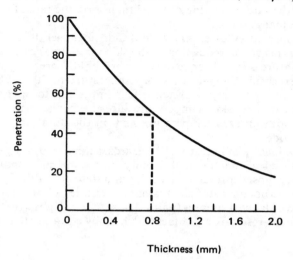

Source: Hardy et al. (36).

In view of the foregoing, it appears that there are two spectral regions with different modes of action: the first is essentially confined to the near-IR with a peak at approximately 1.2µm; the second is wavelength-independent beyond 2µm and causes surface heating. The near-IR region is potentially more hazardous than the middle- and far-IR because of the ability to penetrate significantly into the dermal area of the skin and thus to cause greater injury. Caution should therefore be exercised in using high-intensity sources of near-IR. Because of the strong absorption of middle- and far-IR by skin, occupational exposure to high levels of radiant energy (e.g. from furnaces) may be hazardous if the heat load exceeds the capacity of the body's thermoregulatory mechanism.

The manifestation of excessive IR exposure and its effects on skin seem to be confined to the near-IR. The most obvious effects include burns, increased vasodilation of the arteriolar system, and a gradual increase in pigmentation. This increase may be due to chronic exposure, since it seems to persist for an extensive period of time. Moreover, the development of an "erythematous-like" appearance among certain occupational groups (glass workers, furnacemen, etc.) exposed to high intensities of IR may also be viewed as a chronic effect. Chronic exposure to low levels of IR has been reported to cause blepharitis (37). Matelsky (38) reported that exposure of the eyelids to intense IR produced erythema, oedema and blistering similar to that from an ordinary burn. Tissue damage (denaturation) has been found to occur at skin temperatures of approximately 46–47 °C (39). However,

103

pain is induced at a mean skin temperature of 44.5 ± 1.3 °C *(40,41)*. Physiologically, the pain threshold is dependent on skin temperature alone and not on the rate of heating of the skin, nor on the rate of change of an internal thermal gradient.

Skin temperatures lower than 44–45 °C do not generally produce a burn *(37)*. Above this value, reddening of the skin (erythema) can occur. As the skin temperature is raised above the pain threshold, increasing intensities of pain are perceived. Raising the skin temperature to 70 °C will result in destruction of most enzyme systems. It is important to note that pain is primarily related to skin temperature alone, whereas tissue damage is dependent both on skin temperature and on the duration of the hyperthermic episode.

In the evaluation of industrial IR hazards, a number of variables must be considered, such as individual variability, environmental conditions, body surface area exposed, protective clothing and state of health. Typical industrial sources do not have sufficient radiant intensities to cause injury because of normal avoidance responses (pain). Those thermal burns that do occur in industry usually result from contact with hot objects rather than exposure to IR sources.

Other hazards
A number of reports have described specific effects of IR on man, animals and isolated cells. Some of these, however, can be explained as the result of associated environmental effects rather than of heat. Although in some cases these reports may not have been confirmed by independent workers, they are included here for completeness.

Studies by Krivobok *(42)* indicated that IR produced a limited level of organ and tissue degradation in areas remote from the eye. Changes such as vascular congestion in the spleen and kidneys were reported. A long-term decrease in the immunological reactivity (i.e. phagocyte count, phagocytic index and bactericidal properties of the skin) of foundry workers exposed to high-intensity IR at an irradiance of 0.2–0.7 kW/m² was reported by Zelencova *(43,44)*. However, these reports also indicated that, at lower intensity levels, IR stimulated the body's protective mechanisms.

Lehmann et al. *(45)* demonstrated that, when IR was applied to the ulnar nerve area at the elbow, an analgesic effect was found distally in the area supplied by this nerve. Lehmann states that this finding is in agreement with previous experimental evidence that nerve conduction can be temporarily blocked by application of IR.

A study by Borneff & Blumlein *(46)* indicated that the upper respiratory passages of iron foundry workers were damaged by exposure to intensive near-IR over many years. Chronic rhinitis (in most instances with polyps and hyperplasia of the mucous membranes), chronic laryngitis and sinus troubles were prevalent in almost 50% of the exposed workers. These disorders were claimed to be 5–10 times more frequent in the exposed group than in the control group.

Another organ located near the body surface and very sensitive to thermal insult is the testicle. The principal effect encountered from thermal insult in this organ is a temporary reduction in the sperm count. While little information is available on the relationship between IR and decreased sperm count, this interaction potential should not be completely overlooked. Episodes of apnoea (a transient suspension of respiration) have been reported in infants exposed to radiant warmers *(47)*. The relationship between changes in irradiance levels and the incidence of apnoea is not known.

Arima & Fonkalsrud *(48)* studied the incidence of post-operative intestinal adhesions and microscopic intestinal injury resulting from the use of overhead IR heating lamps. The safety of such lamps has been questioned because of their widespread use in neonatal operating rooms. A total of 45 rabbits were used in the study, the intestines being exposed to IR for two hours or longer. At a distance of 45.8 cm from the source, 87% of the rabbits developed adhesions; at a distance of 91.6 cm, 37% developed adhesions. The incidence and severity of the adhesions correlated directly with the period of intestinal exposure to IR. No histological evidence of intestinal injury was apparent in intestines exposed to the radiation under the conditions of this study. Although the results obtained may not be definitely applicable to the human infant, they do suggest that precautions should be taken when viscera are exposed to such external IR sources.

The genetic effects of IR have been studied by Gordon & Surrey *(49)* using rat liver mitochondria. They postulate that the sites of ATP production were the primary target of IR, the occurrence of chromosome aberrations being a secondary phenomenon. In 1971, Gordon et al. *(50)* again demonstrated such effects using pig kidney cells irradiated with near-IR. Significant increases in chromatid breaks and exchanges were observed. Summarizing the work of other investigators in the field, Krell et al. *(51)* stated that near-IR acts to inhibit repair of spontaneous aberrations by interfering with chromosome hydrogen bonding by base pairing, either in the double-stranded DNA or in the complex structure of histone, enzymes and DNA or in both. The energy required to dissociate hydrogen bonds is 0.06 eV on average, which is within the capability of IR. If the radiation can cause such effects, mutagenesis can perhaps also occur. Inheritable changes due to IR, however, have not been demonstrated to date.

The significant number of reports in the literature of genetic effects following exposure of insects, plants and animal cells to IR must give rise to some concern that the radiation may be capable of affecting human cells.

As far as carcinogenesis is concerned, very little direct or indirect evidence exists at present to show that IR may be a causative agent *(52)*.

EXISTING STANDARDS

There are at present no standards for exposure of the skin or eye to IR non-laser "extended" sources. The only one available is standard Z-136.1 of the American National Standards Institute *(53)* which was developed for laser sources. It may be used for evaluating other sources, but it should be noted that the use of laser standards for setting "broad-band" IR industrial

105

standards can be regarded only as a temporary measure. Since lasers operate with very narrow wavelength bands, comparison with the broad-band industrial sources is difficult. In addition, the lack of industrial exposure data makes it extremely difficult to determine conclusively the occupational thresholds for ocular damage. Fairly large safety margins are therefore used to compensate for this uncertainty.

Safe IR exposure levels in the eye have been suggested by a number of investigators. In 1968, Matelsky *(38)* stated that acute ocular damage from incandescent "hot" bodies can occur with radiant exposure of 40–80 kJ/m² incident on the cornea; however, this single radiant exposure concept ignores exposure duration. Based on the threshold data previously discussed, he recommended that a maximum permissible radiant energy of 4–8 kJ/m² would probably prevent the occurrence of chronic effects on the intraocular tissues. Sliney *(7)* has used experimental laser injury data to recommend that a safe level for chronic exposure to industrial IR sources should be limited to an average ocular irradiance of approximately 0.1 kW/m² with allowable incidental exposure for several minutes up to 1 kW/m².

The International Organization for Standardization (ISO) has proposed an IR exposure standard and the American Conference of Governmental Industrial Hygienists (ACGIH) has issued a notice of intent to establish a threshold limit value (TLV) for near-IR from broad-band sources *(54)*. The TLV for occupational exposure to IR for the eye applies to exposure in any eight-hour working day and requires a knowledge of the spectral radiance (L_λ) and total irradiance of the source as measured at the position(s) of the eye of the worker. The proposed TLV are:

   1. To protect against retinal thermal injury, the spectral radiance of a lamp source weighted against the burn hazard function (Table 2) should not exceed:

$$\sum_{\lambda = 400}^{1400} L_\lambda R_\lambda \Delta\lambda \leq \frac{1}{\alpha t^{1/2}}$$

where L_λ is the spectral radiance measured in W/(m²·sr), R_λ is the dimensionless burn hazard function that changes with each increment of the band $\Delta\lambda$, t is the viewing duration in seconds, and α is the subtense of the source measured in radians. $\alpha = l/r$, where l is the length of the lamp and r is the viewing distance, both measured in metres.

   2. To avoid possible delayed effects on the lens of the eye (cataracto-genesis) the IR radiation ($\lambda > 780$ nm) should be limited to 100 W/m². For an IR heat lamp or any near-IR source where a strong visual stimulus is absent, the near-IR (780–1400 nm) radiance as viewed by the eye over an extended period should be limited to:

$$\sum_{\lambda = 780}^{1400} L_\lambda \Delta\lambda = \frac{0.6}{\alpha}$$

This limit is based on a pupil diameter of 7 mm.

   It should be noted that both formulae are empirical and are not dimensionally correct. (To make them dimensionally correct, it would be necessary to

Table 2. Spectral weighting function for assessing
retinal hazards from broad-band optical sources

Wavelength, λ (nm)	Burn hazard function, R_λ
400	1.0
405	2.0
410	4.0
415	8.0
420	9.0
425	9.5
430	9.8
435	10.0
440	10.0
445	9.7
450	9.4
455	9.0
460	8.0
465	7.0
470	6.2
475	5.5
480	4.5
485	4.0
490	2.2
495	1.6
500– 600	1.0
600– 700	1.0
700–1050	$10^{(700-\lambda)/505}$
1050–1400	0.2

Source: ACGIH *(54)*.

insert a meaningless dimensional correction factor k in the right-hand numerator in each formula. In all cases, the numerical value of k will be unity.)

CONTROL MEASURES

The major areas of concern regarding occupational exposure to IR, especially in the near-IR region, are protection of the skin and eyes. This can best be accomplished by engineering means, i.e. by controlling the emission from the source. At distances comparable with the largest source dimension, the irradiance is inversely proportional to the distance from the source. With increasing distance there is a gradual change to the well known inverse square law. In this respect IR behaves in a similar fashion to all other radiation. Measures for controlling emissions have not been extremely sophisticated, yet their effectiveness, if used properly, can be very good. The high surface temperature of hot equipment can be reduced by insulation. However, this method is usually practicable only for equipment with low surface temperatures because of the thickness of the insulating material needed if it is to be effective. Engineering control measures involve the placing of an object between the source and the receiver to reflect and reduce the transmission of IR. Aluminium, because of its high reflectance, is widely used in the form of foil, sheeting or corrugated siding (sheets) as screening for the control of IR from furnaces and other IR sources in the industrial environment. The thickness of the material is not important and only the reflectance needs to be considered. If used, such a surface must be well polished and kept clean for maximum effectiveness. Other types of shield, such as glass, heat-exchanging aluminium cloth and absorbing plastic, use absorptive material to accept IR and give up heat by convection. Several shields may be used, cooled either by special ventilation or by water circulation. For convenience, such shields should probably be portable so that they can be removed in case of emergency repair or maintenance of machinery. Enclosures, e.g. baffled lamp housings, are employed to control the potential hazards from laboratory arc sources, such as spectroscopic equipment and optical calibration sources. Reflective booths and curtains have been used in welding operations to protect passers-by. Although the majority of welding curtains have been fairly effective in reducing the transmission of ultraviolet and visible radiation, recent investigations have shown that some transparent curtains transmit up to 80% of the IR *(55)*. The spectral transmission properties of such curtains should be examined before they are used for protection against IR exposure. In paint and enamel drying operations, where an intense array of heating lamps may be used, such enclosures may be provided with glass or metal doors and an interlock system. In addition, it is advisable to post warning signs.

Pre-employment medical examinations, with particular attention to eye and skin lesions, can be an important means of preventing the assignment of susceptible workers to work associated with intense IR.

If engineering controls are absent or inadequate an alternative method, but one which is probably the least effective, is the use of personal protective devices. For protection of the skin, lightweight cotton clothing is recommended. Additional protection can be afforded by the wearing of reflective aluminized aprons, coats or gloves. For excessive radiant heat loads, reflective suits may be necessary.

As regards protection of the eye, the filters used in welding goggles, and in furnace inspection goggles for glass, steel and foundry workers, were originally developed empirically. However, optical transmission characteristics for the near-IR region are specified by various standards *(56–58)*.

The various types of goggles and spectacles fitted with suitable filters are shown in Fig. 13. In a recent study, Campbell *(59)* found that only one of 55 shade models of welding filter plates failed to meet the ANSI specifications for IR transmittance up to 2.6μm. However, in absorbing the incident radiation, filter plates may be heated to such an extent that corneal heating with accompanying pain occurs, prompting the worker to remove

Fig. 13. Recommended eye protectors

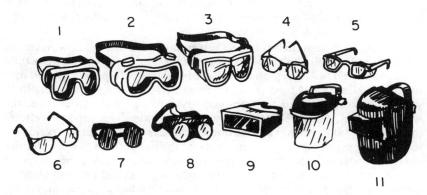

Note. 1. Goggles, flexible fitting, regular ventilation. 2. Goggles, flexible fitting, hooded ventilation. 3. Goggles, cushioned fitting, rigid body. 4. Spectacles, metal frame, with sideshields. 5. Spectacles, plastic frame, with sideshields. 6. Spectacles, metal-plastic frame, with sideshields. 7. Welding goggles, eyecup type, tinted lenses. 8. Welding goggles, coverspec type, tinted lenses. 9. Welding goggles, coverspec type, tinted plate lens. 10. Face shield (available with plastic or mesh window). 11. Welding helmet.

the goggles or spectacles. Reflective metal coatings deposited on the front surface of the absorbing filter should not aggravate this situation if the incident energy is really reflected. Fortunately, electric welding arcs emit relatively little radiation in the far-IR region. Hubner et al. *(28)* recommended that filter specifications for the degree of transmittance in the near-IR region should be made more severe; they could be less stringent for the middle- and far-IR regions (except for high-temperature welding). As far as furnace inspection goggles are concerned, however, they recommended that the eye protection specifications should be made considerably more stringent under all circumstances. In foundries where cobalt-blue glass is used when estimating the temperature of the melting metal, it is advisable to use heat-absorbing glasses as well.

An important factor in the design of eye protection devices is comfort. No matter how effective the filter may be, if goggles are not reasonably comfortable they will not be worn. In reality, a compromise must be reached between comfort and safety. However, in the future, safety may have to be emphasized more strongly because of the rapid increase in new, near-IR radiation emitting sources, plus the rise in the number of potentially exposed workers.

PROBLEMS AND RECOMMENDATIONS

In general, the data presented in the early literature, which comprises the greater part of that available on the bioeffects of IR radiation, are of

109

doubtful value due to lack of information on instrument sensitivity and source spectrum. There is therefore a need for further research to determine threshold limits, irradiance values, the effect of time, and the epidemiology of effects due to IR radiation exposure.

Skin hazards seem to be confined essentially to radiation of wavelength 1.1–1.2μm. Possibly the near-IR region is potentially more hazardous than the middle- and far-IR regions because of the ability to penetrate well into the dermis. Caution should therefore be exercised in situations where individuals may be exposed to high-intensity occupational sources, i.e. the newer sources, which emit in the near-IR region. The reported effects include acute skin burn, vasodilation of capillary beds, and an increased, long-lasting pigmentation. Extensive research on the effects of low-intensity chronic exposure is lacking. The research reported, however, has revealed only an erythematous-like appearance and some eyelid inflammation in certain occupations. It is self-evident that the skin has an inherent natural protective mechanism that enables it to "sense" the warming effect of IR before the skin temperature reaches the pain or burn threshold. In addition, the body has a fairly effective heat dissipation mechanism for protection against the potentially damaging effects of IR. Some injury, however, may occur in the future from excessive exposure to high-intensity near-IR sources, but at present such cases have been infrequent. The effect of middle- and far-IR as a factor in causing heat stress may decrease in the future because of the use of newer sources that emit predominantly in the near-IR region.

There are infrequent references to the occurrence of discomfort similar to that of chronic conjunctivitis as a result of exposure to IR. The increased evaporation of the tear film could, of course, result in an abnormal state of the conjunctiva and cornea. Where the tear film is reduced as a result of malfunction, mainly of the accessory lachrymal glands, a state known as "dry-eye" is produced. In cases of "dry-eye", it may be assumed that IR exposures would aggravate the disease; exposures of such persons to IR should therefore be avoided.

Threshold limits for the occupational exposure of the cornea, lens and retina to IR do not exist at present. Although limited experimental thresholds for damage to ocular structures have been established, there are still insufficient data for comparative purposes.

There is a great need for research to establish exposure standards, determine irradiance levels, and evaluate the role that exposure duration plays in causing damage to the eye and skin. Data collected should include not only the area and spectral distribution of the source but also the working conditions, protective measures and working distance, so that a more precise estimation of the radiation dose incident on the eye and skin can be made.

Although present knowledge of IR bioeffects is insufficient for the conclusive determination of threshold levels, it appears that a more conservative standard of 0.1 kW/m^2 would be prudent, especially when a comparison is made with man's natural environment.

110

CONCLUSIONS AND RECOMMENDATIONS[a]

Conclusions

All people are exposed to IR radiation from sunlight, artificial light and radiant heating. Exposures to IR are quantified by irradiance (W/m^2) and radiant exposure (J/m^2) to characterize biological effects on the skin and cornea. However, near-IR exposure to the retina requires knowledge of the radiance (W/(m^2·sr)) of the IR source. With most IR sources in everyday use the health risks are considered minimal; only in certain high radiant work environments are individuals exposed to excessive levels.

Comprehensive measurements for hazard evaluation should include (a) spectral irradiance and radiance of the source, (b) exposure duration, and (c) frequency of repeated exposures. However, if the source is not too intense, it may suffice to measure only the total irradiance and radiance to assess the hazard.

The interaction of IR radiation with biological tissues is mainly thermal. IR radiation may augment the biological response to other agents. The major health hazards are thermal injury to the eye and skin, including corneal burns from far-IR, heat stress, and retinal and lenticular injury from near-IR radiation.

Recommendations

Control measures should include education and training of all personnel working with IR sources. Engineering controls include the proper layout of working areas, construction of baffles and enclosures, a separation distance from the source, and good ventilation to reduce the risk of heat stress. Personal protection may consist of special eyewear and appropriate clothing and protective garments.

Individuals chronically exposed to high levels of IR, such as glass workers, should wear IR-absorbing protective lenses.

Further research on the health hazards of exposure to IR does not seem urgent at present, but this position may change following understanding of the fundamental mechanisms of the interaction of heat with biological systems.

REFERENCES

1. *International lighting vocabulary,* 3rd ed. Paris, International Commission on Illumination, 1970 (Publication No. 17 (E-1.1)).
2. **Keyes, R.J., ed.** *Optical and infrared detectors.* Berlin, Springer-Verlag, 1977, Vol. 19.

---

[a] These conclusions and recommendations are those pertaining to infrared radiation made by the WHO Working Group on Health Implications of the Increased Use of NIR Technologies and Devices, Ann Arbor, USA, October 1985.

3. **Wolfe, W. & Zissis, G.J., ed.** *The infrared handbook*. Washington, DC, Government Printing Office, 1981.

4. *Criteria for a recommended standard: occupational exposure to hot environments*. Cincinnati, OH, National Institute for Occupational Safety and Health, 1972 (Publication No. HSM 72-10269).

5. **Boettner, E.A. & Wolter, J.R.** Transmission of the ocular media. *Investigative ophthalmology,* **1**: 766–783 (1962).

6. **Geeraets, W.J. & Berry, E.R.** Ocular spectral characteristics as related to hazards from lasers and other light sources. *American journal of ophthalmology,* **66**: 15–20 (1968).

7. **Sliney, D.H.** Nonionizing radiation. *In:* Cralley, L.V., ed. *Industrial environmental health*. New York, Academic Press, 1972, Vol. 1, pp. 171–241.

8. **Fischer, F.P. et al.** Über die zur Schädigung des Auges nötige Minimalquantität von ultraviolettem und infrarotem Licht [On the minimum amount of ultraviolet and infrared light required to damage the eye]. *Archiv für Augenheilkunde,* **109**: 462–467 (1935).

9. **Franke, W.** Der Feuerstar in meßtechnischer Beziehung und Gewerbehygienischer Bedeutung [Glassblowers' cataract with regard to the measuring technique and occupational hygiene]. *Archiv für Gewerbepathologie und Gewerbehygiene,* **16**: 539–554 (1958).

10. **Ruth, W.** *A method to evaluate occupational hazards from infrared radiation*. Thesis, Loughborough, United Kingdom, 1975.

11. **Turner, H.S.** *The interaction of infrared radiation with the eye: a review of the literature*. Columbus, OH, Ohio State Research Foundation, 1970.

12. **Duke-Elder. W.W.** The pathological action of light upon the eye, II: Action upon the lens; theory of the genesis of cataract. *Lancet,* **1**: 1250–1254 (1926).

13. **MacKenzie, W.A.** *A practical treatise on diseases of the eye*. Philadelphia, PA, Blanchard & Lee, 1985.

14. **Meyhofer, W.** (1886). Quoted by Turner, H.S. *The interaction of infrared radiation with the eye: a review of the literature*. Columbus, OH, Ohio State Research Foundation, 1970.

15. **Lydahl, E.** Infrared radiation in cataracts. *Acta ophthalmologica,* Suppl. 166, pp. 1–63 (1984).

16. **Goldmann, H.** The origin of glassblowers' cataract. *Annales d'oculistique,* **172**: 13–41 (1935).

17. **Goldmann, H. et al.** The permeability of the eye lens to infrared. *Ophthalmologica,* **120**: 198–205 (1958).

18. **Robinson, W.** Glass-workers' cataract. *Ophthalmology,* **13**: 353–554 (1915).

19. **Verhoeff, F.H. et al.** The pathological effects of radiant energy on the eye: an experimental investigation with a systematic review of the literature. *Proceedings of the American Academy of Science,* **51**: 630–818 (1916).

20. **Vogt. A.** [Fundamental investigations of the biology of infrared]. *Klinisches Monatsblatt für Augenheilkunde,* **89**: 251–263 (1932) (in German).

21. **Goldmann, H.** Genesis of heat cataract. *Archives of ophthalmology,* **9**: 314–316 (1933).
22. **Pitts, D.G. & Cullen, A.P.** Determination of infrared radiation levels for acute ocular cataractogenesis. *Von Graefes Archiv für klinische und experimentelle Ophthalmologie,* **217**: 285–297 (1981).
23. **Wolbarsht, M.L.** Damage to the lens from infrared. *Proceedings of the Society for Photo-Optical and Instrumentation Engineers,* **229**: 121–142 (1980).
24. **Jacobsen, J.H. et al.** *The effects of thermal energy on anterior ocular tissues.* Wright-Patterson Air Force Base, Ohio, 1963 (AMRL-TDR-63-53).
    Dunn, K.L. Cataract from infrared rays (glass-workers cataract). *Archives of industrial hygiene and occupational medicine,* **1**: 166–180 (1950).
26. **Keatings, G.F. et al.** Radiation cataracts in industry. *Archives of industrial health,* **11**: 305–314 (1955).
27. **Wallace, J. et al.** An epidemiological study of lens opacities among steel workers. *British journal of industrial medicine,* **28**: 265–271 (1971).
28. **Hubner, H.J. et al.** Measurement of radiant power at welding processes and consequences for eye protection against IR radiation. *Optik,* **31**: 462–476 (1970).
29. **Hager, G. et al.** Fire cataracts among locomotive firemen. *Verkehrsmedizin,* **18**: 443–449 (1971).
30. **Sensintaffar, E.L. et al.** An analysis of a reported occupational exposure to infrared radiation. *American Industrial Hygiene Association journal.* **39**: 63–69 (1978).
31. **Medvedovskaya, T.P.** [Data on the condition of the eye in workers at a glass factory]. *Gigiena i sanitarija,* **35**: 105–106 (1970) (in Russian).
32. **Clark, B.A.J.** Welding filters and thermal damage to the retina. *Australian journal of optometry,* **51**: 91–98 (1968).
33. **International Commission on Radiological Protection.** *Report of the Task Group on Reference Man.* Oxford, Pergamon Press, 1975 (ICRP Publication No. 23).
34. **Bloom, W. & Fawcett, D.W.** *A textbook of histology.* Philadelphia, PA, Saunders, 1968.
35. **Jacques. J.A. et al.** Spectral reflectance of human skin in the region 0.7–2.6μm. *Journal of applied physiology,* **8**: 297–299 (1955).
36. **Hardy, J.D. et al.** Spectral transmittance and reflectance of excised human skin. *Journal of applied physiology,* **9**: 257–264 (1956).
37. **Michaelson, S.M.** Human exposure to non-ionizing radiant energy — potential hazards and safety standards. *Proceedings of the Institute of Electrical and Electronics Engineers,* **60**: 389–421 (1972).
38. **Matelsky, I.** The non-ionizing radiations. *In: Industrial hygiene highlights.* Pittsburgh, PA, Industrial Hygiene Foundation of America, 1968, Vol. 1, p. 140.
39. **Henriques, F.C., Jr.** Studies of thermal injury: V. Predictability and significance of thermally induced rate processes leading to invisible thermal injury. *Archives of pathology,* **43**: 489–502 (1947).

40. **Hardy, J.D. et al.** Influence of skin temperature upon pain threshold as evoked by thermal radiation. *Science,* **114**: 149–150 (1951).
41. **Hardy, J.D.** Thermal radiation, pain and injury. *In:* Licht, S., ed. *Therapeutic heat.* New Haven, CT, 1958, Vol. 2, p.157.
42. **Krivobok, V.T.** [Effect of infrared rays on eyes]. *Eksperimental'naja medicina,* **8**: 53–55 (1941) (in Russian).
43. **Zelencova, S.P.** [Immunological reactivity of the organism under the effects of interrupted infrared radiation]. *Vračebnoe delo,* **12**: 88–91 (1968) (in Russian).
44. **Zelencova, S.P.** [The effect of intermittent infrared radiation on the status of natural immunological reactivity of workers]. *Gigiena truda i professional'nye zabolevanija,* **14**: 22–26 (1970) (in Russian).
45. **Lehmann, J.F. et al.** Pain threshold measurements after therapeutic application of ultrasound, microwaves and infrared. *Archives of physical medicine and rehabilitation,* **39**: 560–565 (1958).
46. **Borneff, J. & Blumlein, H.** Damages of the respiratory passages in occupational exposure to heat. *Medizinische Klinik,* **13**: 494–497 (1960).
47. Infrared radiant warmers. *Health devices,* November 1973, p. 4.
48. **Arima, E. & Fonkalsrud, E.W.** The relationship of intestinal adhesions to infrared heating lamp exposure. *Journal of pediatric surgery,* **10**: 231–234 (1975).
49. **Gordon, S.A. & Surrey, K.** Red and far-red action on oxidative phosphorylation. *Radiation research,* **12**: 325–339 (1960).
50. **Gordon, S.A. et al.** The induction of chromosomal aberration in pig kidney cells by far-red light. *Radiation research,* **45**: 274–287 (1971).
51. **Krell, K. et al.** Effects of red, far-red and infra-red radiant energy. *In: Symposium on Biological Effects and Measurement of Light Sources.* Washington, DC, US Food and Drug Administration, 1977, pp. 247–254 (Publication No. 77-8002).
52. **Cunningham-Dunlop, S. & Kleinstein, B.H.** *A current literature report on the carcinogenic properties of ionizing and non-ionizing radiation, Part I. Optical radiation.* Cincinnati, OH, National Institute for Occupational Safety and Health, 1977 (Publication No. 78-122).
53. *Safe use of lasers.* New York, American National Standards Institute, 1980 (ANSI Z-136.1).
54. *Threshold limit values for chemical substances and physical agents in the workroom environment.* Cincinnati, OH, American Conference of Governmental Industrial Hygienists, 1985.
55. **Moss, E.E. & Gawenda, M.C.** *Optical radiation transmission levels through transparent welding curtains.* Cincinnati, OH, National Institute for Occupational Safety and Health, 1978 (Publication No. 78-176).
56. *American national standard practice for occupational and educational eye and face protection.* New York, American National Standards Institute, 1980 (ANSI Z-87.1).
57. *Green protective spectacles and screens for steelworks operatives.* London, British Standards Institution, 1952 (BS 1729).

58. *Welding, cutting and brazing.* Washington, DC, US Department of Labor, 1976 (Code of Federal Regulations, Title 29, Chapter XVII, part 1910.252).
59. **Campbell, D.L.** *Report on tests of welding filter plates.* Cincinnati, OH, National Institute for Occupational Safety and Health, 1976 (Publication No. 76-198).

4

Radiofrequency radiation

J.A. Elder, P.A. Czerski,
M.A. Stuchly, K.H. Mild & A.R. Sheppard

CONTENTS

	Page
Introduction	117
Quantities and units	118
Exposure sources and devices	121
Ambient environment	122
High-power sources	122
Low-power sources	122
Occupational exposure	123
Medical exposure	124
Instrumentation and measurement	125
Calibration techniques and accuracy	125
Survey of potentially hazardous fields	126
Dosimetry	128
Interaction mechanisms	134
Thermal mechanisms of interaction	134
Athermal mechanisms of interaction	136
Electric shock and burn	137
Biological effects	138
Human studies	138
Non-human studies	148
Protective measures and standards	151
Conclusions and recommendations	154
Conclusions	154
Recommendations	157
References	158

INTRODUCTION

The radiofrequency (RF) fields considered in this chapter extend from
300 kHz to 300 GHz. Natural levels of these fields in the environment are
low. Numerous uses are made of this part of the electromagnetic spectrum in

various areas of human activity, such as communications, navigation, industry and medicine, and have resulted in workers and the general population being exposed to RF fields exceeding the naturally occurring levels. As with any form of energy, RF energy has the potential to interact with biological systems, and the outcome may be of no significance, may cause different degrees of harm, or may be beneficial.

There are comprehensive reviews providing detailed information on the biological effects of RF fields on animals and man. It is the purpose of this chapter to give a brief overview of the biological effects pertinent to a health professional who is not a specialist in the field of bioelectromagnetics. Furthermore, information is given on sources and levels of RF exposure, and on methods of measuring exposure fields. Concepts of dosimetry employed in quantifying the interactions of RF fields with biological systems and interaction mechanisms are also outlined. Finally, advice is given on protective measures which should prove useful in practical situations involving human exposure to RF fields.

QUANTITIES AND UNITS

Fields at RF are quantified in terms of electric field strength (E) and magnetic field strength (H). In the SI system, E is expressed as volts per metre (V/m) and H as amperes per metre (A/m). Both are vector fields, i.e. they are characterized by magnitude and direction at each point. Comprehensive reviews of concepts, quantities, units and terminology for nonionizing radiation protection, including radiofrequency radiation, are available (1,2). More information on electric and magnetic fields can be found in Chapter 5.

Electromagnetic waves are waves of electric and magnetic forces, where a wave motion is defined as propagation of disturbances in a physical system. A change in the electric field is accompanied by a change in the magnetic field, and vice versa.

Electromagnetic waves are characterized by a set of parameters that include frequency (f), wavelength (λ), electric field strength (E), magnetic field strength (H), electric polarization (P) (the direction of the E field), velocity of propagation (c) and Poynting vector (S). Fig. 1 illustrates the propagation of an electromagnetic wave in free space and some of the parameters describing the wave. The frequency is defined as the number of complete changes of the electric or magnetic field at a given point per second, and is expressed in hertz (Hz). The wavelength is defined as the distance between two consecutive crests or troughs of the wave (maxima and minima). The frequency, wavelength and wave velocity (v) are interrelated as follows:

$$v = f\lambda \tag{1}$$

In free space, wave velocity is equal to the velocity of light. In materials, the wave velocity and wavelength depend on the electrical properties of the material, i.e. permittivity (ε) and permeability (μ). The permittivity defines

118

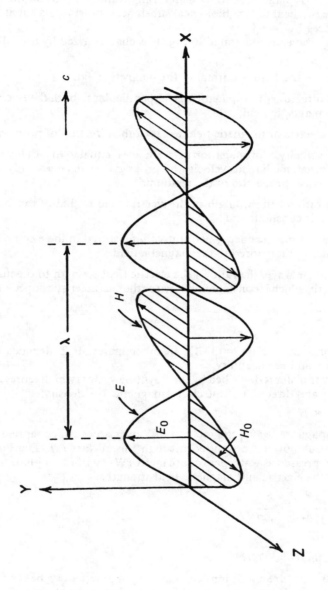

Fig. 1. A plane wave propagating in the direction X with a velocity c equal to the speed of light

119

the material interactions with the electric field, and the permeability expresses the interactions with the magnetic field. Biological substances have a permittivity that differs vastly from that of free space, changes by many orders of magnitude over the RF range, and depends on the type of tissue. The permeability of biological substances, however, is equal to that of free space *(3)*.

A plane wave, as illustrated in Fig. 1, is characterized by the following features:

— the electric field is normal to the magnetic field;

— the direction of propagation is perpendicular to both the electric and the magnetic field;

— no electric or magnetic field exists in the direction of propagation;

— the velocity of propagation in free space is equal to that of light, while in other media the velocity of propagation depends only on the electrical properties of the medium;

— the ratio of the strength of the electric field to that of the magnetic field is constant; and

— at any time the energy per unit volume stored in the electric field is equal to that stored in the magnetic field.

For a plane wave, the ratio of the electric field strength to the magnetic field strength, which is constant, is known as the characteristic impedance (Z):

$$Z = \frac{E}{H} \tag{2}$$

In free space, $Z = 120\pi \simeq 377\Omega$; in other materials Z depends on the permittivity and permeability.

Energy transfer is described by the Poynting vector, which represents the magnitude and direction of the electromagnetic flux density:

$$S = E \times H \tag{3}$$

For a propagating wave, the integral of S over any surface represents the instantaneous power transmitted through this surface *(2)*. The Poynting vector is expressed as watts per square metre (W/m^2) and for plane waves is related to the electric and magnetic field strengths:

$$S = \frac{E^2}{120\pi} = \frac{E^2}{377} \tag{4}$$

and

$$S = 120\pi H^2 = 377H^2 \tag{5}$$

Not all exposure conditions encountered in practice can be represented by plane waves. Close to sources of RF radiation the relationships characteristic of plane waves are not satisfied. The electromagnetic field radiated by an antenna can be divided into two regions: the near-field zone and the far-field zone. Frequently, the near-field is subdivided further. In the immediate vicinity of an antenna is the reactive zone, where the energy is nearly fully stored and not radiated. Next to it is the radiation near-field. The criterion most frequently used to divide the space between the far- and near-fields is that the phase of the fields from all of the points of the radiating antenna aperture does not differ by more than one sixteenth of the wave-length. The distance from the antenna corresponding to this criterion is:

$$r = \frac{2a^2}{\lambda} \tag{6}$$

where a is the greatest dimension of the antenna.

In the near-field, exposure has to be characterized by both the electric and the magnetic fields. In the far-field one of these suffices, as they are interrelated by equation (2). In practice, exposures in the near-field often occur at frequencies below 300 MHz.

Exposures to RF fields are further complicated by interactions of electro-magnetic waves with objects. In general, when electromagnetic waves encounter an object a portion of the incident energy is reflected, some is absorbed by the object, and some is transmitted. The proportion of energy transmitted, absorbed or reflected depends on the frequency and polar-ization of the field, and the electrical properties and shape of the object. A superimposition of the incident and reflected waves results in standing waves, and spatially non-uniform field distribution. Since waves are totally reflected from metallic objects, standing waves are formed close to such objects.

Since interactions of RF fields with biological systems depend on many different field characteristics, and because of the complexities of exposure fields encountered in practice, the following factors should be considered in describing exposures to RF fields:

— near- or far-field;

— electric and magnetic field strengths for the near-field, or one of these for the far-field;

— spatial variations of the magnitude of the field(s);

— field polarization, i.e. the direction of the electric field with respect to the direction of wave propagation.

EXPOSURE SOURCES AND DEVICES

There are numerous and continually increasing uses of RF energy, some of which result in the exposure of workers or the general population to RF fields. In the evaluation of importance and impact of RF sources on human health the following factors should be taken into consideration:

— the potential for producing hazardous levels of RF radiation under normal operating conditions and under conditions of possible malfunction;

121

— the number of sources in use; and

— the number of people who might be exposed and the exposure duration.

Furthermore, hazard awareness and training in safety rules on the part of the personnel operating or servicing a potentially harmful device, and others who work in the vicinity of such a device, are very important.

Ambient environment

The urban RF radiation environment is dominated by radio and television transmissions. The estimated cumulative median exposure power density in the United States due to FM radio and VHF and UHF television is $50 \mu W/m^2$. About 0.6% of those living in large American cities may be exposed to general environmental levels equal to or greater than $10 \, mW/m^2$ (4). Similar estimates have not been made in other countries.

High-power sources

A high-power source is arbitrarily defined as producing a main-beam density of $1 \, W/m^2$ at a distance of 100 m from the source. Low-power sources may produce equivalent main-beam power densities very close to the source, but the levels decrease very rapidly with distance and are much less than $1 \, W/m^2$ at distances of the order of 100 m. Systems belonging to the high-power source category are generally characterized by high transmitter power and high antenna gain, so that the $1 \, W/m^2$ at 100 m criterion is satisfied. This category, in addition to radio and television transmitters, includes tracking and acquisition radars (including air traffic control radar, weather radar and aircraft acquisition radar) as well as communication systems, such as satellite communications earth terminals and tropo-scatter systems (5).

The main-beam average power densities generated by some high-power radar systems can be greater than $1 \, kW/m^2$. However, the opportunity for exposure to the primary beam close to the source is remote.

Low-power sources

Common RF sources, such as traffic radar used by police, microwave relay systems used in telephone communications, and cable television distribution, are considered to be low-power sources. Irradiation by the main beam from a typical traffic radar results in exposure levels of the order of $10 \, mW/m^2$ or less at distances closer than 10 m. Exposure of persons to radiation produced by microwave relay systems used in telephone or television communications is much less than $10 \, mW/m^2$. Since the latter are generally mounted on tall towers or buildings and have highly collimated main beams, irradiation of people by the main beam is extremely unlikely.

In microwave ovens, the energy is enclosed and emission of radiation is not intended. Technological limitations, however, usually result in some small leakage of energy. Personal exposure from microwave ovens is very small because of the rapid decrease in power density with distance. For

122

instance, for a leakage of 50 W/m² at 5 cm from the oven surface, the power density at 0.3 m is less than 15 W/m², and at 1 m it is about 0.1 W/m².

Occupational exposure

Induction heating
Conductive materials can be heated by eddy currents induced by applying alternating magnetic fields to the material, and such heating is used mainly for forging, annealing, tempering, brazing and soldering. The operating frequency can reach a few MHz, but this type of equipment usually operates from 50 or 60 Hz up to about 10 kHz. Since the dimensions of coils producing the magnetic fields are often small there is seldom high-level whole-body exposure, but rather local exposure of, for instance, the hands. Magnetic flux density to the hands of the operator may reach 25 mT or 20 kA/m *(6,7)*. In most cases the flux density is less than 1 mT or about 800 A/m. The field strength for other parts of the body of the operator is considerably lower. The electric field strength near the induction heater is usually low.

Dielectric heating
RF energy is used to heat dielectric materials in a number of industrial processes, the most common application being sealing of plastics. Other uses include glue drying, curing particle boards and panels, and heating fabrics and paper. These RF heaters usually operate at one of the industrial, scientific and medical (ISM) frequencies, namely 13.56, 27.12 or 40.68 MHz.

The relatively high output power, from a few kW to a few tens of kW, and the use of unshielded electrodes can produce relatively high fields around RF heaters. Surveys in the United States have indicated that 60% of the devices measured exposed the operator to E fields greater than 200 V/m, and 29% to H fields greater than 0.5 A/m *(8)*. Similar results were obtained in a Canadian study *(9)*. As part of an epidemiological study of RF sealer workers in Sweden field strengths were measured for various anatomical positions, and in 55% of the cases they were in excess of either 300 V/m or 0.8 A/m at one or more locations *(10)*.

Clearly, there is major concern for operators working with RF heaters, particularly since such heaters operate at frequencies in or near the region of human resonant absorption for a person in electrical contact with the ground. Hence, workers near some of these devices absorb RF energy at rates above those recommended by, for example, the American National Standards Institute (ANSI) *(11–13)*. To reduce stray fields to acceptable levels it is necessary to shield the electrode system, divert the RF energy, or reduce exposure by administrative measures *(14)*.

Communication systems
While most workers in the fields of communication and radar are exposed only to low-strength fields, a few situations exist in which workers can potentially be exposed to high levels of RF radiation *(5,15–17)*. Field

strengths are usually very low (below 1μW/cm^2) in radio transmitter rooms, near the bases of transmitter towers, and in adjacent high-rise buildings and surrounding areas. High-level exposures occur in the vicinity of antennas of high power transmitters. Workers climbing FM television towers may be exposed to high field strengths with a local E field of up to about 1000 V/m and a local H field of about 5 A/m (16). Potentially high exposure also exists when the interlocks of the transmitter cabinet are defeated and the door is open; the equivalent power density can then be up to 2000 W/m^2. Owing to coupling from other transmitters through the antenna feeders, high power densities may even be found in the cabinet when the transmitter is switched off.

Other sources of exposure to the general population and workers are portable and mobile radio transmitters. Strengths of the electric and magnetic fields close to the antennas are quite high, even for low output powers (5,18). However, the rate of energy deposition or specific absorption rate (SAR) in the operator is a more appropriate measure for evaluating hazards. Recently, measurements of SARs have been made in the near-field of typical antennas, and these data can be used for assessing exposure hazards from portable transmitters (18).

Medical exposure

The earliest therapeutic application of electromagnetic fields was in diathermy. Two types of diathermy are commonly used, short-wave (usually at about 27 MHz) and microwave (usually at 915 or 2450 MHz). Only a part of the patient's body is exposed to RF energy and exposure duration is limited (typically 15–30 minutes). However, exposure intensity is high and sufficient to cause a sustained increase in tissue temperature. Exposures to operators of short-wave diathermy devices may exceed 60 V/m and/or 0.16 A/m for operators standing in their normal positions (in front of the diathermy console) for some treatment regimes. Stronger fields are encountered close to the electrodes and cables (19).

Recently, electromagnetic fields have been also successfully used in inducing local hyperthermia for cancer therapy (17). As in diathermy, the patient is exposed to intense fields for a short time. There is relatively little information on operator exposure. One of the devices around which exposure fields have been measured operates at 13.56 MHz and employs coils for heating the torso, neck or thigh. Depending on the power to the coil and the coil type, exposure may exceed the ANSI guidelines (11–13) when the operator is 25 cm–1 m from the coil (20).

In medical diagnosis, RF fields are used in conjunction with static and slowly varying magnetic fields in magnetic resonance imaging (MRI). Systems used at present operate at frequencies ranging from about 6 to 100 MHz. In Canada, the Federal Republic of Germany, the United Kingdom and the United States guidelines have been issued that limit patient exposure to the dose rates close to those recommended by ANSI (11–13). The RF field in MRI devices is almost fully contained within the patient enclosure and exposure of the operators is negligible.

124

INSTRUMENTATION AND MEASUREMENT

No single measurement technique or instrumentation arrangement is capable of operating over the entire frequency range dealt with in this chapter, i.e. 300 kHz–300 GHz. In general, techniques and instruments for use in the microwave range (above about 1 GHz) are not suitable for use at lower frequencies, and vice versa.

Instrumentation can be divided into three basic parts: sensor, leads and meter. The sensor consists of an antenna combined with a detector, and provides a signal that is carried back by the leads to the metering instrumentation. To accomplish this without causing perturbation of or coupling to the field, the leads are made of high-resistance wires, or are shielded and oriented in such a manner that they do not couple to the fields. Optical fibre links are also used. Metering instrumentation consists of signal conditioning circuitry and a display device.

No existing instrument actually measures power density directly; all measure one or more components of the electric field or the magnetic field, or both. The power density can then be calculated from the far-field, plane wave relationships set out in equations (4) and (5). Loop antennas are used as magnetic field sensors, and commercially available instruments cover a frequency range up to about 300 MHz. Dipole antennas are used as electric field sensors. A single dipole or loop gives information on the polarization of the field, but the antenna must be aligned with the field direction. Three dipoles or loops in an orthogonal array provide isotropic (non-directional) response.

The two main types of detector are thermocouples and diodes. Bolometric devices are rarely used. All three types respond to the square of the voltage induced in the antenna, and the outputs can be converted directly to equivalent power density. Diode detectors become inaccurate in fields of high strengths, such as those associated with very short pulse radiation having moderate average power densities. Thermocouple and bolometric devices utilize heating by RF currents and are therefore true square-law detectors even at very high instantaneous field strengths. The bolometric instruments tend to be very robust as they are inherently self-protected against overload. The sensitivity of thermocouple instruments does not significantly depend on temperature and drift is minimal. In the thin-film form, however, they are prone to burn-out in high instantaneous fields.

Most potentially hazardous electromagnetic fields are close to radiation sources where the field strength changes markedly in space. To make meaningful measurements, especially in the near-field, it is important that the probe responds to the designated field parameter only and does not have spurious responses (e.g. it measures only H field but not E field). Probe response may be isotropic or, if nonisotropic, the probe has to be oriented to produce maximum readings. For a more detailed description of the principles of instrumentation and techniques the reader is referred to other sources (11-13,21-23).

Calibration techniques and accuracy

Several methods have been developed to calibrate RF instruments for hazard assessment under conditions simulating free space with an overall uncertainty of less than about 1 dB. These methods permit traceability to primary standards. One technique, suitable above 500 MHz, involves the use of an anechoic chamber and a waveguide horn antenna whose gain is precisely calibrated. Below 500 MHz, the preferred technique involves use of a transverse electromagnetic cell (24). With use of such a cell only a moderate amount of RF power is needed to generate fields of sufficient strength to calibrate probes over their intended range of operation. Tests that must be made in addition to absolute calibration are probe performance with respect to temperature, directional response, modulation, and RF interference with the probe output cable and readout instrumentation.

Probe sensitivity varies with frequency. For high frequency probes variations of about 1 dB are typical, and for lower frequency probes about 2 dB. At the calibration frequency an accuracy of about 0.5 dB is achievable. Nonlinearity of the response over the measured range can in some cases be large, up to about 1 dB. The error in isotropicity of response is often within about 0.5 dB for microwaves and about 1 dB in the short wave region.

For hand-held meters, the presence of the person performing the survey can significantly affect the reading. An indication of the magnitude of this effect can be obtained by remote readout techniques to compare readings with and without an operator in the vicinity.

The overall uncertainty caused by the above factors varies with the type of monitor. However, for quality instruments and good measurement procedures, accuracies of about 2 dB can be achieved in far-field measurements. For near-field measurements, a rapid spatial field variation is an additional source of error. For instance, near an RF sealer, E^2 and H^2 have been found to vary as $1/r^4$ with distance (14), and thus a small error in the distance determination results in a large error in the equivalent power density determination. A total error of some 3 dB is not uncommon in such situations.

A comparative study of a large number of hazard instruments showed a large variation in the performance of the instruments (25). The study indicates the need for regular and quite frequent calibration of instruments used to monitor potentially hazardous RF fields. It is also good practice to make a functional check of the instruments before and after each survey. This does not replace the need for calibration; it merely verifies that the instruments behave as previously and that, for instance, all antenna elements are intact.

Survey of potentially hazardous fields

Before a survey of potentially hazardous electromagnetic fields is made, it is important to determine as many of the unknown characteristics of the sources of these fields and their likely propagation characteristics as possible. This facilitates a better estimate of expected field strengths and more appropriate selection of instruments and procedures. A check list may include:

— output power for all the sources;

126

— nominal frequency (or frequencies) and spurious frequencies including harmonics;

— modulation characteristics: type, peak and root mean square values;

— for deliberate radiators, type of antenna and its beamwidth, polarization and scanning mode; for other devices, type of applicator, product to be manufactured, and duty cycle of the operation;

— polarization of the exposure field (which may be different from that of the antenna);

— distance of the sources to a test site; in a workplace, operator position.

A review of such check lists, though rather elementary, is necessary if the surveyor is to avoid simple but often surprising situations. For leakage fields it is possible that significant levels of spurious frequency fields may exist and produce RF interference for which the instrument has not been designed. If the exposure guideline is frequency-dependent, e.g. ANSI *(12)*, the harmonics may in some cases be important for compliance. Some instruments are sensitive to capacitive coupling with low frequency fields, so that the instrument becomes responsive to the RF space potential between the probe and the electronic readout device.

In the far-field, under free space conditions, on-axis power density S can be calculated from the equation:

$$S = \frac{GP_T}{4\pi d^2} = \frac{A_e P_T}{\lambda^2 d^2} \tag{7}$$

where G is the antenna gain (power ratio), P_T is the net power delivered to the antenna, d is the distance to the antenna, λ is the wavelength and A_e is the effective area of the antenna. If G is not known, a useful approximation of S can be obtained by substituting A, the physical aperture area of an antenna, for A_e; the estimated value of S will be somewhat larger than the actual value. Equation (7) can be used to estimate S at distances greater than about $a^2/2$, where a is the largest aperture dimension. At closer distances, the values given by the equation are too large and the near-field approximate estimate should be used:

$$S = \frac{4P_T}{A_e} \tag{8}$$

Ground reflections could increase S by a factor of four, and even higher factors apply if focusing structures such as corner reflectors are present nearby. On the other hand, it should be recognized that fields measured in the absence of a person may in some cases be misleading with regard to the hazard. For example, a person exposed in front of a reflecting plane reduces the magnitude of the standing wave. Area scanning must therefore be used to obtain spatial averages of exposure fields.

In the case of frequencies below 100 MHz or large aperture antennas, the existence of potentially hazardous reactive near-fields becomes relevant. These fields are not readily calculable; their contribution to apparent power density will vary very rapidly with distance — as $(1/d^3)^2$ and $(1/d^2)^2$ — so care must be exercised to avoid instrument overload as the source is approached. Furthermore, all measurements in the near-field should be made without the presence of an operator.

Measurements to evaluate fields at a workplace should be performed immediately after installation of RF-emitting equipment, and then after any major repair or other changes (e.g. in the device design, protection equipment or mode of operation of the source) that may influence the RF exposure fields.

In some countries special rules may apply to certain equipment for which there are emission standards, such as microwave ovens, RF sealers and diathermy equipment.

DOSIMETRY

Exposure of organisms to RF fields results in induction of RF fields and currents inside the body. The internal rather than the external fields and currents are responsible for interactions with biological systems independently of whether these interactions are thermal or athermal.

A dosimetric quantity that is widely used in quantifying interactions of RF fields with biological systems is the specific absorption rate, SAR. It is defined as the rate of energy transfer to a unit mass of the body *(2)*.

$$\text{SAR} = \frac{d}{dt}\left(\frac{\Delta W}{\Delta m}\right) = \frac{d}{dt}\left(\frac{\Delta W}{\rho \Delta V}\right) \tag{9}$$

where ΔW is the energy transferred to a mass Δm, ΔV is the volume containing mass Δm and ρ is the density. For sinusoidally varying fields, the SAR in a small volume of tissue throughout which the electric field is constant is:

$$\text{SAR} = \frac{\sigma E_{\text{in}}^2}{\rho} \tag{10}$$

where σ is the tissue conductivity and E_{in}^2 is the root mean square magnitude of the electric field in the tissue. The SAR is expressed in units of watts per kilogram (W/kg) or derived units, e.g. mW/g.

The initial rate of temperature increase when heat losses are neglected is also directly proportional to the SAR:

$$\frac{dT}{dt} = \frac{\text{SAR}}{c} \tag{11}$$

where T is temperature, t is time and c is the specific heat capacity.

128

Two SARs are frequently used: the average SAR, usually the whole-body average, defined as the total energy transferred to the body per unit time divided by the total mass; and the local SAR corresponding to a small volume *(26)*.

The SAR in a biological body depends on several exposure parameters such as frequency, intensity, polarization, radiation source-body configuration (the far-field and the near-field) and the presence of reflecting surfaces nearby. The SAR also depends on the size, shape and electrical properties of the body. The spatial distribution of the SAR inside the body is usually highly non-uniform and depends on all of the above parameters.

Considerable progress has been made in theoretical and experimental dosimetry. Several general reviews *(26–28)* and comprehensive descriptions of theoretical *(29–31)* and experimental methods and results *(32)* are available. Theoretical methods are generally reliable for calculations of whole-body average SAR, but at present experimental methods are more viable for determination of spatial distributions of SAR within exposed biological bodies *(33–36)*.

The whole-body average SAR for exposures in the far-field is a function of frequency and polarization. Fig. 2 illustrates a typical dependence, in this case for a spheroidal model of an average man (1.75 m, 70 kg) exposed in the far-field to 10 W/m^2. The E polarization corresponds to the electric field parallel to the main body axis, the H polarization to the magnetic field parallel to the main body axis, and the K polarization to the wave propagation from head to toe. The maximum absorption occurs at about 70–80 MHz for E polarization. This frequency is referred to as the resonant frequency for man. It should be noted that at this frequency the power absorbed is a few times greater than that obtained by multiplying the surface area of the body cross-section by the incident power density.

For a man standing in contact with RF ground, the resonant frequency shifts to about 30–40 MHz, and the SAR increases by a factor of about two.

The average SAR also depends on the size and shape of the body, as illustrated in Fig. 3, which shows the SAR curves for three species for E polarization and 10 W/m^2 *(29)*.

The spatial distribution of the SAR within an exposed body is highly non-uniform. Fig. 4 shows local SARs in the mid-plane of a model of man exposed in the far-field in the E polarization *(34)*. At all frequencies the maximum SAR is produced in the neck, although its magnitude is greater at lower frequencies. In the near-field even greater non-uniformities in the SAR are produced; Fig. 5 shows local SARs in the body mid-plane for exposure to dipole antennas close to the body *(34)*. For exposures in the near-field, such as those resulting from use of portable transmitters, ratios of the local to average SAR as high as 200 may occur *(34)*.

Dosimetric data on average SARs are useful in extrapolating observed effects in animals to those expected in man. For example, if a mouse and a man are exposed at 1 GHz (Fig. 3) the average SAR in the mouse is about ten times greater than that in the man; hence, if a biological effect is observed in a mouse at a given power density, a greater power density would be expected to be necessary to cause the same effect in man. Such extrapolation, however,

129

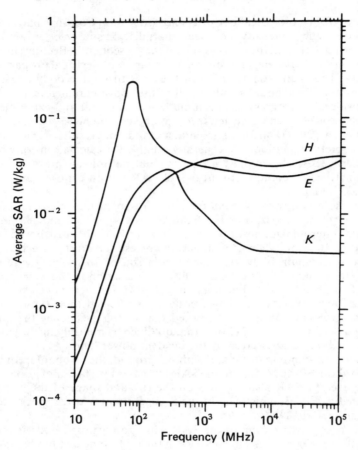

Fig. 2. The average specific absorption rate (SAR)
in a spheroidal model of an average man (1.75 m, 70 kg)
exposed to a plane wave of 10 W/m²

does not take account of differences in spatial distributions of the SAR. For man at frequencies above about 5 GHz the energy is deposited close to the body surface. Even at lower frequencies, i.e. above 900 MHz, the SAR in a human torso decreases exponentially with distance from the surface *(34,36)*. It is at frequencies between about 30 and 300 MHz that high local SARs occur in the head and torso. At frequencies below about 50 MHz for a man in contact with RF ground very high local SARs are produced in the ankles *(37)*.

Fig. 3. The average absorption rates (SAR)
in spheroidal models of a man, monkey and mouse
exposed to a plane wave of 10 W/m² in the E polarization

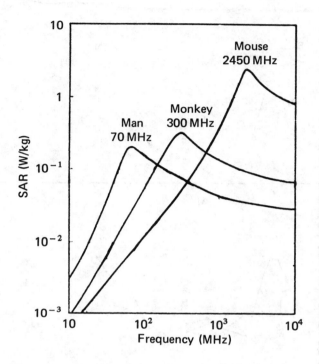

In experiments with laboratory animals the whole-body average SAR is
frequently determined by measuring the total absorbed power as a difference between the incident, reflected and transmitted powers to an enclosed
chamber (the chamber's losses should also be considered). Average SARs
are frequently measured by calorimetric methods, and local SARs by im-
plantable, nonperturbing thermometers. A detailed review of experimental
dosimetric methods can be found elsewhere (32).

131

Fig. 4. Local values of the SAR in the mid-section of a full-scale model of a man exposed to a plane wave at three frequencies

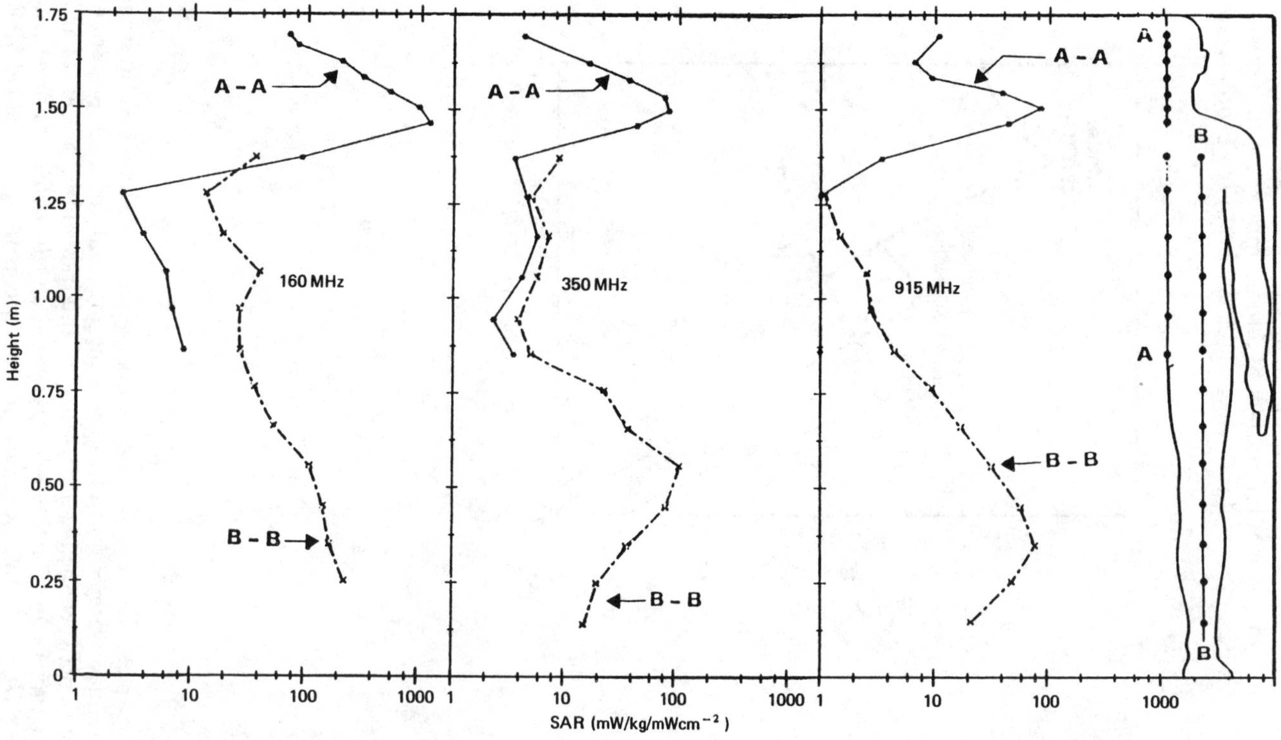

Fig. 5. Local values of the SAR in the mid-section of a full-scale model
of a man exposed in the near-field of resonant dipoles at three frequencies

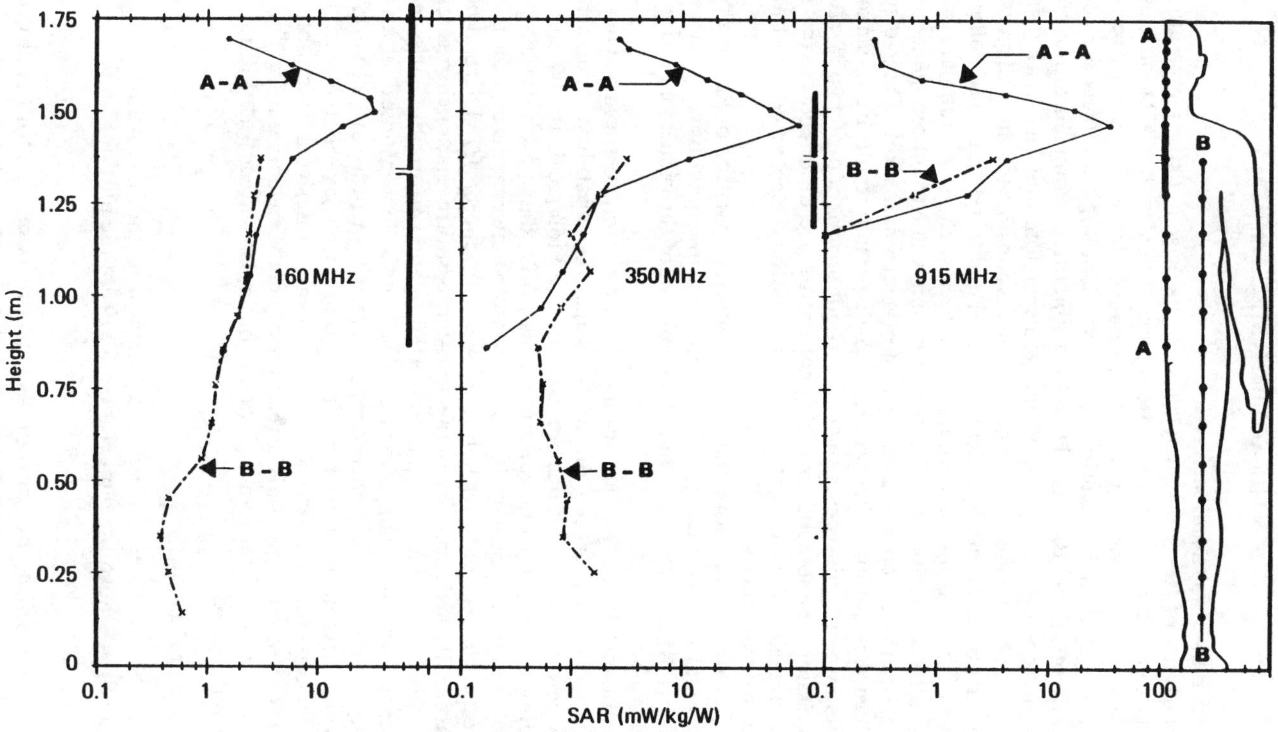

133

INTERACTION MECHANISMS

Thermal mechanisms of interaction

Absorption of RF energy may cause an increase in tissue temperature. The initial rate of temperature increase is proportional to the SAR (equation (11)). Molecular phenomena involved in a conversion of RF energy to thermal energy and their implications in terms of biological effects are reviewed elsewhere *(38,39)*. Biological tissues exhibit three strong relaxation phenomena responsible for the α-, β- and γ-dispersion, and one weak dispersion known as δ. The molecular phenomena responsible for the α-dispersion are the least understood, appearing to be related to relaxation of counterions, but in some tissues intracellular structures such as the tubular apparatus may also contribute *(38)*. The β-dispersion is due to inhomogeneous structures such as membranes and their interfacial polarization. The γ-dispersion is due to relaxation of free water, and δ-dispersion results from relaxation of bound water, amino acids and charged side groups of proteins *(38)*. All these dispersion phenomena in biological tissues are described on the macroscopic scale by the dielectric permittivity (the dielectric constant and loss factor or conductivity).

Because of thermoregulatory responses, deposition of RF energy in the body may not necessarily lead to an increase in temperature. A review of the heat transfer mechanisms and development of the bioheat equation can be found elsewhere *(40)*.

When RF energy deposition and conversion to thermal energy in a biological body exceed the heat dissipation capabilities, an increase in temperature occurs. It has been shown that biological effects, such as heat killing of cells *(41)* or heat mutagenesis *(42)*, depend on the temperature profile in time. To quantify the relationship between this and the biological effect a concept of "thermal dose" was introduced *(41)*. Thermal dose is a mathematical description of the time–temperature relationship as related to the biological endpoint, and converted to an exposure time at some reference temperature. This is expressed as "equivalent-minutes" at an empirically determined reference temperature. From plots of cell survival following heating *in vitro* for different times at several temperatures, an Arrhenius-like plot as shown in Fig. 6 can be derived. This plot shows a break at about 43 °C, and the relationship between temperature (T) and time (t) needed to obtain an equivalent biological effect (cell killing) can be formulated as:

$$t_1 = t_2 R \ (T_1 - T_2) \tag{12}$$

where R can be calculated as a function of the inactivation energy, ΔH (J/mol) and absolute temperature, T. From the plot in Fig. 6:

$$R = e^{-\Delta H/R(T^2 + T)} = e^{-\Delta H/8T(T + 1)} \tag{13}$$

In this equation, the constant 8 is an approximation for the gas constant (8.314 J(mol⋅K)). The numerator of the exponent contains a constant of 1 K, which cancels the unit of K in the denominator, thus providing the correct dimensions. Obviously this is an approximation, though the

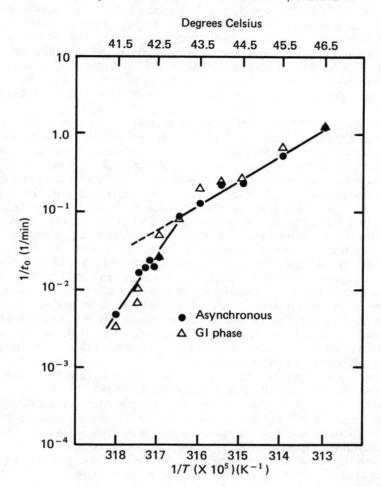

Fig. 6. An Arrhenius plot of heat inactivation showing the inverse of time $1/t_0$ as a function of absolute temperature T

relationship is supported by many *in vitro* and *in vivo* data. A convenient reference temperature for heat killing of cells is 43 °C, and thus the thermal dose may be expressed in minutes equivalent to heating at 43 °C. However, this reference temperature may vary, particularly for different chosen end-points, and may be relative to the normal physiological temperature of the tissue *(43)*.

135

Athermal mechanisms of interaction

A well established athermal mechanism of interaction at frequencies below a few tens of MHz is through electrical stimulation of excitable membranes of nerve and muscle cells (44,45). RF fields can induce currents sufficient to stimulate excitable tissue for frequencies below 1 MHz (44). The threshold current densities for stimulation, and approximate relationships between the strengths of the RF exposure fields and the induced current densities are available (44,45). Very high field strengths are necessary to reach the stimulation thresholds (electric field strength above 100 kV/m, magnetic field strength above 1.5 kA/m) (45).

Some experimental results can be interpreted to mean that effects occurred without significant changes in temperature and well below the stimulation threshold, and should be attributed to non-thermal mechanisms at the molecular level. At low levels, experimental demonstrations include behavioural changes in continuous-wave or modulated fields (46–50); altered efflux of calcium from brain tissue exposed in vitro to extremely low frequency modulated waves (51–53); changes in EEG waves (47,54); impaired killing ability of lymphocytes (55); chromosomal changes in developing mouse sperm cells; changes in intracellular enzyme activity (56); altered firing rates of molluscan pacemaker neurons (57,58); and changes in growth rate of yeast cells (59,60). The calcium efflux studies show effects that are specific to modulation frequency and field amplitude, thus introducing the concepts of frequency and power windows in the biological response to electromagnetic fields. From some of these studies it is quite clear that temperature elevation was not important (e.g. those in which modulation frequency was found to be a critical variable), but for others athermal mechanisms are implied though not conclusively demonstrated by the experimental conditions and data.

In very strong fields, interactions such as pearl chain formation and orientation occur through field-induced forces (38,39).

Furthermore, athermal mechanisms are proposed for rectification by cell membranes (61), which has been observed in plant cells for frequencies below about 10 MHz (62–64); vibrational resonances in DNA molecules (65); and structural transformations from random to coiled configurations in polymeric proteins (66). Cell membrane rectification may be important for fields somewhat greater than 10^2 V/m (in situ) at frequencies below 10 MHz. Careful measurements recently performed in various laboratories have not confirmed the earlier claims of resonance in DNA molecules (67,68).

In addition, the RF hearing phenomenon may be considered as athermal or microthermal since the temperature rise is very small (as low as 10^{-5} °C). However, the effects on the auditory system by pulsed microwaves are described by a model in which rapid thermal expansion launches an acoustic wave in the head (69–72).

A number of untested theoretical models have been proposed (73–75). Cooperative interactions among cell surface charges, leading to a low frequency collective mode in the EEG range, have been proposed by Grodsky (76). This mode could exchange energy with a weak, low-frequency field and thus provide understanding of the observations of calcium efflux, but the

136

model is not quantitatively predictive. A model was proposed by Fröhlich *(77)* for enzyme activity whereby long-range cooperative forces act together with high-frequency molecular vibrations ("elastic modes") and provide a source of energy to be coupled to the molecules. In this model, an electromagnetic field may couple strongly to one of the vibrational energy stages of biomolecules in the range of 10^{13}–10^{15} Hz. Effects may not be observed for frequencies below tens or hundreds of GHz, although Fröhlich has suggested that the very slow rate constants for enzymatic processes may lead to effects at very greatly reduced frequencies. Adey *(73)* proposed that electromagnetic fields couple to biological systems at the cell membrane, where cooperative processes among surface molecules (glycoproteins) amplify small changes in ion binding. The non-equilibrium nature of the process was emphasized to indicate how the system could obtain biologically and energetically significant changes. In turn, electrochemical signals are thought to communicate across the membrane and thereby call into play all the cellular apparatus for response to the cell environment, including membrane-related enzymes.

Davidov *(78)* proposed that the longitudinal vibrations along the axis of alpha helical molecules could form solitary wave quasiparticles (solitons) as a means of energy transport over molecular distances. In application of the soliton concept, Lawrence & Adey *(79)* proposed the existence of solitons in the membrane-spanning channel proteins. Moreover, they proposed that interaction of the solitons with electromagnetic waves was a mechanism for bioelectromagnetic effects.

Electric shock and burn
Electromagnetic fields at frequencies below a few MHz also interact with biological bodies through electrical charges induced on ungrounded or poorly grounded metallic objects such as cars, trucks, cranes, wires and fences. When a person comes in contact with such objects, current passes to ground through the person. This current depends on the total charge on the object, which in turn depends on the frequency and electric field strength, as well as the object geometry and capacitance and the person's impedance to ground.

Above a certain threshold the current to ground is perceived by the person as a tingling or prickling sensation in the finger or hand touching the charged object for frequencies below about 100 kHz, and as heat at higher frequencies. A severe shock is experienced above a much higher threshold. Both thresholds depend on frequency, the surface area of contact and the individual. The thresholds are generally higher for men than for women and children, and there are also individual differences *(80)*.

Burns occur when the RF current entering the body through contact between a small cross-section of the body, such as a finger, and an electrically charged object exceeds 200 mA. Detailed data on electric shock and burn thresholds for various RF frequencies can be found elsewhere *(80)*.

BIOLOGICAL EFFECTS

This review of biological effects of RF radiation is divided into two sections. The first, on human studies, is a detailed summary of the literature to emphasize the state of knowledge of the effects of RF radiation on the human body. The second section, on non-human studies, gives only a brief summary of the effects in biological systems ranging from *in vitro* preparations to live animals.

Human studies

Present knowledge can be summarized as follows.

● For the broad range of frequencies between 300 kHz and 300 GHz, cutaneous perception of heat and thermal pain may be an unreliable sensory mechanism for protection against potentially harmful RF radiation exposure levels. This is because (*a*) RF energy can be absorbed in tissue below the cutaneous thermal receptors, and (*b*) adverse effects occur at temperatures below the threshold (45 °C) of thermal pain *(3,81)*.

● Although some studies have associated lens defects with microwave radiation exposure, the present view is that low-level, chronic exposure to microwave radiation does not induce cataracts in man *(81,82)*.

● When the human head is exposed to pulsed radiation such as radar, an audible sound described as a click, buzz, chirp or knocking sensation is perceived by some individuals; the sound appears to originate from within or behind the head. This phenomenon is called "RF sound" or "RF hearing", and varies with pulse duration and pulse repetition frequency *(81,83)*.

● Numerical models of thermoregulatory responses in man can be used to predict RF exposure conditions that may cause undesirable temperatures in the body *(84,85)*.

● Currently available human data are very limited and not useful for deriving exposure limits for RF radiation. This conclusion is based on various problems with such data, such as exposure assessment and documentation of methods, including statistical procedures *(86–89)*.

The following is a review of the literature on which the summary statements listed above were based.

Cutaneous perception

Exposure of the human body to microwave radiation can cause heating that is detectable by the temperature-sensitive receptors in the skin. Several investigators have experimentally determined the microwave intensities that cause sensations of warmth and thermal pain.

138

Hendler and colleagues *(90,91)* showed that the sensation of warmth in an exposed circular area (37 cm²) of the forehead was greatest for infrared (IR) radiation and less for radiation at 10 GHz. The reason is that the outer skin layers contain the thermal sensors, and therefore the more superficial the energy deposition the lower the threshold.

Justesen et al. *(92)* confirmed the earlier work on perception of microwave radiation in a study of human adults whose forearms were exposed to 2450 MHz or to IR radiation. A 15-fold difference between microwave and IR thresholds for detectable warning was attributed to (*a*) different scattering (nearly two thirds of the incident microwave energy is scattered and not absorbed); and (*b*) the different depths of penetration of the two forms of energy. None of the experimental subjects who experienced both types of radiation could distinguish a difference in sensory quality. The authors concluded that the same set of superficial thermoreceptors was being stimulated but less efficiently so by the more deeply penetrating, more diffusely absorbed RF energy.

For both RF and IR radiation at intensities producing sensations of warmth, a threshold of warmth was experienced when the temperature of tissue 0.2 mm below the skin surface was increased by 0.01–0.02 °C over the temperature of a deeper layer in the skin. It was also noted that there was a persistent sensation of warmth for 0.7 seconds after exposure ceased, indicating the continued presence of an effective temperature difference between the subcutaneous tissue layers *(90)*.

Schwan & Foster *(38)* exposed a 7 cm diameter area of the forehead (equivalent to the area exposed in Hendler's studies) to 2.88 GHz radiation, and measured the length of time that elapsed before the person was aware of a sensation of warmth. The authors found that the reaction times were not linearly proportional to the reciprocal of the incident power density, and concluded that subjective awareness of warmth was not a reliable indication of microwave hazard.

Vendrik & Vos *(93)* exposed a 13 cm² area of the inner forearm to 3 GHz of pulsed radiation and found the threshold for temperature changes to be 0.4–1.0 °C. Skin temperature increases that were kept below 1 °C were linear with the power density for six exposure durations. In contrast to the regularity of skin temperature changes induced by RF, the reports of temperature sensations were variable. Sensations of warmth occurred from less than 0.5 seconds to 3.5 seconds after rapid rises in skin temperature. The sensations did not cease when the skin temperature began to drop. In this study, RF radiation at 3 GHz was found to be a factor of 10 less effective in producing a rise in temperature than IR radiation at a similar intensity.

Cook *(94)* determined the pain threshold in six subjects who were exposed to 3 GHz radiation at five different sites on the body surface. The initial skin temperature ranged from 31.5 °C to 33.5 °C. Pain resulted when a critical skin temperature (46 °C) was reached. The skin temperature corresponding to burning pain was found to be independent of the area of exposure, radiation intensity, exposure time and anatomical site. At high intensities, the exposure time needed to produce pain was an inverse function of the power density. The sensations of warmth and pain with microwave heating differed little from those resulting from IR radiation.

In summary, the few studies on thermal pain and warmth sensations in human beings exposed to frequencies in the range 3–10 GHz show that cutaneous perception may be a reliable indicator of an unsafe exposure level only at RF frequencies of the order of tens of GHz or more with wavelengths comparable to or smaller than the thickness of the skin. Under these conditions most of the energy is absorbed in the outer tissue layers containing thermal sensors. At lower frequencies, much of the energy is absorbed within the body below the superficial skin layers. In all mammals tested, the threshold temperature (42 °C) for cellular injury for sustained elevations (seconds to tens of seconds) is below the threshold (45 °C) of pain(95). These results strongly indicate that cutaneous perception of RF energy is not a reliable protection against potentially harmful levels of RF radiation over the broad frequency range of 300 kHz–300 GHz (80).

Ocular effects
The potential of RF radiation to induce cataracts and less significant lens defects and opacities has been the subject of several studies. In a case-control study, Cleary et al. (96) found no difference in cataract formation among US Army and Air Force veterans. Later, Cleary & Pasternack (97) studied subclinical or minor lens changes in 736 microwave workers and 559 controls. The number of defects was found to increase significantly in both groups with increasing age, and microwave workers were found to have more lens changes than controls. The authors suggested that the subclinical lens changes observed with greater frequency in the microwave workers may indicate accelerated aging of lens tissue. The types of lens defect that accounted for the differences in eye scores between the two groups were posterior polar defects and opacification. In an expanded analysis, the increase in lens defects was correlated with duration of microwave work, exposure score, and duration of exposure interaction. The group of microwave workers was further characterized by the identification and evaluation of five occupational subgroups. The subgroups were found to differ with respect to the average values for age, duration of work, exposure score and eye score. The subgroup involved in research and development of microwave equipment had the greatest evidence of lens defects. The authors also examined the possibility that ionizing radiation rather than microwave radiation was responsible for the observed differences in lens defects, but concluded that there was no evidence for this.

Majewska (98) studied the eyes of 200 Polish workers employed for from 6 months to 12 years at installations with microwave-generating equipment that operated between 600 MHz and 10.7 GHz. Although cited as "high intensity" by the author, intensity levels were not specified. Two hundred age-matched controls were also examined. Lens changes noted in 168 of the microwave workers as against 148 of the controls were stated to be statistically significant. In the same study, the effects of longer-term exposure were evaluated by comparing 100 controls with 102 employees, drawn from the original group, who had worked with high-frequency electromagnetic wave generators for more than four years. The extent of opacities in the exposed group was greater than in the controls in each 5-year age group for ages ranging from 20 to 50 years. Among microwave workers, lens changes uncontrolled for age also showed an increase with length of employment.

A survey of ocular anomalies in 377 exposed personnel and 320 controls from eight military installations was carried out by Odland *(99)*. Exposed personnel were defined as those whose primary duties involved the operation or maintenance of radar equipment, while the duties of the controls did not permit actual or potential exposure to radar. The actual work assignments were not stated. The frequency of lens anomalies was similar in the two groups; however, the frequency of anomalies between control and exposed groups was different for individuals with a family history of diabetes, nontraumatic cataract, glaucoma or defective vision. Lens changes were noted in 29% of the exposed individuals with such a history, compared with 17% in controls with a family history of eye problems.

Shacklett et al. *(100)* reported the results of eye examinations of military and civilian personnel, consisting of 477 with a history of microwave exposure and 340 controls without exposure. The authors stated that detailed work histories were recorded (including time spent with different types of equipment) but information on typical exposure settings was not given. Differences between the two groups in the frequency of opacities, vacuoles and posterior subcapsular iridescence were reported as not being statistically significant, but the type of statistical test was not stated. An age-dependent increase in lens changes was noted in both groups.

Siekierzynski et al. *(101)* reported no differences in lens opacities in two groups exposed, respectively, to $2 W/m^2$ and $2-60 W/m^2$.

Zaret *(102)* has attributed a number of cases of cataract development to microwave exposure. These instances have generally been related to acute exposure to high-intensity fields in the workplace. For most cases, work history information sufficient to estimate dose rates or exposure conditions is lacking *(103)*.

Although some studies have associated lens defects with microwave radiation exposure, no data at present would support a conclusion that low-level chronic exposure to microwave radiation induces cataracts *(81,82,104,105)*.

Numerical modelling of thermoregulatory responses in man
Based on heat-balance considerations alone, Durney et al. *(29)* calculated approximate whole-body SARs that produce a rectal temperature of 39.2 °C in 60 minutes of RF exposure in hot and humid environments. The model can best be applied to radiation conditions producing primarily deep heating and not surface heating. At 70 MHz (the resonant frequency for the electrically ungrounded adult human being) and at 39.4 °C and 80% relative humidity, exposure at an SAR of 4 W/kg was calculated to increase rectal temperature to 39.2 °C in 60 minutes. A healthy person exposed to these conditions would experience some degree of discomfort along with the rise in body temperature, a rise in heart rate, and profuse sweating. All responses would increase with time, and after about 60 minutes a healthy person

would be on the verge of collapse, with unpleasant but reversible symptoms such as those reported in experiments on human heat tolerance. At an ambient temperature of 41.1 °C and 80% relative humidity the calculations predict that an SAR of 1 W/kg would raise rectal temperature to 39.2 °C in 60 minutes.

Thermal models range from a simple cylindrical model of the human arm to one of the entire human body *(106)*. Stolwijk *(107–109)* and Stolwijk & Hardy *(110)* developed a human model composed of six segments (head, trunk, arms, hands, legs and feet), each of which had four layers (skin, fat, muscle and core). The geometry of each was cylindrical except for the head, which was spherical. Analysis by digital computer included metabolic heat production, sweating, blood flow to all layers, and convective and radiant exchange with the environment.

Emery et al. *(111)* described the thermal effects of a uniform deposition of RF energy for a one-dimensional model of heat conduction. This model tends to overestimate the temperature profile in the body because heat flow is assumed to be only from the core to the skin.

Spiegel et al. *(85,112,113)* combined a block model with a two-dimensional extension, in which heat flows from the core to the skin as well as along the major axis of the body. The temperature distribution was calculated in a resting, nude, 70 kg, 170 cm tall man (air temperature 30 °C, relative humidity 30%) exposed to a plane wave at 80 and 200 MHz *(113)*. The electric field vector was oriented parallel to the major axis of the body. Whole-body resonance occurred for the 80 MHz field, and partial-body resonance occurred in the arms for the 200 MHz field. For the 80 MHz field an incident power density of 100 W/m^2 was used (SAR 2.25 W/kg). For the 200 MHz field incident power densities were 100 and 325 W/m^2 (SAR 0.58 and 1.9 W/kg, respectively). At 100 W/m^2 the whole-body SAR at 80 MHz was about four times larger than that at 200 MHz. The distribution of temperatures for the two frequencies was quite different: for the 80 MHz field the highest temperature of 41.6 °C was in the lower thigh; for the 200 MHz field (SAR 1.9 W/kg) the highest temperature (42.9 °C) occurred in the arms and, in addition, the temperature in the neck rose to 40.6 °C.

Spiegel *(84)* modelled the thermal response of the human being in the near-field of a 45 MHz and 200 MHz antenna. The lower-frequency antenna was 18.6 cm in front of the human body and the other antenna was placed 10 cm in front of the face. The 45 MHz frequency represents the resonant frequency for an electrically grounded adult human being and the 200 MHz frequency is near the resonant frequency (375 MHz) of the head *(114)*. Negligible heating occurred at antenna input power levels less than 50 W, but higher levels produced temperature increases in various body regions that were clearly hazardous *(84)*. For example, at 200 MHz (400 W antenna input power) a whole-body SAR of 2.44 W/kg caused a rise in neck temperature to 42 °C, where protein denaturation and other adverse cellular effects occur. At 45 MHz (600 W antenna input power) a whole-body SAR of only 0.64 W/kg produced 42 °C in the ankles.

Although the combined RF heat transfer model reported here is realistic enough to predict gross effects and trends, the model could be further

142

refined from two-dimensional to three-dimensional. In addition, altered blood flow in tissues at temperatures in excess of 40°C has not been incorporated into the model, and it therefore probably overestimates the magnitude of the temperature rise. Furthermore, little is known about the effects of RF energy on the predominant heat-dissipating mechanisms of sweating and vasodilation (85). The physiological responses incorporated into the models are those of humans exposed to high ambient temperatures and/or exercise, but not to RF radiation. The extent of the similarity of response to RF radiation exposure and heat stress is debatable because of the unique characteristics of RF energy absorption within the body. Nevertheless, the models are useful approximations for predicting hazardous core and localized temperatures in the human body exposed to ambient conditions, both in the workplace and in the general environment.

Auditory effects ("RF hearing")

The earliest report on the auditory perception of pulsed microwaves appeared in 1956 as an advertisement of the Airborne Instruments Laboratory (115). The advertisement described observations made in 1947 on the hearing of sounds that occurred at the repetition frequency of a radar while the listener stood close to an antenna.

Frey (116–118) systematically studied the human auditory response to pulse-modulated radiation. The subjects, who were more than 30 m from an enclosed antenna, reported hearing a transient buzzing sound on exposure to the intermittent rotating beam. The apparent location of the sound was described as a short distance behind the head and was independent of orientation. The peak power density thresholds for RF hearing were 2.66 kW/m^2 for 1310 MHz and 50 kW/m^2 for 2982 MHz fields, and the average power density thresholds were 4 and 20 W/m^2, respectively.

The highest effective frequency of RF hearing is between 6.5 and 8.9 GHz, and the lowest effective frequency is 216 MHz (116,117,119).

Several other studies have been performed on auditory RF effects (69–72,120–126) and a comprehensive description has been published (72). At 2450 MHz, a threshold energy density of 0.4 J/m^2 per pulse, an energy absorption per pulse of 16 mJ/kg, was calculated (123). It has been shown both theoretically and experimentally that radiation-induced pressure changes result from the absorption of RF pulses and could produce significant acoustic energy in solution (71). Audible sounds are produced by rapid thermal expansion, resulting from a rise of only 5×10^{-6} °C.

Evidence from many studies, such as measurement of acoustic transients in water, potassium chloride solution, tissues (71) and muscle-simulating materials (124), investigations of the relationship of pulse duration and threshold (71,121-123,125), characteristics of field-induced cochlear microphonics in laboratory animals (69,70) and theoretical calculations (72), indicates that thermoelastic expansion in the head is the mechanism that best explains the characteristics of the RF hearing phenomenon.

It is well documented that some people can hear pulsed RF radiation as a buzz, click or knock. Furthermore, the thermoelastic expansion mechanism explains how the pulse of RF energy can be transformed to an acoustic

143

impulse in the head, although it is not known what structure(s) in the head transduce the microwave energy to acoustic energy. A very low average power density can cause an acoustic response in the head, and there is the potential for exposure of the public and workers to pulsed fields that induce the effect. There is therefore a need to inform people of the RF hearing phenomenon because of the potential psychological effects of RF hearing, particularly in those who may have no knowledge as to the origin of the sounds.

Occupational surveys/clinical studies
The majority of reports in the literature concern people occupationally exposed in industrial or military settings. Barron et al. *(127)* conducted a study to evaluate changes in various physical characteristics of radar personnel employed by an airframe manufacturer. A total of 226 radar workers were grouped according to duration of exposure (0–2, 2–5 or 5–13 years). The radar exposure included S-band (2880 MHz) and X-band (9375 MHz) frequencies. Exposure zones at various distances from the antenna were specified and used to estimate three ranges of power density ($>$ 131, 39–131 and $<$ 39 W/m^2). Because of the relatively low power densities, those exposed to $<$ 39 W/m^2 were eliminated from the study. The 88 controls had not been occupationally exposed to radar. The age distribution of all subjects ranged from 20 to more than 50 years, with the majority under 40 years of age. However, the controls were older: 46% as against 17% of the exposed group were 40 years of age or older. Platelet counts and urinalyses were similar in the two groups. Ocular anomalies of several different types were found in twelve of the exposed group compared with only one of the controls. The medical surveillance programme was extended to permit periodic re-examination. No significant differences in physical health status were noted, but none of the results was tested statistically *(128)*.

The Soviet and Eastern European literature describes a collection of symptoms in personnel industrially exposed to RF energy *(88)*. These collective symptoms, which have been variously called "neurasthenic syndrome", "chronic overexposure syndrome" or "microwave sickness", are based on subjective complaints that include headaches, sleep disturbances, weakness, lessened libido, impotence, pains in the chest, and a general feeling of being unwell *(129)*. Also described are labile functional cardiovascular changes including bradycardia (or occasional tachycardia), arterial hypertension (or hypotension) and changes in cardiac conduction; this form of neurocirculatory asthenia is also attributed to nervous influence *(130,131)*.

In Poland, the health status of 841 men occupationally exposed to pulse-modulated microwave radiation was evaluated *(101,132,133)*. The men had been exposed for various periods of time, some for over 10 years. They were placed in one of two groups on the basis of exposure level. One group consisted of 507 men exposed to mean power densities greater than 2 W/m^2, with short-term exposures estimated to reach 60 W/m^2. The other group was 334 men exposed to mean power densities less than 2 W/m^2. The health conditions evaluated covered three main categories: neurasthenia, functional disturbances of the digestive tract and cardiovascular disturbances with abnormal electrocardiogram (ECG) findings. According to Polish occupational exposure criteria, these conditions are considered contraindications for work in a microwave environment. The neurasthenia was defined by a variety of symptoms such as fatigue, headaches, sleep disturbances and difficulties in memorizing and concentrating. Psychological examinations were made. Comparisons were made both between and within exposure groups according to age and duration of occupational exposure. The two groups were found to be similar with respect to the distribution of these symptoms and conditions, indicating no dependence on level of exposure.

Djordjevic et al. (134) reported on medical evaluations of radar workers aged 25–40 years with a work history of 5–10 years. Evaluation of the work environment led to the conclusion that the workers were exposed to pulsed microwaves within a wide range of intensities but generally at levels less than 50 W/m². The control group consisted of 220 persons reported to be similar in age, work regime and socioeconomic status, and having never worked with microwave sources. The two groups did not differ with respect to nervous and cardiovascular system function, ECG, multiple biochemical and haematological indicators, frequency of sleep disturbance, inhibition of sexual activity and impairment of memory. Radar workers, however, reported more subjective complaints of headache, fatigue and irritability. Based on their survey of the working conditions, the authors attributed the latter result to such factors as poor lighting and ventilation and high noise levels in the environment of the radar workers, as well as the need to concentrate on the radar screen. If true, the work environments for the two groups were evidently not similar in all respects.

More recently, an exploratory study was published of male physiotherapists using RF, infrared and ultrasound diathermy equipment (135). The study population consisted of 3004 men who responded to a mail questionnaire survey. The cohort was divided into subgroups according to their use of ultrasound, microwave, short wave and infrared diathermy equipment. High and low exposures were approximated by considering length of employment and the number of treatments given per week. An association between heart disease (primarily ischaemic heart disease) and exposure to short wave (27 MHz) radiation was the only consistently significant finding in comparisons between high- and low-exposure groups. In general, heart disease was less prevalent than in a general population of comparable sex, age and race, perhaps due to higher socioeconomic status in the study group, to an occupation in a medical setting, and to a "healthy worker effect".

Mortality studies
Lilienfeld et al. (136) assessed the potential health consequences of microwave irradiation of the American Embassy in Moscow. The health of about 1800 employees and 3000 dependants who had served in the Embassy between 1943 and 1976 was compared with that of about 2500 employees

and 5000 dependants at eight other embassies or consulates in Eastern Europe over the same period. The microwave irradiation of the Moscow Embassy was first detected in 1953, and subsequently varied in intensity, direction and frequency over time. The highest intensity was 0.18 W/m² and the frequencies ranged from 0.6 to 9.5 GHz *(137,138)*. Information on illness and other conditions or symptoms was sought from employment medical records, and from a self-administered health history questionnaire and death certificates.

No convincing evidence was found at the time of the analysis to implicate microwaves in the development of adverse health effects. Limitations inherent in the study were: uncertainties associated with the reconstruction of the employee populations and dependants; difficulties of obtaining death certificates; the low level of response to the questionnaire; the statistical power of the study; and, most critically, ascertainment of exposure. No records were available on where employees lived or worked, so one had to rely on questionnaire responses to estimate an individual's potential for exposure. The highest exposure level (0.18 W/m²) was recorded for only 6 months during 1975/1976; thus, the group exposed to the most intense fields had the shortest cumulative time of exposure and of observation.

Robinette & Silverman *(139)* and Robinette et al. *(140)* examined mortality and morbidity among American naval personnel occupationally exposed to radar. The exposure group (probably highly exposed) consisted of technicians involved in repair and maintenance of radar equipment. The controls (probably minimally exposed) were involved in the operation of radar or radio equipment. It was estimated that radio and radar operators (in the low-exposure group) generally received less than 10 W/m², whereas gunfire control and electronics technicians (in the high-exposure group) received higher levels during their duties. Over 40 000 retired personnel were included in the study, with about equal numbers in the two exposure classifications. The mean age in 1952 of the low-exposure group was 20.7 years, and that of the high-exposure group 22.1 years. An effort was also made with naval personnel to develop an index of potential exposure, the "hazard number". This number was based on months of duty multiplied by the sum of the power ratings (equipment output power of ship gunfire-control radars or aircraft search radars) where technicians were assigned.

Information was obtained from medical records of four main types: death certificates, inservice hospital admissions, admissions to hospital of retired personnel, and disability compensation records. There were no statistically significant differences in strokes, cancers of the digestive tract and respiratory system, and leukaemias, although these diseases occurred slightly more often in the high-exposure group. The reported mortality ratio of 1.64 for malignant neoplasms of the lymphatic and haematopoietic system was not statistically significant. Comparisons made within the high-exposure group across hazard number categories revealed that two were statistically significant: (*a*) the difference in respiratory tract cancer between those with a hazard number smaller than 5000 as against those with a number larger than 5000; and (*b*) the test for trend for all diseases combined. These results may be misleading, however, since one or two false positive

146

findings might be expected when many statistical comparisons are made. In addition, information relevant to the development of lung cancer, such as a history of smoking, could not be obtained.

Letters in the *New England journal of medicine* suggest both an association and a lack of association of leukaemia with occupational exposure to a variety of electric and magnetic fields, including radio, television and other electronic devices *(141,142)*. Reviews of the relevant literature prior to 1978 provide little evidence that RF exposures are carcinogenic *(143–145)*. However, the subject remains controversial for many reasons, one being the lack of well designed studies on man or animals with adequate exposure data and that are statistically adequate for drawing reliable conclusions.

Congenital anomalies and reproductive effects
In 1965 Sigler et al. *(146)* studied occupational exposure to radar and length of military service among fathers of children with Down's syndrome. The fathers of the children with Down's syndrome had more military service experience than control fathers, but the difference was not statistically significant. More frequent exposure to radar was reported by case fathers, a difference that was statistically significant. Exposure to radar occurred primarily in assignments as radar operators or technicians. Cohen et al. *(147)* expanded the study from 216 to 344 verified cases and their matched pairs; the previously noted differences disappeared in the extended analysis.

Lancranjan et al. *(148)* studied 31 adult males with an average age of 33 years and a mean exposure of 8 years (range 1–17 years) to electromagnetic fields that "frequently were in the range of tens to hundreds of mW/cm^2" at frequencies between 3.6 and 10 GHz. A group of 30 men of similar average age and with no known exposure to microwaves served as a control. Statistical analysis of the results showed no differences in urinary content of 17 ketosteroids or total gonadotropin between the exposed and control groups. Slight but statistically significant decreases were reported for exposed personnel in the number of sperm cells per millilitre of semen, the percentage of motile sperm cells in the ejaculate, and the percentage of normal sperm cells. Spermatogenesis improved in two thirds of the subjects after exposure ceased, which the investigators felt supported the argument for a microwave effect. However, exposures were poorly defined, and the number of men evaluated was very small.

Kallen et al. *(149)* examined the outcome of pregnancy in 2018 Swedish physiotherapists, to whom 2043 infants were born, including 25 pairs of twins. The overall results of pregnancy were better than expected in this cohort, as measured by eight different criteria. Finally, 33 cases of perinatal death and major malformation in the study group were compared with 63 controls. Of several criteria explored, the only positive finding was that the women who had a dead or malformed infant used short-wave equipment more frequently than the control physiotherapists. No information was presented on short-wave radiation levels typical of the various physiotherapy work environments, on the types of equipment most commonly used in Sweden, or on X-ray exposures.

147

At present, the data on humans exposed to RF radiation are not adequate or sufficiently developed to be useful in defining risk or determining exposure limits.

Non-human studies
The following summary statements for non-human studies are based on a recent comprehensive review *(150)*.

Thermoregulatory responses to RF radiation
Thermoregulatory effectors such as peripheral vasodilation, evaporation, metabolism and behaviour may be activated during exposure to RF radiation *(151,152)*. Many responses are activated in the absence of any measurable change in deep-body temperature during RF radiation exposure *(152)*.

In general, the SAR required to increase the activity of a thermoregulatory effector or to raise body temperature in animals decreases with increasing body mass *(153)*. For example, four different responses in the squirrel monkey *(154–157)* were activated by SARs of 0.6–1.5 W/kg, whereas 5.3–29 W/kg were required for the mouse *(158–161)*. Thermoregulatory responses by mammals during RF radiation exposure at levels that produce heat stress are similar to those to high ambient temperature *(151,152)*.

Most thermoregulatory responses, as well as other biological effects, have been recorded in laboratory animals exposed to RF radiation at normal room conditions of temperature, humidity, air flow, etc. Thus, it is reasonable to predict that the threshold for effects due to RF heating are lower at ambient conditions that exacerbate thermal effects *(150)*.

Due to heat stress from absorbed RF energy, dose rates of 3.6–7 W/kg are lethal to rats, rabbits, dogs and rhesus monkeys exposed for 1–4 hours at normal laboratory conditions of temperature and humidity *(162–164)*.

Reproductive effects
RF radiation was found to be teratogenic at exposure conditions that approached lethal levels for the pregnant animal *(165)*. For example, SARs of 11.1–12.5 W/kg were teratogenic in rats exposed for 20–40 minutes *(166)*. Reduced fetal weight seemed to occur consistently in rodents exposed gestationally to SARs greater than 4.8 W/kg *(167)*. Temporary sterility occurred in male rats at an SAR of 5.6 W/kg, which caused a significant increase in testicular temperature *(168)*.

Effects on the blood-forming and immune systems
Partial or whole-body exposure of animals to RF radiation may lead to a variety of changes in the blood-forming and immune systems; stimulatory or suppressive changes have occurred depending on the exposure conditions, species, and biological parameters *(89,169–173)*. In those cases where the reversibility of RF radiation effects on the immune system was examined, the effects proved transient *(169,174)*. Effects of RF radiation on the blood-forming and immune systems similar to those caused by glucocorticoid-mediated stress responses have been reported *(175–179)*.

148

Effects on the blood-forming and immune systems have been reported at SARs greater than 0.4 W/kg; however, there is a lack of convincing evidence for RF radiation effects on these systems without some form of thermal involvement. Many reports, particularly those describing effects of acute exposure, show an association between RF-induced thermal loading or increased body temperature and changes in the blood, blood-forming and immune systems (169).

Nervous system

Acute or chronic continuous- or pulsed-wave irradiation of animals at SARs greater than 2 W/kg can produce morphological alterations in the central nervous system. These changes were qualitatively similar after acute or chronic exposure and at different SARs, but quantitatively more alterations occurred in the affected neuronal structure at higher SARs and after chronic exposure. The changes were found less frequently in animals allowed to survive several days to weeks after exposure ceased (180).

There is no conclusive evidence that RF radiation affects the blood–brain barrier at SARs below 2 W/kg (181,182).

RF radiation, especially pulse-modulated radiation, appeared to have a potentiating effect on drugs that affect nervous system function (50,129,183–186).

RF fields sinusoidally modulated at extremely low frequencies, especially 16 Hz, caused central nervous system changes in vitro (52,73,187) and in the live animal (188). The physiological significance of these effects is not established. The same type of field was also shown to change EEG patterns in the brain (46,54).

Behaviour

Changes in locomotor behaviour occurred after continuous-wave exposures at an SAR as low as 1.2 W/kg (189).

Reductions in conditioned behaviour were reported during exposures at an SAR of 2.5 W/kg (190) and such behaviour ceased at an SAR of 10 W/kg (191). Alterations in conditioned performance measured following exposure also occurred at SARs of 2.5 W/kg or more (192).

The threshold for detection of microwaves by a laboratory animal such as the rat may be as low as 0.6 W/kg (193). However, it is not certain that animals avoid or attempt to escape from continuous-wave microwaves, even at very high power levels.

The effects of drugs on behaviour were augmented after pulsed-wave radiation exposures of 30 minutes at an average SAR of 0.2 W/kg (50).

All of the above statements are based on studies of the rat; few behavioural studies have used other species.

Behavioural thermoregulation was altered after only several minutes of exposure at an SAR of 1 W/kg in the rat (194) and the squirrel monkey (156). In general, behaviour was modified by microwave energy input approximately equal to one quarter to one half of the resting metabolic rate of the animal (195). Behavioural alterations were reported to be reversible with time after termination of exposure (195).

Cataracts

RF radiation was found cataractogenic if localized exposure of the eye was of sufficient intensity and duration *(196,197)*. For single acute exposures, the threshold intensity for cataract production exceeded $1000 \, W/m^2$, equivalent to about $92 \, W/kg$ *(197)*. Multiple exposures at intensities near threshold values for single acute exposures resulted in lens opacities *(196)*. The most effective frequencies appear to be microwave frequencies in the range $1-10 \, GHz$ *(81)*.

In contrast to the above conclusions, which are based on acute, near-field exposures to the rabbit eye or head, only one *(198)* of five studies *(199–202)* reported cataracts in unrestrained animals after far-field exposure to near-lethal values. No cataracts occurred in rabbits exposed chronically at $100 \, W/m^2$ (maximal SAR in head of $17 \, W/kg$) for 180 days *(201)*. Although the threshold RF power density for cataract induction for long-term (days) exposure of animals (including man) has not been defined, the threshold is probably significantly higher than that required to induce many other physiological changes.

Endocrine and other physiological effects

In general, responses due to acute exposure of the thyroid gland have been consistent with the known effects of heat *(203)* and changes in corticosterone have occurred in association with increases in body temperature *(204–206)*. Changes in clinical chemistry and cardiovascular effects are consistent with those due to heating *(187)*.

Molecular, subcellular and cellular effects

For most molecular or subcellular systems exposed *in vitro,* no consistent biological effects have been demonstrated that can be attributed to RF-specific interactions. Exceptions include chain-length-dependent microwave absorption by DNA *(207)*, conformational transitions in a model protein *(66)* and sodium and potassium ion transport across red blood cell membranes *(208–212)*.

No consistent effects have been demonstrated on the growth and colony-forming ability of single cells such as bacteria that can be attributed to RF-specific interactions *(213,214)*; however, there are reports *(59,60)* of frequency-specific alterations in growth rates of yeast cells exposed at $41-42 \, GHz$.

The electrophysiological properties of single cells, especially the firing rate, membrane potential and resistance of neurons in isolated preparations, may be affected by RF radiation in a manner different from that caused by generalized heating *(57,58)*.

Genetics and mutagenesis

In general, RF radiation of low to moderate intensity did not cause mutations in biological systems unless temperature increased considerably *(215)*.

The only exposures that are potentially mutagenic are those at high power densities, i.e. those that result in substantial thermal loading at sensitive sites or result in extremely high E field forces. An exception is a

150

study describing chromosomal changes in sperm cells of mice exposed to pulsed microwave radiation at high peak but low average field strengths *(216)*.

Life span and carcinogenesis
Few studies have specifically addressed longevity or cancer incidence *(217)*. There is no conclusive evidence of the effect of long-term exposure to RF radiation on the life span of experimental animals or on cancer *(218)*.

There is evidence from one group of investigators that chronic exposure to RF radiation (SAR of 2–3 W/kg) resulted in cancer promotion or cocarcinogenesis in three different tumour systems in mice *(219,220)*.

PROTECTIVE MEASURES AND STANDARDS

The aim of protective measures is to eliminate or to reduce, as far as possible, unnecessary exposure of humans to RF fields, and to keep all exposure below applicable limits for the general public or occupationally exposed people. Unnecessary exposure is defined as that which is unrelated and incidental to the intended purpose and use of the equipment.

A distinction can be made between unconfined sources or deliberate emitters, such as radar, telecommunications or broadcasting installations, and confined sources which by design and intended purpose emit radiation into a confined space, such as microwave ovens. In the latter case, emission of stray fields into the surrounding area does not serve any useful purpose, and should be kept to a minimum. In all cases, sound engineering design which takes safety aspects into account is a basic protective measure. Engineering controls and means of reducing exposure levels and preventing accidental exposure, such as safety interlocks, screening and shielding materials, are the subject of engineering textbooks.

As a rule, stray fields and leakage should be reduced to levels well below the limits applicable to the general public or for occupational exposure. In several countries this problem is addressed by voluntary or mandatory equipment performance standards, which specify permissible emission levels for a particular type of equipment (e.g. microwave oven standards). In some RF protection recommendations such as ANSI C95.1-1982 *(12)* a class of low power equipment is defined and is exempt from safety rules.

Another basic rule of safety is to use RF equipment according to its intended purpose and operating instructions. This implies that an operating manual addresses safety aspects and the users are properly trained in safe operation of the equipment. It should be stressed that awareness of potential hazards, together with professional and safety training of operators, are essential requirements for protection against RF radiation. The code of practice on protection of workers against RF radiation prepared by the International Radiation Protection Association and the International Labour Office *(221)* provides detailed information on protective measures. The code deals with the assignment of responsibilities to the employer, the responsible user, the safety inspector and radiation workers.

151

Various forms of protective garments and eyewear have been developed to allow people to work in areas where they could otherwise receive unacceptably high levels of radiation. In general, the properties of the garments and eyewear that have been developed up to now are not satisfactory, and their use is not recommended at present. However, the use of non-reflective materials for reducing levels of exposure from secondary sources, i.e. reflections from objects and walls in the vicinity of RF emitters, is effective.

Radiation surveys around RF sources are a basic control measure. The aim of such surveys is to determine areas where human presence should be restricted or prohibited. Procedures for such surveys and measuring equipment should be standardized to achieve reliable and comparable results (see the section on instrumentation and measurement, p. 125). Several national standards give recommended procedures (11–13,221,222).

In the case of unconfined RF sources, particularly deliberate emitters, siting and installation should be based initially on a computation of field levels in areas intended to be restricted or for human occupancy. Such computations should be verified by measurements and radiation surveys. When conducting such surveys it should be noted that:

— radiation exposure in uncontrolled areas should not exceed the limits recommended;

— radiation levels should be known in controlled areas; these areas should be designated and the maximum occupancy time posted;

— the immediate vicinity of unmanned, high-power sources of RF radiation (e.g. some radio or television transmitting systems) should be fenced off to prevent unauthorized access to places where overexposure could occur;

— RF devices should be positioned as far away as is necessary from areas normally occupied by non-radiation workers;

— the siting of an RF device should take into account the possibility of multiple exposures from radiation fields and leakage from other radiation sources in the vicinity;

— there should be no unnecessary metallic objects near any radiating RF device, as the presence of such objects may result in high field strengths in some locations;

— shielding or screening should be achieved by use of absorptive rather than reflective materials;

— all RF devices should comply with applicable standards of design, construction and performance as specified in the appropriate regulations, where such regulations exist; and

152

— some high-voltage RF devices may emit X-rays, and appropriate protection should be provided.

Ancillary hazards should also be considered. RF burns or shock may result from charges induced on metallic objects situated in the vicinity of RF sources *(80)*. Care should be taken to eliminate situations in which RF shock or burns may occur.

Sparking on metallic objects may ignite or detonate flammable gases or vapours. Care should be taken to ensure that electro-explosive devices are not placed in RF fields of sufficiently high level to cause detonation. Wires of electric blasting caps, under certain conditions, may pick up sufficient energy from RF fields to cause the caps to explode. The susceptibility of blasting caps to these fields depends on the frequency, polarization and strength of the field, and various factors in the design of the detonator, such as its immunity from RF interference. The hazardous field strength depends on the field frequency: the lower the frequency, the more susceptible the detonators are. More information, including recommended distances from various sources, can be found in the American National Standards Institute safety guide *(13)*.

Apart from the primary biological hazards of direct exposure to RF energy, there are the more subtle influences that can affect users of electronic prosthetic devices, such as life support systems and diagnostic medical devices. Such electromagnetic interference may cause a variety of malfunctions. The offending signals can be a product of intended radiation, such as radio, television and radar transmission, or unintended signals generated by microwave or short-wave diathermy equipment, microwave hyperthermia equipment, engine ignition systems, electric razors or electromechanical relays. Electromagnetic interference can be diminished or eliminated by separation, shielding, filtering, and proper installation.

Improvements in cardiac pacemaker design have largely eliminated problems with electromagnetic interference. Pacemakers are typically tested for such interference at a frequency of 450 MHz, where susceptibility is generally considered to be the greatest.

Protective measures may be summarized as follows.

- Control of RF emissions from industrial and medical sources should be achieved by proper equipment design, appropriate shielding, interlocks and proper location of equipment.

- All personnel should receive both verbal and written information on techniques and work practices, so as to avoid unnecessary exposure.

- Warning signs should be located in all areas where exposure to RF fields could potentially exceed adopted limits.

- Exposure levels should be monitored by conducting surveys at regular intervals or immediately following physical or electronic alterations to

153

equipment that could affect the exposure levels. Records should be maintained of surveys conducted, monitoring instrumentation used, names of survey personnel and exposed workers involved, and any action taken.

Appropriate protective measures require a well organized RF radiation protection programme that by necessity must be based on a well formulated set of standards. At the national level, several mandatory or voluntary standards have been introduced and are periodically modified and revised. Comprehensive reviews are available of standards introduced or revised over the past 25 years (223–227). Although initially national standards differed widely, recently some trend towards standardization has emerged. This emerging international consensus is reflected in the guidelines for exposure limits to RF fields issued by the International Radiation Protection Association (228). The exposure limits proposed in these guidelines are presented in Tables 1 and 2.

RF exposure limits are specified in terms of electric or magnetic field strengths or power density. In recent standards, the numerical values of exposure limits are derived from consideration of effects induced at various SAR levels. For example, the IRPA recommendations (228) adopted a whole-body average SAR of 0.4 W/kg as a basic limit for occupational exposure and 0.08 W/kg for exposure of the general public. Derived exposure limits, specified in V/m, A/m or W/m², are averaged over six minutes, and it is recommended that the instantaneous exposure should not exceed the time-averaged values by more than a factor of 100. Other recommendations, such as ANSI C95.1-1982 (12), do not address exposure limits averaged over periods shorter than the specified time-averaging period. The longer such a period is, the higher the implied instantaneous exposure limits. Thus, the duration of the time-averaging period adopted in recommendations is a point of practical importance. The biological basis for the determination of an instantaneous exposure limit is as yet incomplete, available data are controversial, and the subject requires further study.

CONCLUSIONS AND RECOMMENDATIONS[a]

Conclusions
Fields at radiofrequencies from 300 kHz to 300 GHz are quantified in terms of electric field strength (V/m) and magnetic field strength (A/m). At frequencies above 300 MHz, power density (W/m²) is usually used in hazard evaluation. Instrumentation to measure RF fields is commercially available. Except for fields far from sources, both the electric and the magnetic fields

---

[a] These conclusions and recommendations are those pertaining to radiofrequency radiation made by the WHO Working Group on Health Implications of the Increased Use of NIR Technologies and Devices, Ann Arbor, USA, October 1985.

154

Table 1. Occupational exposure limits to radiofrequency electromagnetic fields

Frequency range (MHz)	Unperturbed RMS electric field strength (V/m)	Unperturbed RMS magnetic field strength (A/m)	Equivalent plane wave power density	
			W/m²	mW/cm²
0.1–1	614	$1.6/f$	—	—
>1–10	$614/f$	$1.6/f$	—	—
>10–400	61	0.16	10	1
>400–2 000	$3f^{\frac{1}{2}}$	$0.008f^{\frac{1}{2}}$	$f/40$	$f/400$
>2 000–300 000	137	0.36	50	5

[a] These values are provided for information only, and are not to be considered for determining compliance.

Note. Hazards of RF burns should be eliminated by limiting currents from contact with metal objects. In most situations this may be achieved by reducing the *E* values from 614 to 194 V/m in the range from 0.1 to 1 MHz and from $614/f$ to $194/f^{\frac{1}{2}}$ in the range from >1 to 10 MHz. f = frequency in MHz.

Source: IRPA *(228).*

Table 2. Exposure limits to radiofrequency electromagnetic fields for the general public

Frequency range (MHz)	Unperturbed RMS electric field strength (V/m)	Unperturbed RMS magnetic field strength (A/m)	Equivalent plane wave power density	
			W/m²	mW/cm²
0.1–1	87	$0.23/f^{\frac{1}{2}}$	—	—
>1–10	$87/f^{\frac{1}{2}}$	$0.23/f^{\frac{1}{2}}$	—	—
>10–400	27.5	0.073	2	0.2
>400–2 000	$1.375f^{\frac{1}{2}}$	$0.0037f^{\frac{1}{2}}$	$f/200$	$f/2 000$
>2 000–300 000	61	0.16	10	1

Note. f = frequency in MHz.

Source: IRPA *(228).*

155

may have to be measured for complete specification. Contact with large, ungrounded metal objects in the presence of strong RF fields may be hazardous.

At any given frequency, natural background radiation is very low compared with levels produced by man-made sources. The urban population is exposed to power densities of the order of hundreds of mW/m^2 (hundreds of mV/m and tenths of mA/m). Higher exposures may occur close to radiation sources. Levels of occupational exposure vary considerably with frequency and application. In some cases, exposures of the order of $100 W/m^2$ (hundreds of V/m and $1-10 A/m$) occur for short durations, but in most cases exposures are much lower. In therapeutic applications exposures are high enough to cause an elevation of tissue temperature by a few degrees.

The specific absorption rate, SAR (W/kg), is widely used as a dosimetric quantity and exposure limits can be derived from SARs. The SAR in a biological body depends on several exposure parameters such as frequency, intensity, polarization, radiation source-body configuration, reflecting surfaces, and body size, shape and electrical properties. Furthermore, the spatial distribution of the SAR inside the body is highly non-uniform. Non-uniform energy deposition results in non-uniform deep-body heating and may produce internal temperature gradients. At frequencies above 10 GHz, the energy is deposited close to the body surface. The maximum SAR occurs at about 70 MHz for the standard man, and at about 30 MHz when the person is standing in contact with RF ground. At extreme conditions of temperature and humidity, whole-body SARs of 1–4 W/kg at 70 MHz are expected to cause a core temperature rise of about 2 °C in healthy human beings in one hour.

High-level RF radiation is a source of thermal energy that carries all of the known implications of heating for biological systems, including burns, temporary and permanent changes in reproduction, cataracts, and death. For the broad range of radiofrequencies, cutaneous perception of heat and thermal pain is unreliable for the detection of RF radiation, because at many frequencies much of the energy is deposited below the level of the thermal receptors in the skin. RF heating is an interaction mechanism that has been studied extensively. Thermal effects have been observed at less than 1 W/kg, but temperature thresholds have not been determined for most effects. The time–temperature profile must be considered in assessing biological effects.

Biological effects also occur where RF heating is neither an adequate nor a possible mechanism. These effects often involve modulated RF fields and millimetre wavelengths. Various hypotheses have been proposed but have not yet yielded information useful for deriving human exposure limits. There is a need to understand the fundamental mechanisms of interaction, since it is not practical to explore each RF field for its characteristic biophysical and biological interactions.

There is very little information on the effects of RF radiation on man, and limited data on responses of animals exposed to frequencies above 10 GHz and below 10 MHz; most animal research has been carried out at frequencies near 3 GHz.

156

In general, no changes in chromosomes, DNA or the reproductive potential of animals exposed to RF radiation have been reported in the absence of significant rises in temperature, though there are limited data on DNA and chromosomal changes at non-thermal levels. Other effects reported in the absence of a significant rise in temperature include changes in ion transport across cell membranes, electrophysiological properties of neurons and mobilization of calcium ions, especially in brain tissue.

Effects on the blood-forming and immune systems have been reported at SARs greater than 0.4 W/kg.

Elevated maternal body temperature is known to be associated with birth defects, and RF radiation is teratogenic at exposure conditions that approach lethal levels for the pregnant animal. Synergistic effects with chemical teratogens have been reported.

Neurons in the central nervous system of experimental animals can be affected by acute high-level and by chronic low-level exposures at SARs greater than 2 W/kg. Modulated RF fields can alter EEG rhythms. Pulsed radiation may have a potentiating effect on drugs that affect nervous system function. Some types of animal behaviour are disrupted at SARs of 1– 4 W/kg; these behavioural alterations appear to be reversible with time following termination of exposure.

A single acute exposure of the eye to high-intensity RF radiation, if applied for a sufficient time, can cause cataracts and retinal changes in experimental animals. No epidemiological data at present support a conclusion that low-level, chronic exposure to RF radiation induces cataracts in man, although some studies have associated ocular lens defects with RF exposure.

Pulsed RF radiation can be heard by some human beings as a click, buzz or chirp, but this effect has no known medical significance.

Human data are currently limited and do not provide adequate information about the relationship between prolonged low-level RF radiation exposure and increased mortality or morbidity, including cancer incidence. In epidemiological studies and clinical reports of RF effects in man, the problems of quantification are numerous and include uncertainties about "dose", health effects, latent periods, dose–response relationships, and interactions with other physical or chemical agents.

Recommendations

The progress made in understanding energy transfer from RF fields to biological systems permits recommendations for health protection. The available data also indicate the need for further research.

1. Limiting the exposure of people to RF radiation by means of voluntary or mandatory standards is desirable. To promote better understanding and to avoid misinterpretation, the scientific basis and rationale for the limits set out in such standards should be made available. It is hoped that this will lead to convergence in the limits recommended by various national bodies. The emerging international consensus is reflected in the 1984 IRPA recommendations on RF exposure limits (228).

157

2. Protecting workers and the public requires evaluation of human exposures from any RF source. Limiting the exposure of people should be a consideration in the purchase of equipment. For equipment in use, particularly at frequencies of the order of tens of MHz and where workers are in contact with RF ground, efforts should be made to reduce exposure by shielding or by other means. Appropriate education and training are essential for personnel occupationally exposed to RF radiation. In the kHz and MHz range, preventive measures should be employed to avoid hazards from contact with large, ungrounded metal structures. An international repository should be established for reports that include medical and exposure records on cases of human injury and disease attributed to RF exposure.

3. Assessment should be made of the energy deposition from portable and mobile transmitters and other sources that may produce high local SARs. Guidelines for limits on such exposures should be developed.

4. Human exposure and dosimetric data are urgently needed for use in epidemiological studies. Data are particularly needed in the areas of malignancies, visual system effects, reproductive effects, and functional disturbances of the central nervous system.

5. The distribution, quantification and physiological significance of energy absorption in human and animal tissues require further studies, particularly of hazardous increases in core and localized temperature and the consideration of ambient conditions. Information is also needed on the influence of exposure to RF radiation on the course of human disease and on thermal tolerance in individuals with impaired physiological states.

6. The responses of biological systems and their components (including responses to amplitude- and pulse-modulated fields) need to be further evaluated with respect to:
   — in vivo and in vitro tests of mutagens, carcinogens and teratogens;
   — dysfunction of the central nervous system;
   — synergistic effects; and
   — biophysical, electrochemical and cooperative molecular mechanisms.

7. Research is needed at frequencies where few data exist, particularly below 10 MHz.

REFERENCES

1. **International Non-Ionizing Radiation Committee of the International Radiation Protection Association.** Review of concepts, quantities, units, and terminology for non-ionizing radiation protection. *Health physics,* **49**: 1329–1362 (1985).

2. *Radiofrequency electromagnetic fields. Properties, quantities, and units, biophysical interaction and measurements.* Washington, DC, National Council on Radiation Protection and Measurements, 1981 (NCRP Report No. 67).

3. **Schwan, H.P. & Foster, K.R.** RF-field interactions with biological systems. *Proceedings of the Institute of Electrical and Electronics Engineers,* **68**: 104–113 (1980).

4. **Tell, R.A. & Mantiply, E.D.** Population exposure to VHF and UHF broadcast radiation in the United States. *Proceedings of the Institute of Electrical and Electronics Engineers,* **68**: 6–12 (1980).

5. **Stuchly, M.A.** Potentially hazardous microwave radiation sources — a review. *Journal of microwave power,* **12**: 370–381 (1977).

6. **Lovsund, P. & Mild, K.H.** [Low frequency electromagnetic field near some induction heaters]. Stockholm, National Board of Occupational Health and Safety, 1978 (in Swedish).

7. **Stuchly, M.A. & Lecuyer, D.W.** Induction heating and operator exposure to electromagnetic fields. *Health physics,* **49**: 693–700 (1985).

8. **Conover, D. et al.** Measurement of electric- and magnetic-field strengths from industrial radiofrequency (6–38 MHz) plastic sealers. *Proceedings of the Institute of Electrical and Electronics Engineers,* **68**: 17–20 (1980).

9. **Stuchly, M.A. et al.** Radiation survey of dielectric heaters in Canada. *Journal of microwave power,* **15**: 113–121 (1980).

10. **Mild, K.H. et al.** [Effects on humans of high levels of exposure to radiofrequency radiation. A study on the health and exposure of plastic welding machine operators]. *Arbete och hälsa,* **12**: 1–66 (1987) (in Swedish).

11. *Recommended practice for the measurement of hazardous electromagnetic fields — RF and microwave.* New York, American National Standards Institute, 1981.

12. *Safety levels with respect to human exposure to radiofrequency electromagnetic fields, 300 kHz to 100 GHz.* New York, American National Standards Institute, 1982 (ANSI C95.1-1982).

13. *Safe distances from radiofrequency transmitting antennas for electric blasting operations.* New York, American National Standards Institute, 1985 (ANSI C95.5-1985).

14. **Eriksson, A. & Mild, K.H.** Radiofrequency electromagnetic leakage fields from plastic welding machines. Measurements and reducing measures. *Journal of microwave power,* **20**: 95–107 (1985).

15. **Mild, K.H.** Occupational exposure to radiofrequency electromagnetic fields. *Proceedings of the Institute of Electrical and Electronics Engineers,* **68**: 12–17 (1980).

16. **Mild, K.H.** Radiofrequency electromagnetic fields in Swedish radio stations and tall FM/TV towers. *Bioelectromagnetics,* **2**: 61–69 (1981).

17. **Stuchly, M.A. & Stuchly, S.S.** Industrial, scientific, medical and domestic applications of microwaves. *IEE proceedings, A,* **130**: 467–503 (1983).

18. **Stuchly, S.S. et al.** Energy deposition in a model of man in the near field. *Bioelectromagnetics,* **6**: 115–129 (1985).

19. **Stuchly, M.A. et al.** Exposure to the operator and patient during short wave diathermy treatments. *Health physics,* **42**: 341–366 (1982).

20. **Stuchly, M.A. et al.** Operator exposure to radiofrequency fields near a hyperthermia device. *Health physics,* **45**: 101–107 (1983).

21. **Eggert, S. et al.** Near-zone field-strength meter for measurement of RF electric fields. *Radio science,* **14**(65): 9–14 (1979).

22. *Instrumentation for nonionizing radiation measurement.* Rockville, MD, US Department of Health and Human Services, 1984 (Publication (FDA) 84-8222).

23. **Tell, R.A.** Instrumentation for measurement of electromagnetic fields: equipment, calibrations and selected applications. *NATO Advanced Study Institutes series A: life sciences,* **49**: 95–162 (1983).

24. **Crawford, M.** Generation of standard EM fields using TEM transmitted cells. *IEEE transactions on electromagnetic compatibility,* **EMC-16**: 189 (1974).

25. **Bostrom, R. et al.** Calibration of commercial power density meters at RF and microwave frequencies. *IEEE transactions on instrumentation and measurement,* **IM-35**: 111–115 (1986).

26. **Durney, C.H.** The physical interactions of radiofrequency radiation fields and biological systems. *In:* The impact of proposed radio-frequency radiation standards on military operations. Springfield, VA, National Technical Information Service, 1985 (AGARD Lecture Series No. 138).

27. **Durney, C.H.** Electromagnetic dosimetry for models of humans and animals: a review of theoretical and numerical techniques. *Proceedings of the Institute of Electrical and Electronics Engineers,* **68**: 33–40 (1980).

28. **Gandhi, O.P.** State of knowledge for electromagnetic absorbed dose in man and animals. *Proceedings of the Institute of Electrical and Electronics Engineers,* **68**: 24–32 (1980).

29. **Durney, C.H. et al.** *Radiofrequency radiation dosimetry handbook,* 2nd ed. Brooks Air Force Base, Texas, USAF School of Aerospace Medicine, 1978 (Report SAM-TR-78-22).

30. **Durney C.H. et al.** *Radiofrequency radiation dosimetry handbook,* 3rd ed. Brooks Air Force Base, Texas, USAF School of Aerospace Medicine, 1980 (Report SAM-TR-80-32).

31. **Durney C.H. et al.** *Radiofrequency radiation dosimetry handbook,* 4th ed. Brooks Air Force Base, Texas, USAF School of Aerospace Medicine, 1986).

32. **Stuchly, M.A. & Stuchly, S.S.** Experimental radio and microwave dosimetry. *In:* Polk, C. & Postow, E., ed. *Handbook of biological effects of electromagnetic radiation.* Boca Raton, FL, CRC Press, 1986.

33. **Guy, A.W. et al.** Average SAR and distribution in man exposed to 450 MHz radiofrequency radiation. *IEEE transactions on microwave theory and techniques,* **MTT-32**: 752–762 (1984).

34. **Stuchly, S.S. et al.** Energy deposition in a model of man: frequency effects. *IEEE transactions on biomedical engineering,* **BME-33**: 702–711 (1986).

35. **Stuchly, M.A. et al.** Exposure of man in the near-field of a resonant dipole: comparison between theory and measurements. *IEEE transactions on microwave theory and techniques,* **MTT-34**: 26–31 (1986).

36. **Stuchly, M.A. et al.** Exposure of human models in the near and far field — a comparison. *IEEE transactions on biomedical engineering,* **BME-32**: 609–616 (1985).

37. **Gandhi, O.P. et al.** Likelihood of high rates of energy deposition in the human legs at the ANSI recommended 3–30 MHz RF safety levels. *Proceedings of the Institute of Electrical and Electronics Engineers,* **73**: 1145–1147 (1985).

38. **Schwan, H.P. & Foster, K.R.** RF-field interactions with biological systems: electrical properties and biophysical mechanisms. *Proceedings of the Institute of Electrical and Electronics Engineers,* **68**: 104–113 (1980).

39. **Stuchly, M.A.** Interaction of radiofrequency and microwave radiation with living systems. A review of mechanisms. *Radiation and environmental biophysics,* **16**: 1–14 (1979).

40. **Bowman, H.D.** Heat transfer and thermal dosimetry. *Journal of microwave power,* **16**: 121–133 (1981).

41. **Sapareto, S.A. & Devey, W.C.** Thermal dose determination cancer therapy. *International journal of radiation: oncology — biology — physics,* **10**: 787–800 (1984).

42. **Lingren, D.** The temperature influence on the spontaneous mutation rate. I. Literature review. *Hereditas,* **10**: 265–278 (1972).

43. **Lan, M.P.** Induced thermal resistance in the mouse ear. The relationship between heating time and temperature. *International journal of radiation biology,* **35**: 481–485 (1979).

44. **Bernhardt, J.** The direct influence of electromagnetic fields on nerve and muscle cells of man within the frequency range of 1 Hz to 30 MHz. *Radiation and environmental biophysics,* **16**: 309–323 (1979).

45. **Bernhardt, J.H. et al.** *Gefährdung von Personen durch electromagnetische Felder* [Hazards to human health from electromagnetic fields]. Neuherberg, D. Reimer Verlag, 1983 (STH-Berichte No. 2).

46. **Bawin, S.M. et al.** Effects of modulated very high frequency fields on specific brain rhythms in cats. *Brain research,* **58**: 365–384 (1973).

47. **Dumanskiy, J.D. & Shandala, M.G.** The biological action and hygienic significance of electromagnetic fields of superhigh and ultrahigh frequencies in densely populated areas. *In:* Czerski, P. et al., ed. *Biological effects and health hazards of microwave radiation.* Warsaw, Polish Medical Publishers, 1974, pp. 289–293.

48. **Rudnev, M.I. et al.** The use of evoked potential and behavioral measures in the assessment of environmental insult. *In:* Otto, D.A., ed. *Multidisciplinary perspectives in event-related brain potential research.* Research Triangle Park, NC, US Environmental Protection Agency, 1978, pp. 444–447.

49. **Shandala, M.G. et al.** Patterns of change in behavioral reactions to low power densities of microwaves. *Abstracts, International Symposium on the Biological Effects of Electromagnetic Waves (URSI), Airlie, VA, 1977.*

50. **Thomas, J.R. et al.** Microwave radiation and chlordiazepoxide: synergistic effects on fixed-interval. *Behavioral science,* **203**: 1357–1358 (1979).

51. **Bawin, S.M. et al.** Effects of modulated VHF fields on the central nervous system. *Annals of the New York Academy of Sciences,* **247**: 74–81 (1975).

52. **Blackman, C.F. et al.** Induction of calcium ion efflux from brain tissue by radiofrequency radiation: effects of modulation frequency and field strength. *Radio science,* **14**(65): 93–98 (1979).

53. **Blackman, C.F. et al.** Induction of calcium ion efflux from brain tissue by radiofrequency radiation: effect of sample number and modulation frequency on the power-density window. *Bioelectromagnetics,* **1**: 35–43 (1980).

54. **Takashima, S. et al.** Effects of modulated RF energy on the EEG of mammalian brains. *Radiation and environmental biophysics,* **16**: 15–27 (1979).

55. **Lyle, D.B. et al.** Suppression of T-lymphocyte cytotoxicity following exposure to sinusoidally amplitude-modulated fields. *Bioelectromagnetics,* **4**: 281–292 (1983).

56. **Byus, C.V. et al.** Alterations in protein kinase activity following exposure of cultured human lymphocytes to modulated microwave fields. *Bioelectromagnetics,* **5**: 341–351 (1984).

57. **Seaman, R.L. & Wachtel, H.** Slow and rapid responses to CW and pulsed microwave radiation by individual aplysia pacemakers. *Journal of microwave power,* **13**: 77–86 (1978).

58. **Wachtel, H. et al.** Effects of low-intensity microwaves on isolated neurons. *Annals of the New York Academy of Sciences,* **247**: 46–62 (1975).

59. **Grundler, W. & Keilmann, F.** Sharp resonances in yeast growth prove nonthermal sensitivity to microwaves. *Physical review letters,* **51**: 1214–1216 (1983).

60. **Grundler, W. et al.** Resonant growth rate response of yeast cells irradiated by weak microwaves. *Physical letters, A,* **62**: 463–466 (1977).

61. **Pickard, W.F. & Rosenbaum, F.J.** Biological effects of microwaves at the membrane levels: two possible athermal electrophysiological mechanisms and a proposed experimental test. *Mathematical biosciences,* **39**: 235–253 (1978).

62. **Barsoum, Y.H. & Pickard, W.F.** The vacuolar potential of characean cells subjected to electromagnetic radiation in the range 200–8200 MHz. *Biolectromagnetics,* **3**: 393–400 (1982).

63. **Barsoum, Y.H. & Pickard, W.F.** Effects of electromagnetic radiation in the range 20–400 MHz on the vacuolar potential of characean cells. *Bioelectromagnetics,* **3**: 193–201 (1982).

64. **Barsoum, V.H. & Pickard, W.F.** Radio-frequency rectification in electrogenic and nonelectrogenic cells of *Chara* and *Nitella*. *Journal of membrane biology,* **65**: 81–87 (1982).

65. **Edwards, G.S. et al.** Resonant microwave absorption of selected DNA molecules. *Physical review letters,* **53**: 1284–1287 (1984).

66. **Schwartz, G. & Seelig, J.** Kinetic properties and the electric field effect of the helix-coil transition of poly (γ-benzyl L-glutamate) determined from dielectric relaxation measurements. *Biopolymers,* **6**: 1263–1277 (1968).

67. **Foster, K.R. et al.** "Resonances" in the dielectric absorption of DNA? *Journal of biophysics,* **52**: 421–425 (1987).

68. **Gabrill, C. et al.** Microwave absorption in aqueous solutions of DNA. *Nature,* **328**: 145–146 (1987).

69. **Chou, C.K. et al.** Cochlear microphonics generated by microwave pulses. *Journal of microwave power,* **10**: 361–367 (1975).

70. **Chou, C.K. et al.** Characteristics of microwave-induced cochlear microphonics. *Radio science,* **12**(6): 221–227 (1977).

71. **Foster, K.R. & Finch, E.D.** Microwave hearing: evidence for thermoacoustic auditory stimulation by pulsed microwaves. *Science,* **185**: 256–258 (1974).

72. **Lin, J.C.** *Microwave auditory effects and applications.* Springfield, IL, Charles C. Thomas, 1978.

73. **Adey, W.R.** Tissue interactions with nonionizing electromagnetic fields. *Physiological reviews,* **61**: 435–514 (1981).

74. **Swicord, M. & Postow, E.** *In:* Polk, C. & Postow, E., ed. *Handbook of biological effects of electromagnetic radiation.* Boca Raton, FL, CRC Press, 1986.

75. **Taylor, L.S.** The mechanisms of athermal microwave biological effects. *Bioelectromagnetics,* **2**: 259–267 (1981).

76. **Grodsky, I.T.** Possible physical substrates for the interaction of electromagnetic fields with biologic membranes. *Annals of the New York Academy of Sciences,* **247**: 117 (1975).

77. **Fröhlich, H.** The extraordinary dielectric properties of biological materials and the action of enzymes. *Proceedings of the National Academy of Sciences of the United States of America,* **72**: 4211 (1975).

78. **Davidov, A.S.** *Biology and quantum mechanics.* New York, Pergamon Press, 1982.

79. **Lawrence, A.F. & Adey, W.R.** Nonlinear wave mechanisms in interactions betweeen excitable tissue and electromagnetic fields. *Neurological research,* **4**: 115–153 (1982).

80. **Guy, A.W. & Chou, C.K.** *Hazard analysis: very low frequency through medium frequency range.* Brooks Air Force Base, Texas, USAF School of Aerospace Medicine, 1982.

81. **Elder, J.A.** Special senses. *In:* Elder, J.A. & Cahill, D.F., ed. *Biological effects of radiofrequency radiation.* Washington, DC, US Environmental Protection Agency, 1984 (Publication EPA-600/8-83-026F).

82. **Cleary, S.F.** Microwave cataractogenesis. *Proceedings of the Institute of Electrical and Electronics Engineers,* **68**: 49–55 (1980).

163

83. **Chou, C.K. et al.** Auditory perception of radio-frequency electromagnetic fields. *Journal of the Acoustical Society of America,* **71**: 1321–1334 (1982).

84. **Spiegel, R.J.** The thermal response of a human in the near-zone of a resonant thin-wire antenna. *IEEE transactions on microwave theory and techniques,* **MTT-30**: 177–185 (1982).

85. **Spiegel, R.J.** A review of numerical models for predicting the energy deposition and resultant thermal response of humans exposed to electromagnetic fields. *IEEE transactions on microwave theory and techniques,* **MTT-32**: 730–746 (1984).

86. **Hill, D.** Human studies. *In:* Elder, J.A. & Cahill, D.F., ed. *Biological effects of radiofrequency radiation.* Washington, DC, US Environmental Protection Agency, 1984 (Publication EPA-600/8-83-026F).

87. **Roberts, N.J., Jr. & Michaelson, S.M.** Epidemiological studies of human exposures to radiofrequency radiation. *International archives of occupational and environmental health,* **56**: 169–178 (1985).

88. **Shandala, M.G. et al.** [Principal directions of research on biological effects of microwaves in the USSR]. *Gigiena i sanitarija,* **10**: 4–7 (1981) (in Russian).

89. **Smialowicz, R.J.** Hematologic and immunologic effects. *In:* Elder, J.A. & Cahill, D.F., ed. *Biological effects of radiofrequency radiation.* Washington, DC, US Environmental Protection Agency, 1984 (Publication EPA-600/8-83-026F).

90. **Hendler, E.** Cutaneous receptor response to microwave irradiation. *In:* Hardy, J.D., ed. *Thermal problems in aerospace medicine.* London, Unwin, 1968, pp. 149–161.

91. **Hendler, E. et al.** Skin heating and temperature sensation produced by infra-red and microwave irradiation. *In:* Herzfeld, C.M., ed. *Temperature: its measurement and control in science and industry. Part 3. Biology and medicine.* New York, Reinhold, 1963, pp. 211–230.

92. **Justesen, D.R. et al.** A comparative study of human sensory thresholds: 2450-MHz microwaves vs far-infrared radiation. *Bioelectromagnetics,* **3**: 117-125 (1982).

93. **Vendrik, A.J.H. & Vos, J.J.** Comparison of the stimulation of the warmth sense organ by microwave and infrared. *Journal of applied physiology,* **13**: 435–444 (1958).

94. **Cook, H.F.** The pain threshold for microwave and infra-red radiations. *Journal of physiology,* **118**: 1–11 (1952).

95. **Hardy, H.D. et al.** *Pain sensations and reactions.* New York, Hafner Publishing Co., 1967.

96. **Cleary, S.F.** Cataract incidence in radar workers. *Archives of environmental health,* **11**: 179–182 (1965).

97. **Cleary, S.F. & Pasternack, B.S.** Lenticular changes in microwave workers: a statistical study. *Archives of environmental health,* **12**: 23–29 (1966).

98. **Majewska, K.** Investigations on the effect of microwaves on the eye. *Polish medical journal,* **7**: 989–994 (1968).

99. **Odland, L.T.** Radiofrequency energy: a hazard to workers? *Industrial medicine and surgery*, **42**: 23–26 (1973).

100. **Shacklett, D.E. et al.** Evaluation of possible microwave-induced lens changes in the United States Air Force. *Aviatiation, space, and environmental medicine*, **46**: 1403–1406 (1975).

101. **Sierkierzynski, M. et al.** Health surveillance of personnel occupationally exposed to microwaves. III. Lens translucency. *Aerospace medicine*, **45**: 1146–1148 (1974).

102. **Zaret, M.M.** Selected cases of microwave cataract in man associated with concomitant annotated pathologies. *In:* Czerski, P. et al., ed. *Biological effects and health hazards of microwave radiation.* Warsaw, Polish Medical Publishers, 1974, pp. 294–301.

103. **Zaret, M.M.** Cataracts following use of microwave oven. *New York State journal of medicine*, **74**: 2032–2048 (1974).

104. **Appleton, B.** Microwave lens effects in humans. II. Results of five year survey. *Archives of ophthalmology*, **93**: 257–258 (1975).

105. **Appleton, B. & McCrossan, G.C.** Microwave lens effects in humans. *Archives of ophthalmology*, **88**: 259–262 (1972).

106. **Wissler, E.H.** A mathematical model of the human thermal system. *Bulletin of mathematical biophysics*, **26**: 147–166 (1964).

107. **Stolwijk, J.A.J.** *Expansion of a mathematical model of thermoregulation to include high metabolic rates.* Washington, DC, National Technical Information Service, 1969.

108. **Stolwijk, J.A.J.** *A mathematical model of physiological temperature regulation in man.* Washington, DC, National Technical Information Service, 1971.

109. **Stolwijk, J.A.J.** Mathematical model of thermal regulation. *Annals of the New York Academy of Science*, **33**: 309–325 (1980).

110. **Stolwijk, J.A.J. & Hardy, J.D.** Control of body temperature. *In:* Douglas, H. & Lee, K., ed. *Handbook of physiology. Reactions to environmental agents.* Baltimore, MD, Williams & Wilkins, 1977, pp. 45–88.

111. **Emery, A.F.** The numerical thermal simulation of the human body when undergoing exercise or nonionizing electromagnetic irradiation. *Journal of heat transfer*, **98**: 284–291 (1976).

112. **Spiegel, R.J.** Numerical modeling of thermoregulatory systems in man. *In:* Elder, J.A. & Cahill, D.F., ed. *Biological effects of radiofrequency radiation.* Washington, DC, US Environmental Protection Agency, 1984 (Publication EPA-600/8-83-026F).

113. **Spiegel, R.J. et al.** A thermal model of the human body exposed to an electromagnetic field. *Bioelectromagnetics,* **1**: 253–270 (1980).

114. **Hagmann, M.J. et al.** Heat resonance: numerical solutions and experimental results. *IEEE transactions on microwave theory and techniques,* **MTT-27**: 809–813 (1979).

115. **Airborne Instruments Laboratory.** An observation on the detection by the ear of microwave signals. *Proceedings of the IRE,* **44**(Oct.): 2A (1956).

116. **Frey, A.H.** Auditory system response to radio frequency energy. *Aerospace medicine,* **32**: 1140–1142 (1961).

117. **Frey, A.H.** Human auditory system response to modulated electromagnetic energy. *Journal of applied physiology,* **17**: 689–692 (1962).

118. **Frey, A.H.** Some effects on human subjects of ultra-high-frequency radiation. *American journal of medical electronics,* **2**: 28–31 (1963).

119. **Constant, P.C. Jr.** Hearing EM waves. *In:* Jacobson, B., ed. *Digest of the Seventh International Conference on Medical and Biological Engineering.* Stockholm, Karolinska Institute, 1967, p. 349.

120. **Cain, C.A. & Rissman, W.J.** Mammalian auditory responses to 3.0 GHz microwave pulses. *IEEE transactions in biomedical engineering,* **BME-25**: 288–293 (1978).

121. **Chou, C.K. & Guy, A.W.** Microwave-induced auditory responses in guinea pigs: relationship of threshold and microwave-pulse duration. *Radio science,* **14**(65): 193–197 (1979).

122. **Frey, A.H. & Messenger, R.** Human perception of illumination with pulsed ultrahigh-frequency electromagnetic energy. *Science,* **181**: 356–358 (1973).

123. **Guy, A.W. et al.** Microwave-induced acoustic effects in mammalian auditory systems and physical materials. *Annals of the New York Academy of Science,* **247**: 194–215 (1975).

124. **Olsen, R.G. & Hammer, W.C.** Microwave-induced pressure waves in a model of muscle tissue. *Bioelectromagnetics,* **1**: 45–54 (1980).

125. **Tyazhelov, V.V. et al.** Some peculiarities of auditory sensations evoked by pulsed microwave fields. *Radio science,* **14**: 259-263 (1979).

126. **Gournay, L.S.** Conversion of electromagnetic to acoustic energy by surface heating. *Journal of the Acoustical Society of America,* **40**: 1322–1330 (1966).

127. **Barron, C.I. et al.** Physical evaluations of personnel exposed to microwave emanations. *Journal of aviation medicine,* **26**: 442–452 (1955).

128. **Barron, C.I. & Baraff, A.A.** Medical considerations of exposure to microwaves (radar). *Journal of the Americal Medical Association,* **168**: 1194–1199 (1958).

129. **Baranski, S. & Edelwejn, Z.** Pharmacologic analysis of microwave effects on the central nervous system in experimental animals. *In:* Czerski, P. et al., ed. *Biological effects and health hazards of microwave radiation.* Warsaw, Polish Medical Publishers, 1974.

130. **Silverman, C.** Epidemiologic studies of microwave effects. *Proceedings of the Institute of Electrical and Electronics Engineers,* **68**: 78–84 (1980).

131. **Silverman, C.** Epidemiology of microwave radiation effects in humans. *In:* Castellani, A., ed. *Epidemiology and quantitation of environmental risk in humans from radiation and other agents.* New York, Plenum, 1985, pp. 433–458.

132. **Czerski, P. et al.** Health surveillance of personnel occupationally exposed to microwaves. I. Theoretical considerations and practical aspects. *Aerospace medicine,* **45**: 1137–1142 (1974).

133. **Siekierzynski, M. et al.** Health surveillance of personnel occupationally exposed to microwaves. II. Functional disturbances. *Aerospace medicine*, **45**: 1143–1145 (1974).

134. **Djordjevic, A. et al.** A study of the health status of radar workers. *Aviation, space and environmental medicine*, **50**: 396–398 (1979).

135. **Hamburger, S. et al.** Occupational exposure to nonionizing radiation and an association with heart disease. An exploratory study. *Journal of chronic disease*, **36**: 791–802 (1983).

136. **Lilienfeld, A.M. et al.** *Foreign Service health status study — evaluation of health status of Foreign Service and other employees from selected Eastern European posts. Final Report.* Washington, DC, Department of State, 1978.

137. **Pollack, H.** Epidemiologic data on American personnel in the Moscow Embassy. *Bulletin of the New York Academy of Medicine*, **55**: 1182–1186 (1979).

138. **United States Senate Committee on Commerce, Science, and Transportation.** *Microwave irradiation of the US Embassy in Moscow.* Washington, DC, US Government Printing Office, 1979.

139. **Robinette, C.D. & Silverman, C.** Causes of death following occupational exposure to microwave radiation (radar) 1950–1974. *In:* Hazzard, D.G., ed. *Symposium on Biological Effects and Measurements of Radiofrequency/Microwaves.* Rockville, MD, US Department of Health, Education, and Welfare, 1977, pp. 338–344 (Publication (FDA) 77-8026).

140. **Robinette, C.D. et al.** Effects upon health of occupational exposure to microwave radiation (radar). *American journal of epidemiology*, **112**: 39–53 (1980).

141. **Calle, E.E. & Savitz, D.A.** Leukemia in occupational groups with presumed exposure to electrical and magnetic fields. *New England journal of medicine*, **313**: 1476–1477 (1985).

142. **Milham, S., Jr.** Mortality from leukemia in workers exposed to electrical and magnetic fields. *New England journal of medicine*, **307**: 249 (1982).

143. **Baranski, S. & Czerski, P.** *Biological effects of microwaves.* Stroudsburg, PA, Dowden, Hutchinson & Ross, 1976.

144. **Dwyer, M.J. & Leeper, D.B.** *A current literature report on the carcinogenic properties of ionizing and non-ionizing radiation. II. Microwave and radiofrequency radiation.* Cincinnati, OH, Department of Health, Education, and Welfare, 1978 (Publication (NIOSH) No. 78-134).

145. **Justesen, D.R. et al.** *Compilation and assessment of microwave bioeffects. A selective review of the literature on biological effects of microwaves in relation to the satellite power system (SPS).* Washington, DC, US Department of Energy, 1978.

146. **Sigler, A.T. et al.** Radiation exposure in parents of children with mongolism (Down's syndrome). *Bulletin of the Johns Hopkins Hospital*, **117**: 374 (1965).

147. **Cohen, B.H. et al.** Parental factors in Down's syndrome: results of the second Baltimore case–control study. *In:* Hook, E.B. & Porter, I.H., ed. *Population cytogenetics: studies in humans.* New York, Academic Press, 1977, pp. 301–352.

148. **Lancranjan, I. et al.** Gonadic function in workmen with long-term exposure to microwaves. *Health physics,* **29**: 381–383 (1975).

149. **Kallen, B. et al.** Delivery outcome among physiotherapists in Sweden: is non-ionizing radiation a fetal hazard? *Archives of environmental health,* **37**: 81–85 (1982).

150. **Elder, J.A.** Summary and conclusions. *In:* Elder, J.A. & Cahill, D.F., ed. *Biological effects of radiofrequency radiation.* Washington, DC, US Environmental Protection Agency, 1984 (Publication EPA-600/8-83-026F).

151. **Adair, E.R., ed.** *Microwaves and thermoregulation.* New York, Academic Press, 1983.

152. **Gordon, C.J.** Effect of RF-radiation exposure on body temperature: thermal physiology. *In:* Elder, J.A. & Cahill, D.F., ed. *Biological effects of radiofrequency radiation.* Washington, DC, US Environmental Protection Agency, 1984 (Publication EPA-600/8-83-026F).

153. **Gordon, C.J. & Ferguson, J.H.** Scaling the physiological effects of exposure to radiofrequency electromagnetic radiation: consequences of body size. *International journal of radiation biology,* **46**: 387–397 (1984).

154. **Adair, E.R.** Microwaves and thermoregulation. *In:* Mitchell, J.C., ed. *USAF Radiofrequency Radiation Bioeffects Research Program — a review.* San Antonio, TX, USAF School of Aerospace Medicine, 1981, pp. 145–158 (Review 4-81).

155. **Adair, E.R. & Adams, B.W.** Microwave induced peripheral vasodilation in squirrel monkey. *Science,* **207**: 1381–1383 (1980).

156. **Adair, E.R. & Adams, B.W.** Microwaves modify thermoregulatory behavior in squirrel monkey. *Bioelectromagnetics,* **1**: 1–20 (1980).

157. **Adair, E.R. & Adams, B.W.** Adjustments in metabolic heat production by squirrel monkeys exposed to microwaves. *Journal of applied physiology,* **52**: 1049–1958 (1982).

158. **Gordon, C.J.** Effects of ambient temperature and exposure to 2450-MHz microwave radiation on evaporative heat loss in the mouse. *Journal of microwave power,* **17**: 145–150 (1982).

159. **Gordon, C.J.** Influence of heating rate on control of heat loss from the tail in mice. *American journal of physiology,* **244**: R778–R784 (1983).

160. **Gordon, C.J.** Effect of 2450 MHz microwave exposure on behavioral thermoregulation in mice. *Journal of thermal biology,* **8**: 315–319 (1984).

161. **Ho, H.S. & Edwards, W.P.** Oxygen-consumption rate of mice under differing dose rates of microwave radiation. *Radio science,* **12**: 131–138 (1977).

162. **Berman, E. et al.** Lethality of mice and rats exposed to 2450 MHz circularly polarized microwaves as a function of exposure duration and environmental factors. *Journal of applied toxicology,* **5**: 23–31 (1985).

163. **Lotz, W.G.** Hyperthermia in radiofrequency-exposed rhesus monkeys: a comparison of frequency and orientation effects. *Radiation research,* **102**: 59–70 (1985).
164. **Michaelson, S.M. et al.** Physiologic aspects of microwave irradiation of mammals. *American journal of physiology,* **201**: 351–356 (1961).
165. **Berman, E.** Reproductive effects. *In:* Elder, J.A. & Cahill, D.F., ed. *Biological effects of radiofrequency radiation.* Washington, DC, US Environmental Protection Agency, 1984 (Publication EPA-600/8-83-026F).
166. **Lary, J.M. et al.** Teratogenic effects of 27.12 MHz radiofrequency radiation in rats. *Teratology,* **26**: 299–309 (1982).
167. **Berman, E. et al.** Fetal and maternal effects of continual exposure to 970-MHz circularly polarized microwaves. (Personal communication).
168. **Berman, E. et al.** Tests of mutagenesis and reproduction in male rats exposed to 2450-MHz (CW) microwaves. *Bioelectromagnetics,* **1**: 65–76 (1980).
169. **Rudnev, M.I. & Goncher, N.M.** Changes in morphofunctional and cytochemical indices of blood leukocytes after exposure to microwave radiation of low intensities. *Radiobiology,* No. 5, pp. 645–649 (1985).
170. **Shandala, M.G. et al.** Effects of microwave radiation on cell immunity in conditions of chronic exposure. *Radiobiology,* No. 4, pp. 544–546 (1983).
171. **Vinogradov, G.I. et al.** The influence of nonionizing microwave radiation on the autoimmune reactions and antigenic structure of proteins. *Radiobiology,* No. 6, pp. 840–843 (1985).
172. **Prausnitz, S. & Susskind, C.** Effects of chronic microwave irradiation on mice. *IRE transactions on biomedical electronics,* **9**: 104–108 (1962).
173. **Susskind, C.** *Nonthermal effects of microwave radiation.* Berkeley, CA, University of California Electronics Research Laboratory, 1962 (Report RADC-TDR-62-624, Annual Scientific Report, 1961–1962 on Contract No. NONR-222(92) and Final Report on Contract AF41(657)-114).
174. **Czerski, P.** Microwave effects on the blood-forming system with particular reference to the lymphocyte. *Annals of the New York Academy of Sciences,* **247**: 232–242 (1975).
175. **Liburdy, R.P.** Radiofrequency radiation alters the immune system: modulation of T- and B-lymphocyte levels and cell-mediated immunocompetence by hyperthermic radiation. *Radiation research,* **77**: 34–36 (1979).
176. **Liburdy, R.P.** Radiofrequency radiation alters the immune system. II. Modulation of *in vivo* lymphocyte circulation. *Radiation research,* **83**: 66–73 (1980).
177. **Rama Rao, G. et al.** Effects of microwave exposure on the hamster immune system. II. Peritoneal macrophage function. *Bioelectromagnetics,* **4**: 141–155 (1983).
178. **Smialowicz, R.J. et al.** Microwaves (2459-MHz) suppress murine natural killer cell activity. *Bioelectromagnetics,* **4**: 371–381 (1983).
179. **Yang, H.K. et al.** Effects of microwave exposure on the hamster immune system. I. Natural killer cell activity. *Bioelectromagnetics,* **4**: 123–139 (1983).

180. **Gage, M.I. & Albert, E.N.** Nervous system. *In:* Elder, J.A. & Cahill, D.F., ed. *Biological effects of radiofrequency radiation.* Washington, DC, US Environmental Protection Agency, 1984 (Publication EPA-600/8-83-026F).

181. **Ward, T.R. et al.** Measurement of blood–brain barrier permeation in rats during exposure to 2450-MHz microwaves. *Bioelectromagnetics,* **3**: 371–383 (1982).

182. **Williams, W.M. et al.** Effect of 2450 MHz microwave energy on the blood–brain barrier to hydrophilic molecules. A. Effect on the permeability to sodium fluorescein. *Brain research reviews,* **7**: 166–170 (1984).

183. **Baranski, S. & Edelwejn, Z.** Studies on the combined effect of microwaves and some drugs on bioelectric activity of rabbit central nervous system. *Acta physiologica polonica,* **19**: 31–41 (1968).

184. **Cleary, S.F. & Wangemann, R.T.** Effect of microwave radiation on pentobarbital-induced sleeping time. *In:* Johnson, C.C. & Shore, M.L., ed. *Biological effects of electromagnetic waves.* Rockville, MD, US Department of Health, Education, and Welfare, 1976, Vol. 1, pp. 311–322 (Publication (FDA) 77-8010).

185. **Lai, H. et al.** Ethanol-induced hypothermia and ethanol consumption in the rat are affected by low-level microwave irradiation. *Bioelectromagnetics,* **5**: 213–220 (1984).

186. **Servantie, B. et al.** Pharmacologic effects of a pulsed microwave field. *In:* Czerski, P. et al., ed. *Biological effects and health hazards of microwave radiation.* Warsaw, Polish Medical Publishers, 1974.

187. **Liddle, C.G. & Blackman, C.F.** Endocrine, physiological and biochemical effects. *In:* Elder, J.A. & Cahill, D.F., ed. *Biological effects of radiofrequency radiation.* Washington, DC, US Environmental Protection Agency, 1984 (Publication EPA-600/8-83-026F).

188. **Adey, W.R. et al.** Effects of weak amplitude modulated microwave fields on calcium efflux from awake cat cerebral cortex. *Bioelectromagnetics,* **3**: 295–307 (1982).

189. **D'Andrea, J.A. et al.** Physiological and behavioral effects of exposure to 2450-MHz microwaves. *Journal of microwave power,* **14**: 351–362 (1979).

190. **de Lorge, J.O. & Ezell, C.S.** Observing responses of rats exposed to 1.28- and 5.62-GHz microwaves. *Bioelectromagnetics,* **1**: 183–198 (1980).

191. **D'Andrea, J.A. et al.** Behavioral effects of resonant electromagnetic power absorption in rats. *In:* Johnson, C.C. & Shore, M.L., ed. *Biological effects of electromagnetic waves.* Rockville, MD, Department of Health, Education, and Welfare, 1976, Vol. 1, pp. 257–273 (Publication (FDA) 77-8010).

192. **Gage, M.I.** Behavior in rats after exposures to various power densities of 2450 MHz microwaves. *Neurobehavioral toxicology,* **1**: 137–143 (1979).

193. **King, N.W. et al.** Behavioral sensitivity to microwave irradiation. *Science,* **273**: 398–401 (1972).

194. **Stern, S. et al.** Microwaves: effect on thermoregulatory behavior in rats. *Science,* **206**: 1198–1201 (1979).
195. **Gage, M.I.** Behavior. *In:* Elder, J.A. & Cahill, D.F., ed. *Biological effects of radiofrequency radiation.* Washington, DC, US Environmental Protection Agency, 1984 (Publication EPA-600/8-83-026F).
196. **Carpenter, R.L.** Ocular effects of microwave radiation. *Bulletin of the New York Academy of Medicine,* **55**: 1048–1057 (1979).
197. **Guy, A.W. et al.** Effect of 2450-MHz radiation on the rabbit eye. *IEEE transations on microwave theory and techniques,* **MTT-23**: 492–498 (1975).
198. **Williams, R.J. et al.** Ultrastructural changes in the rabbit lens induced by microwave radiation. *Annals of the New York Academy of Sciences,* **247**: 166–174 (1975).
199. **Appleton, B. et al.** Investigation of single exposure microwave ocular effects at 3000 MHz. *Annals of the New York Academy of Sciences,* **247**: 125–134 (1975).
200. **Ferri, E.S. & Hagan, G.J.** Chronic low-level exposure of rabbits to microwaves. *In:* Johnsen, C.C. & Shore, M.L., ed. *Biological effects of electromagnetic waves.* Rockville, MD, US Department of Health, Education, and Welfare, 1976, Vol. 1, pp. 129–142 (Publication (FDA) 77-8010).
201. **Guy, A.W. et al.** Long-term 2450-MHz CW microwave irradiation of rabbits: methodology and evaluation of ocular and physiologic effects. *Journal of microwave power,* **15**: 37–44 (1980).
202. **McAfee, R.D. et al.** Absence of ocular pathology after repeated exposure of unanesthetized monkeys to 9.3 GHz microwaves. *Journal of microwave power,* **14**: 41–44 (1979).
203. **Michaelson, S.M. et al.** Biochemical and neuroendocrine aspects of exposure to microwaves. *Journal of the New York Academy of Sciences,* **247**: 21–45 (1975).
204. **Lotz, W.G. & Michaelson, S.M.** Temperature and corticosterone relationships in microwave exposed rats. *Journal of applied physiology,* **44**: 438–445 (1978).
205. **Lotz, W.G. & Michaelson, S.M.** Effects of hypophysectomy and dexamethasone on rat adrenal; response to microwaves. *Journal of applied physiology: respiratory, environmental and exercise physiology,* **47**: 1284–1288 (1979).
206. **Lu, K. et al.** Microwave-induced temperature, corticosterone, and thyrotropin interrelationship. *Journal of applied physiology: respiratory, environmental and exercise physiology,* **50**: 399 (1981).
207. **Swicord, M.L. et al.** Chain-length-dependent microwave absorption of DNA. *Biopolymers,* **22**: 2513-2516 (1983).
208. **Cleary, S.F. et al.** Effects of X-band microwave exposure on rabbit erythrocytes. *Bioelectromagnetics,* **3**: 453–466 (1982).
209. **Fisher, P.D. et al.** Effects of microwave radiation (2450 MHz) on the active and passive components of $^{24}Na^+$ efflux from human erythrocytes. *Radiation research,* **92**: 411–422 (1982).

210. **Ismailov, E.Š.** [Mechanism of effects of microwaves on erythrocyte permeability for potassium and sodium ions]. *Biologija nauki,* **3**: 58–60 (1971) (in Russian).

211. **Liburdy, R.P. & Penn, A.** Microwaves bioeffects in the erythrocyte are temperature and pO_2 dependent: cation permeability and protein shedding occur at the membrane phase transition. *Bioelectromagnetics.* **5**: 283–292 (1984).

212. **Olcerst, R.B. et al.** The increased passive efflux of sodium and rubidium from rabbit erythrocytes by microwave radiation. *Radiation research,* **82**: 244–256 (1980).

213. **Allis, J.W.** Cellular and subcellular effects. *In:* Elder, J.A. & Cahill, D.F., ed. *Biological effects of radiofrequency radiation.* Washington, DC, US Environmental Protection Agency, 1984 (Publication EPA-600/8-83-026F).

214. **Blackman, C.F. et al.** Effects of nonionizing electromagnetic radiation on single-cell biologic systems. *Annals of the New York Academy of Sciences,* **27**: 352–366 (1975).

215. **Blackman, C.F.** Genetics and mutagenesis. *In:* Elder, J.A. & Cahill, D.F., ed. *Biological effects of radiofrequency radiation.* Washington, DC, US Environmental Protection Agency, 1984 (Publication EPA-600/8-83-026F).

216. **Manikowska-Czerska, E. et al.** Effects of 2.45 MHz microwaves on meiotic chromosomes of male CBA/CAY mice. *Journal of heredity,* **76**: 71–73 (1985).

217. **Kirk, W.P.** Life span and carcinogenesis. *In:* Elder, J.A. & Cahill, D.F., ed. *Biological effects of radiofrequency radiation.* Washington, DC, US Environmental Protection Agency, 1984 (Publication EPA-600/8-83-026F).

218. **Guy, A.W. et al.** *Effects of long-term low-level radiofrequency radiation exposure on rats.* Brooks Air Force Base, Texas, USAF School of Aerospace Medicine, 1985, Vol. 9 (document USAFSAM-TR-85-64).

219. **Szmigielski, S. et al.** Acceleration of cancer development in mice by long-term exposition to 2450-MHz microwave fields. *In:* Berteaud, A.J. & Servantie, B., ed. *Ondes Electromagnétiques et Biologie, URSI International Symposium Proceedings, Paris, 1980,* pp. 165–169.

220. **Szmigielski, S. et al.** Accelerated development of spontaneous and benzopyrene-induced skin cancer in mice exposed to 2450-MHz microwave radiation. *Bioelectromagnetics,* **3**: 179–191 (1982).

221. *Code of practice: protection of workers against radiofrequency and microwave radiation in the working environment.* Geneva, International Labour Office, 1986, pp. 41.

222. **Minin, B.Z.** [Microwaves and human safety]. Moscow, Izdatelstvo Sovetskoe Radio Moscow, 1974 (in Russian).

223. **Bowhill, S.A., ed.** *Review of radio science 1978–1980.* Brussels, International Union of Radio Science, 1984.

224. **Bowhill, S.A., ed.** *Review of radio science 1981–1983.* Brussels, International Union of Radio Science, 1984.

225. **Czerski, P.** Radiofrequency radiation exposure limits in Eastern Europe. *Journal of microwave power,* **20**: 233–239 (1985).
226. **Czerski, P.** The development of biomedical approaches and concepts in radiofrequency radiation protection. *Journal of microwave power,* **21**: 9–23 (1986).
227. **Postow, E.** Review of radiofrequency standards. *In:* Polk, C. & Postow, E., ed. *Handbook of biological effects of electromagnetic fields.* Boca Raton, FL, CRC Press, 1986.
228. **International Non-Ionizing Radiation Committee of the International Radiation Protection Association.** Interim guidelines on limits of exposure to radiofrequency electromagnetic fields in the frequency range from 100 kHz to 300 GHz. *Health physics,* **54**: 115–128 (1988).

5

Electric and magnetic fields at extremely low frequencies

L.E. Anderson & W.T. Kaune

CONTENTS

	Page
Introduction	176
Electric and magnetic fields	176
Electric fields	176
Electric field measurements	177
Electric field sources	178
Magnetic fields	181
Magnetic field measurements	182
Magnetic field sources	182
Electric and magnetic field coupling to living organisms	183
Theory of electric field coupling	184
Data on electric field coupling	185
Biophysical analysis of electric field coupling	189
Protective measures for electric field coupling	195
Capacitive discharges and contact currents	195
Protective measures for capacitive discharges and contact currents	198
Theory of magnetic field coupling	203
Data on magnetic field coupling	203
Biophysical analysis of magnetic field coupling	203
Protective measures for magnetic field coupling	205
Review of cellular studies	205
Growth and metabolism	206
Membrane effects and activity	206
Review of animal studies	207
Neural and neuroendocrine systems	207
Reproduction and development	213
Other biological functions	214

Review of human studies 215
    Laboratory investigations 215
    Epidemiological assessments 217
    Cardiac pacemakers 220
Conclusions and recommendations 220
    Conclusions 220
    Recommendations 222
References 224

INTRODUCTION

Owing to the rapidly increasing and varied uses of electric power in industrial society, the likelihood and level of exposure of biological systems to electromagnetic fields for biological systems has increased by orders of magnitude over the past century. Significant increases have occurred in ambient electromagnetic field levels over a frequency range extending from zero to hundreds of GHz. This chapter concentrates on fields with extremely low frequencies (ELF), covering the 30–300 Hz range. Another recent review of the ELF range is given in Environmental Health Criteria No. 35 *(1)*.

It is both appropriate and important to evaluate possible interactions between the man-made electromagnetic environment and living organisms, including man, and whether such interactions are beneficial or detrimental, transient or permanent. In the past two decades, research programmes throughout the world have made significant progress in defining the physical interactions of electric and magnetic fields with living organisms and in describing biological effects resulting from these interactions. Much of this effort has been directed towards electric fields of power frequencies (i.e. 50 and 60 Hz). However, more recently, other ELF frequencies have been examined and research has been expanded to include magnetic fields, both alone and in conjunction with electric fields. Although it is now clear that ELF fields can cause biological effects, the mechanisms of such interactions are largely unknown and the implications for the health of man and animals have yet to be determined.

ELECTRIC AND MAGNETIC FIELDS

Electric fields

Any system of electric charges produces an electric field E at all points in space. Being a vector quantity, E is characterized by both direction and magnitude. An electric charge placed in E will experience a force F, where F and E are related by the following equation:

$$F = qE \tag{1}$$

where q is the size of the charge placed in the field. The force on a positive charge (e.g. a proton) is in the same direction as E, while the force on a negative charge (e.g. an electron) is in the opposite direction.

Except for certain areas of microscopic physics, electric field quantities are expressed in SI units. In this system, the unit of electric charge is the coulomb (C). The smallest quantity of electric charge that has been observed in nature is that of an electron or proton, 1.6×10^{-19} C. It appears that all larger quantities of charge are integral multiples of the electron charge.

The unit of electric field strength is volts per metre (V/m). It is generally much easier to measure the electric potential V than it is to determine the electric field, because the potential is not dependent on the physical geometry of a given system, such as locations and sizes of conductors. The electric potential is usually defined so that the earth's potential is zero.

In many systems, the relationship between V and E can be calculated without too much difficulty. The simplest example, and one of great practical importance, is two large parallel conducting plates that are initially uncharged. Suppose that a charge Q is removed from one of these plates and is placed on the other. (The transfer of charge between the plates could be accomplished using a battery or a transformer connected between them.) A potential and an electric field are then established between the two plates. Except near the edges of the plates, the electric field will be uniform and perpendicular to their surfaces. In this region the magnitude of E, denoted E, is related to V by:

$$E = V/h \qquad\qquad (2)$$

where h is the separation between the plates. Electric fields exert forces on charged particles. In an electrically conductive material, such as a living tissue, these forces will set charges in motion to form an electric current.

The unit of current is the ampere (A). The distribution of current in a three-dimensional volume is frequently specified using the current density vector J, whose direction is the direction of current flow and whose magnitude is equal to dI/dA, were dI is the current crossing a very small surface element of area dA oriented perpendicular to J. The unit of current density is A/m^2. J is directly proportional to E in a wide variety of materials, that is:

$$J = \sigma E \qquad\qquad (3)$$

where the constant of proportionality σ is called the electrical conductivity of the medium. The unit of σ is siemens/metre (S/m). ELF conductivities of various living tissues lie in the approximate range 0.01–1.5 S/m (2–5).

Electric field measurements

Standard procedures exist for the measurement of power frequency electric fields produced by transmission lines (6). These procedures can, in most cases, be directly applied to the measurement of electric fields produced by other ELF sources.

The most common problem is the measurement of an electric field at some point in space. A type of meter, often referred to as a free-body or dipole meter, has been developed for this measurement. A free-body meter consists essentially of two conducting halves and an electronic system that can measure the current induced between the two halves when they are placed in an electric field. These meters perturb electric fields in their

177

vicinity, so that measurements made near conducting surfaces may be less accurate. Commercially available meters, which are about the size of a conventional voltmeter, are adequate for electric field measurements near transmission lines or similar sources. However, because of their size, they may not be suitable for field measurements in laboratory systems, such as electric field exposure systems for laboratory animals. Much smaller meters have been developed for measurements in these kinds of system (7).

Another potential source of error is field perturbations caused by the body of the person making the measurement. To avoid such errors it is recommended that, for measurements under transmission lines, the horizontal distance between the operator and the meter be at least 2.5 m (6).

It is sometimes desirable to measure the electric field acting on a conducting surface, such as the surface of the ground or that of the human body. This measurement can be accomplished easily, using a small surface element which can be located directly over, and very close to, the surface in question. The current induced in this element is directly related to the average electric field acting on it (6,8,9).

Electric field sources
Experiments have shown that there is a vertical electric field present in the lower portion of the earth's atmosphere. The source of this field is positive charge carried from the ground to the upper atmosphere by thunderstorm activity. The mean strength of the ground-level atmospheric electric field is about 130 V/m and its direction is vertically downward (10), but over time the strength of this field is highly variable. Ground-level field strengths in excess of 100 kV/m have been observed on flat, unobstructed surfaces during thunderstorms (10,11).

Electric fields with frequencies in the ELF range result predominantly from man-made sources. The strongest ELF fields to which humans are normally exposed, outside of a few occupational settings, are those produced by electric power generation, transmission and distribution systems. A considerable amount of data has been published on measurements of the electric fields under high-voltage transmission lines (8,12–21). Theoretical work has shown that it is possible to calculate, with good accuracy, the electric fields produced by these sources (12,14,15,19,20,22,23).

Fig. 1 shows electric field strengths for two powerline configurations and system voltages, at ground level and halfway between two of the pylons that support the line. The strength of the field at ground level is generally greatest at this location because the line's conductors are closest to the ground and because shielding from the grounded pylons is at a minimum. The field strength decreases markedly as the pylon is approached (Fig. 2) (19).

Table 1 gives field strengths for different practical powerline configurations; similar data for substations are shown in Table 2. These data show that the largest electric field strengths produced at ground level by transmission lines now in service are about 15 kV/m. Electric field strengths under even higher voltage lines that may be built in the future will probably not significantly exceed this value because of the need to limit shock hazards to personnel in their vicinity.

178

Fig. 1. Calculated electric field strength mid-way between two towers supporting an electric power transmission line

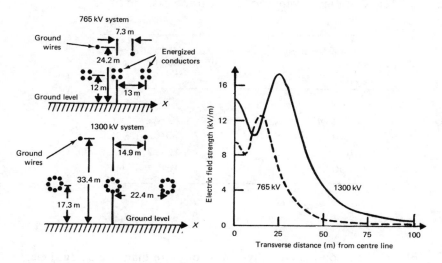

Fig. 2. Measured electric field strength E_0 along the trace of a 380 kV double-circuit line

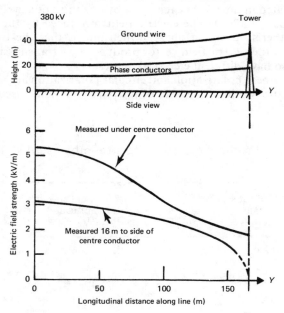

179

Table 1. Maximum electric field strengths at midspan under various configurations and voltages of electric power transmission lines

Highest system voltage (kV)	Electric field strength under line at midspan (kV/m)
123	1–2
245	2–3
420	5–6
800	10–12
1200	15–17

Source: Hauf *(24)*.

Electric field sources typically consist of more than one charged electrode. For example, an alternating current transmission line has conductors or, for higher voltage applications, three bundles of conductors. The sinusoidal voltages of these conductors are approximately equal in magnitude but are separated in time from each other by one third of a cycle (i.e. by an electrical phase of 120°). The electric fields produced by the three conductors are therefore not in phase, which means that the three spatial components of the electric field vector produced by a transmission line will in general also have different electrical phases.

The terms linear polarization and circular polarization are occasionally used to describe two limiting cases of the phase relations discussed in the

Table 2. Maximum electric field strengths in substations

Highest system voltage (kV)	Electric field strength under busbars (kV/m)
123	5–6
245	9–10
420	14–16
800	14–16

Source: Hauf *(24)*.

previous paragraph. An electric (or magnetic) field is said to be linearly polarized if all three spatial components of the field have the same electrical phase. This is the approximate situation for electric fields produced by transmission lines at distances of more than about 15 m from the line's conductors (14,25).

The other limiting case of circular polarization occurs when the two independent components of the field vector are equal in magnitude but are 90° out of phase. This case can be detected with a field meter by noting that there is a plane in which the measured field strength is independent of the orientation of the meter.

Studies are under way at present in Canada, Japan, Sweden, the United Kingdom, the United States and possibly other countries to determine actual exposures to electric fields of people working and living in the vicinity of transmission lines (26–34). These studies have demonstrated that exposure estimates can significantly overstate exposure if they are based on measured unperturbed field values (i.e. fields measured with no people present between the line and the ground) and on estimates of a person's position as a function of time.

Magnetic fields

Like electric fields, magnetic fields are produced by electric charge, but *only* by electric charge in physical motion. Magnetic fields exert forces on other charges, but only on charges in motion. Since the most common manifestation of electric charge in motion is an electric current, it is often said that magnetic fields are produced by electric currents and interact with other electric currents (35).

The magnitude F of the force $\boldsymbol{F}$, acting on an electric charge q, moving with a velocity $\boldsymbol{v}$ perpendicular to a magnetic field $\boldsymbol{B}$, is given by the expression:

$$F = qvB \qquad (4)$$

where v is the magnitude of $\boldsymbol{v}$ and the direction of $\boldsymbol{F}$ is perpendicular to *both* $\boldsymbol{v}$ and $\boldsymbol{B}$. If, instead, $\boldsymbol{v}$ and $\boldsymbol{B}$ were parallel, F would be zero. This illustrates one important characteristic of a magnetic field: it does no physical work because the force, called the Lorentz force, generated by its interaction with a moving charge is always perpendicular to the direction of motion and can, therefore, change only the charge's direction and not its speed. However, magnetic fields can facilitate the transformation of one form of energy into another, as for example in an electric generator where mechanical energy is transformed into electrical energy.

Magnetic fields are specified by two vector quantities, the magnetic flux density, B, and the magnetic field strength, H. B and H have units of tesla (T) and ampere/metre (A/m), respectively. In air, vacuum and, to a lesser but still adequate approximation, in all nonmagnetic materials, B and H are related by the formula:

$$B/H = 4\pi \times 10^{-7} \text{ (T} \cdot \text{m)/A} \qquad (5)$$

Thus only one of the quantities, H or B, needs to be specified.

181

Unfortunately, there is currently no consensus in the reporting of magnetic field levels in the ELF literature related to biological effects, and both flux density and field strength values are in common use. An additional complication is that both the SI unit tesla and the CGS unit gauss (G) have been and are still being used to express flux density values. In this chapter, the primary specification of magnetic field levels is made using flux density (B), but this specification is immediately followed by field strength (H) values. The corresponding values for nonmagnetic media for B expressed in tesla (SI units), B expressed in gauss (CGS units) and H are 1 T, 10 kG and 796 kA/m, respectively.

Magnetic field measurements

An alternating magnetic field induces an electric field and an electromotive force in any conducting circuit exposed to it. This fact is the basis for one type of magnetic field meter that uses coils, often called search coils, to measure the electromotive force (voltage) induced by the magnetic field to be measured. Meters in this class are simple, sensitive, prone to few errors, commercially available, and easily fabricated in the laboratory.

The second type of meter used for the measurement of ELF magnetic fields utilizes the Hall effect, essentially an application of equation (4). Meters in this class are also commercially available but are, because of lower sensitivity, perhaps less well suited to the measurement of ELF magnetic fields from environmental sources. They do have the advantage that they provide measurement at a localized point in space, and they can also be used to measure static magnetic fields.

Standard methods for measuring power frequency magnetic fields under high voltage transmission lines have been proposed by Misakian (6).

Magnetic field sources

Natural phenomena, such as thunderstorms and solar activity, produce time-varying magnetic fields in the ELF range (36). Such fields are generally of low strength, approximately 0.01 mT (8 mA/m). However, during intense magnetic storms, these fields can reach intensities of about 0.5μT (0.4 A/m) (37).

Of greater importance in the context of possible biological effects are the numerous ELF magnetic fields arising from man-made sources. In the lowest intensity range, generally less than 0.3μT (0.24 A/m), are fields found in the home (38,39) and in office environments such as near video display equipment (40). Magnetic fields from ELF communications and power transmission systems are somewhat higher and can approach a level of up to about 15μT (12 A/m) (41–43). Fig. 3 shows magnetic fields under an electric power transmission line carrying currents of 1 kA.

Considerably higher flux densities can occur in the immediate proximity of industrial processes using large induction motors or heating devices. Lovsund et al. (44) documented magnetic fields of from 8 to 70 mT (6.4–56 kA/m) in the steel industry in Sweden. Recently, significant developments in specific areas of medical care have allowed the use of pulsed

182

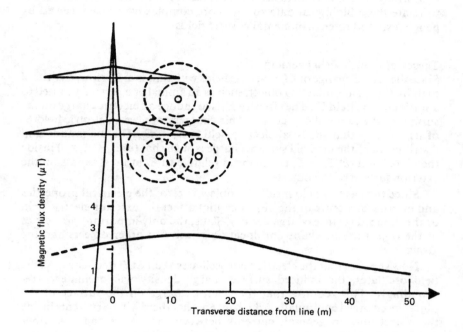

Fig. 3. Magnetic flux density

magnetic fields for various diagnostic and treatment procedures *(45–48)*. Flux densities (field strengths) from these new technologies may range from 1 to 10 mT (0.8–8 kA/m).

ELECTRIC AND MAGNETIC FIELD COUPLING TO LIVING ORGANISMS

Exposure of a living organism to electric and/or magnetic fields is normally specified by the unperturbed field strength, that is, the field strength measured or calculated with the subject removed from the system. The use of this field is convenient because it involves a quantity that is relatively easy to measure or calculate. The fields that actually act on an exposed organism — the fields at the outer surface and inside the body — can be different because of perturbations caused by the body of the exposed subject. These latter fields must, however, be determined in order to specify exposure at the level of living tissues or to relate exposure levels and conditions from one species and/or exposure geometry to another.

For example, most data that can now be used for assessing the risks to humans of exposure to environmental ELF electric fields were obtained in

183

experiments whereby animals housed in highly controlled laboratory conditions were exposed to uniform, vertical ELF electric fields. Estimation of field strengths and current densities at the tissue level will be needed in order to relate these biological data to the more complex exposure received by people exposed to environmental electric fields.

Theory of electric field coupling

Since the rate of change of ELF electric fields with time is quite slow, it is useful to think first of a conducting object, such as an animal or a person, exposed to a static electric field E_0. The effect of E_0 is to induce an electric charge on the surface of the exposed object *(49)*. This charge produces an electric field E_1 of its own, so that the total electric field is $E_0 + E_1$. The magnitude and distribution of the induced surface charge is such that (*a*) $E_0 + E_1 = 0$ inside the object and (*b*) $E_0 + E_1$ is enhanced relative to E_0 at most points on the exterior surface of the object.

Since the electric field inside the object is zero, the electrical properties and internal structure of the object cannot affect the exterior electric field or the induced surface charge density. Thus, the only important properties of the object are its shape and location relative to other objects and the ground.

Now suppose that the electric field oscillates at an ELF frequency. As the field oscillates, the induced surface charge density will oscillate correspondingly. This requires that the electric charge on the surface of the object be changed continually, which means that there will be currents inside the object and, in general, currents between the object and any other conducting objects that are in electrical contact with it. Since living tissues have finite conductivities, internal currents cannot flow without the existence of corresponding electric fields. Thus, both currents and electric fields are induced inside a conducting object by an external ELF electric field.

However, in the ELF range, the variation in surface charge density is so slow that the currents and fields generated inside the object are actually very small. Estimates show that the electric fields induced inside the bodies of humans and animals are generally less than about 10^{-7} of the field outside the body and probably rarely exceed about 10^{-4} of the external field *(50)*. Because these internal fields are so small, the conclusion reached earlier about static electric fields is still valid to a very good approximation; the only properties of a living organism that affect the exterior field and induced surface charge density are its shape and location relative to other objects. Because currents inside the object have as their source induced surface charge, this conclusion can be generalized to total currents passing through any section of the body of a living organism *(50)*.

The argument presented in the preceding paragraph suggests that much useful information can be obtained from experiments with conducting models that simulate the shapes of humans and animals. This technique has been used by Schneider *(19)*, Deno *(28,51)* and Kaune and colleagues *(52–54)*.

Data on electric field coupling

A number of theoretical papers have been published in which exposure of animals and humans has been modelled. Most of these papers *(55–63)* simulated the body of the exposed organism with a sphere, spheroid or ellipsoid, but two used more refined models *(62,63)*. The results of these investigations have been reviewed elsewhere *(64–66)*. Present experimental data exceed in detail and scope the available theoretical data.

Considerable information can be obtained using conducting models that simulate, to a greater or lesser extent, the body shapes of humans and animals. The following paragraphs summarize data obtained using this technique.

Since exposure to electric fields often occurs when the subject is electrically grounded, a parameter of considerable importance is the short-circuit current, that is, the total current that passes between the subject and ground. Published short-circuit current data are summarized in Table 3 for humans *(12,51)*, horses and cows *(8)*, pigs *(67)*, guinea pigs *(53)*, and rats *(54)*. Most of these measurements were taken at only one frequency and body weight. They have been extrapolated to other frequencies, f, and body weights, W, assuming a $fW^{2/3}$ dependence *(50)*. The human data were expressed by Deno *(51)* in terms of body height rather than body weight. In the preparation of Table 3 it was assumed that a body height of 1.7 m is equivalent to a body weight of 70 kg.

The use of the formulae in Table 3 is illustrated by the following example. Consider a 70 kg human exposed to a 10 kV/m, 50 Hz electric field. The top line in Table 3 indicates that the short-circuit current induced in this person will be $(15 \times 10^{-8}) \times 50 \times 70\,000^{2/3} \times 10\,000 = 127\,\mu A$. At 60 Hz this current would be increased to $153\,\mu A$.

Table 3. Short-circuit currents induced in grounded humans and animals by vertical ELF electric fields[a]

Species	Short-circuit current (μA)
Human	$15.0 \times 10^{-8} fW^{2/3}E_0$
Horse	$8.5 \times 10^{-8} fW^{2/3}E_0$
Cow	$8.6 \times 10^{-8} fW^{2/3}E_0$
Pig	$7.7 \times 10^{-8} fW^{2/3}E_0$
Guinea pig	$4.2 \times 10^{-8} fW^{2/3}E_0$
Rat	$4.0 \times 10^{-8} fW^{2/3}E_0$

[a] f = frequency of the applied field (Hz)
E_0 = strength of the applied field (V/m)
W = weight of subject (g)

Deno *(51)* published the first data on induced currents in anatomically detailed human models. Fig. 4 gives data on the distribution of current induced in a grounded man exposed to a vertical electric field. Using the data in this figure, Deno *(28)* estimated average vertical current densities in humans standing on the ground while exposed to vertical, 60 Hz electric fields. These data may be extrapolated to other frequencies in the ELF range by multiplying them by the ratio $f/60$ *(50)*.

Fig. 4. Current induced in a grounded person, of body height h, exposed to a vertical, uniform, ELF electric field

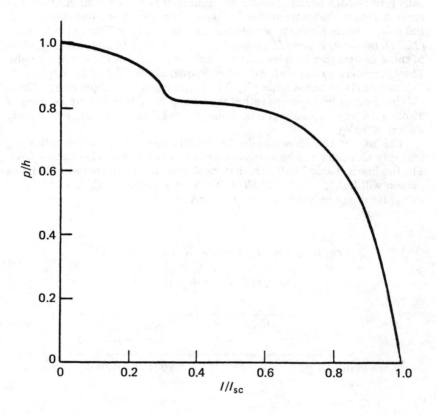

Note. The current I flowing vertically in the body at a height p above ground is expressed as a ratio to the total induced current, I_{sc} (i.e. the short-circuit current). For example, the current passing through the neck ($p/h = 0.8$) is about $0.3 I_{sc}$. I_{sc} can, in turn, be determined using Table 3 in this chapter.

Source: Adapted from Deno *(51)*.

Deno also developed a simple technique for measuring the external electric fields acting directly on the surface of the body. Surface electric fields measured at one frequency can be extrapolated without scaling to any other frequency in the ELF range *(50)*.

Kaune & Phillips *(54)* used Deno's methods to measure surface electric fields and induced-current distributions in grounded rats and pigs exposed to a vertical electric field. Axial (i.e. along the long axis of the body) current densities were estimated from the induced-current data.

Peak surface electric field and current density data, derived from Deno's human measurements and Kaune & Phillips's animal data, are presented in Fig. 5. In this figure, the magnitude and frequency of the unperturbed electric field were assumed to be 10 kV/m and 60 Hz, respectively.

For a vertical external electric field, the axial current densities induced inside a human body are considerably larger than the corresponding quantities for animals, even though the external electric fields are the same (Fig. 5). This conclusion is also true for induced electric fields because of equation (3). These differences mean that the external unperturbed fields, which are almost always used to specify exposure, must be scaled to equalize internal current densities and electric fields in order to extrapolate biological data from one species to another. This process is complicated by the fact that the actual value of the scaling factor depends on which internal quantity is being scaled. For example, a scaling factor for the peak electric field strength acting on the outer surface of the body would be about 4.9 : 1 for humans versus rats, while the scaling factor for *axial* current density in the neck would be about 20 : 1 for the same species comparison. Evidently, knowledge must be obtained about the site of action for a particular biological effect before extrapolation of data across species can be performed.

The surface electric field data given in Fig. 5 are quite limited, in that measurements were made at only a few points on the body. A simple way to calculate electric field strengths averaged over the entire body surface of a grounded subject has been described by Kaune *(52)*. This method requires only that short-circuit current I_{sc} and body surface area A_b be measured or estimated. In terms of these parameters the average electric field E_{avg} acting on the surface of the body is:

$$E_{avg} = \frac{I_{sc}}{2\pi f \varepsilon_0 A_b} \tag{6}$$

where f is frequency and ε_0 the permittivity of space (8.85×10^{-12} F/m). Table 4 gives approximate maximum and average electric fields acting on the surfaces of the bodies of grounded humans, pigs, rats, guinea pigs, horses and cows exposed to a vertical 1 kV/m electric field.

The current density data given in Fig. 5 represent only one (the axial) of three components of the total current density vector. This is a serious limitation in the animal data because it is certain that significant vertical current will also be present. Measurements in three-dimensional models of humans and animals are required to overcome this limitation.

187

Fig. 5. Effects of a vertical electric field of 60 Hz and 10 kV/m on grounded human, pig and rat

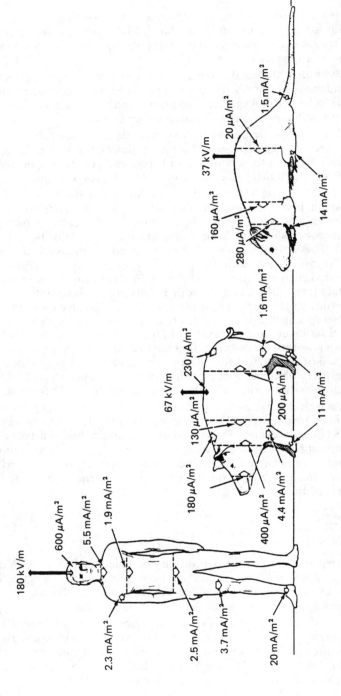

Note. Surface r.m.s. electric field measurements are shown for human and pig, and surface r.m.s. electric field estimates for rats. Estimated axial current densities averaged over selected sections through bodies are shown. Calculated current densities perpendicular to surface of body are shown for human and pig. Relative body sizes are not to scale.

Source: Kaune & Phillips *(54).*

Table 4. Peak and average electric fields
acting on the surfaces of grounded humans and animals
exposed to a vertical 1 kV/m electric field[a]

Species	Electric field (kV/m)	
	Average	Peak
Human	2.7	18
Pig	1.4	6.7
Rat (resting)	0.73	3.7
Rat (rearing)	1.5	—[b]
Horse	1.5	—[b]
Cow	1.5	—[b]

[a] The data are valid for all frequencies in the ELF range.

[b] Data not available.

Guy et al. (68) and Kaune & Forsythe (69) measured induced electric fields and internal current densities in grounded homogeneous models of humans exposed to 60 Hz electric fields. Fig. 6 and 7 summarize Kaune & Forsythe's data for human models, assuming a frequency of 60 Hz and an exposure field strength of 10 kV/m. Note, for example, the enhancements in current density that occur in the armpits, and also in the lower pelvic region when the model is grounded through only one foot.

One limitation of the current density data given in Fig. 6 and 7 is that they pertain only to the exposure of grounded humans. It is more usual for humans to be exposed when they are partially insulated from ground by their shoes. Deno (51) and Kaune et al. (70) have made measurements that enable grounded data to be extrapolated to ungrounded exposure situations. Fig. 8 summarizes the results of Kaune et al.

Data similar to those given in Fig. 6–8 are needed for laboratory animals to enable the determination of quantitative factors to scale animal data to human exposure situations. It is expected that these data will become available in the next few years.

Biophysical analysis of electric field coupling
External electric fields act directly on the surface of the body of an exposed human or animal and they also induce electric currents and fields inside the body. This section presents a discussion of how such fields might affect physiological processes, and summarizes some published estimates of threshold current densities above which biological effects might be expected to occur.

Fig. 6. Current densities measured in the midfrontal plane of a saline model of a standing person exposed to a 60 Hz, 10 kV/m electric field

A. Head and upper torso

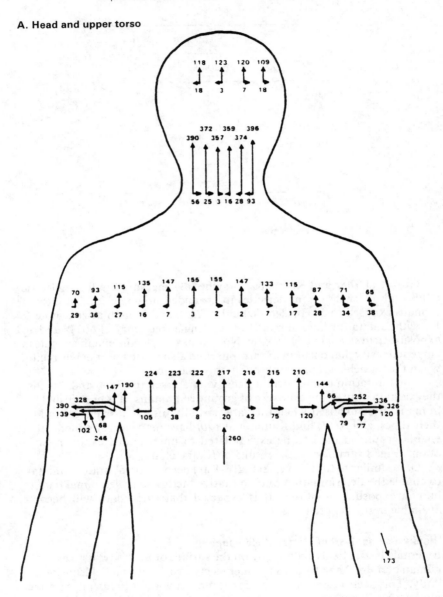

Fig. 6 (contd)

B. Legs and lower torso

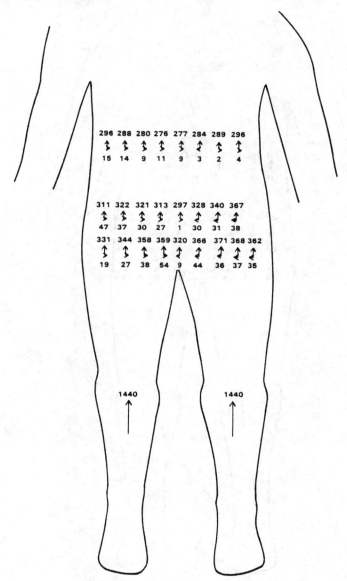

Note. Induced r.m.s. current densities are given in units of nA/cm (1 nA/cm = 10 A/m). The model was grounded with equal currents passing through both feet.

Source: Kaune & Forsythe *(69)*.

Fig. 7. Current densities measured in the midfrontal plane
of a saline model of a standing person
exposed to a 60 Hz, 10 kV/m electric field

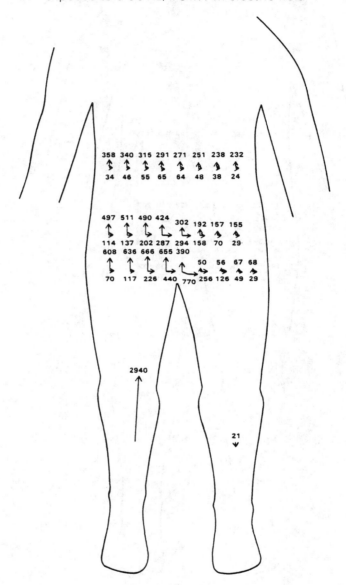

Fig. 8. Average axial current densities induced in a 1.7 m tall person exposed to a vertical 10 kV/m, 60 Hz electric field.

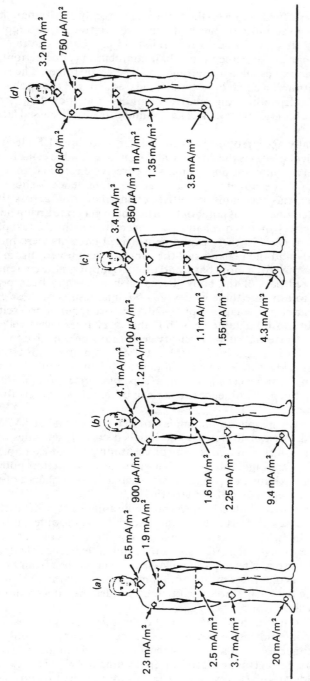

Note. Four positions of the body relative to ground are shown: (a) standing on and in electrical contact with ground; (b) feet elevated 11 mm above the ground to approximately simulate insulating footwear; (c) feet elevated 128 mm above the ground; (d) feet elevated 1.23 m above the ground.

Source: Kaune et al. (70).

The electric field acting on the surface of the body of a human or animal is enhanced over most of the body surface relative to the unperturbed electric field. It is well known that this field can be perceived *(71)*. One known mechanism of perception is hair stimulation (piloerection), that is, the production of oscillatory hair motion by electric forces. The frequency of this vibration can equal or be double the frequency of the applied electric field *(72,73)* depending on relative humidity and, possibly, other parameters. The biological aspects of hair stimulation are discussed later in this chapter.

A generally accepted mechanism of interaction of ELF electric fields with biological tissues is the direct stimulation of excitable cells. It accounts for the ability of humans and animals to perceive electric currents in their bodies and for the possibility that they can experience a shock. At the cellular level, this mechanism consists of the induction across the membranes of excitable cells of potentials sufficient to generate action potentials.

Most research on this mechanism has been done in the context of electric shock hazards *(74)*. In these experiments, fixed currents were introduced into the body and the reactions of the subjects were noted. From a mechanistic point of view, a more basic and useful quantity is the current density in the affected part of the body. Research has recently been performed attempting to estimate these values *(75–78)*. The results of these estimates are that tissue current densities of $10–20 \, A/m^2$ are required to excite action potentials in most excitable cells, but that very long nerve cells aligned parallel to the external current density vector may be sensitive to considerably smaller current densities. Bernhardt *(76)* summarized his estimates by proposing that the threshold current density for the stimulation of action potentials in excitable cells is $1 \, A/m^2$. (It is, of course, possible that lower current densities may influence excitable or other cells through other modes of interaction.)

Fig. 5–8, for people and animals exposed to external 10 kV/m, 60 Hz electric fields, demonstrate that the induced current densities are much smaller than values required for the direct stimulation of excitable cells. It therefore appears highly unlikely that stimulation of action potentials in excitable cells in humans or animals will occur as a consequence of exposure to environmental levels of ELF electric fields.

For electric-field-induced effects on whole animals, it is difficult to know whether a biological effect is due to the field acting at the outer surface of the body or to the fields induced inside the body. No such complication exists for experiments involving cell suspensions, and the existence of effects at low field strengths (e.g. effects on the release of calcium ions from brain tissue *(79–82)*) shows that there must be other mechanisms of interaction between ELF electric fields and biological tissues besides the direct stimulation of excitable cells.

Currently, neurophysiologists and other scientists are considering the view that communication between brain cells occurs not only through synaptic connections but also through each cell's modulation of and sensitivity to the extracellular electrical environment *(83)*. The extracellular electric fields are markedly smaller than those in the membrane, so it is hoped that this mechanistic picture might prove capable of explaining the sensitivity of certain types of tissue to external ELF electric fields. Based on this concept, Bernhardt (76) argues that extracellular electric fields induced by external fields could not, *a priori*, be judged safe unless their strengths were less than the endogenous fields always present in living tissues. This author used electrocardiographic and electroencephalographic data to estimate endogenous fields in tissues in the brain and torso, and arrived at a lower-limit current density of about $1 \, mA/m^2$. Examination of Fig. 5–8 shows that current densities exceeding this level are induced in grounded and ungrounded humans exposed to electric fields characteristic of maximum levels produced at ground level by electric power transmission lines.

There is evidence from *in vitro* studies that brain tissue can be sensitive to induced current densities much smaller than Bernhardt's estimate of minimum endogenous levels. In the calcium-release experiments mentioned above, brain tissue was immersed in a culture medium and was exposed to electric fields in air about the preparation. Effects were seen at a root-mean-square (r.m.s.) field strength of about $3 \, V/m$ and a frequency of $16 \, Hz$ (79,81). Modelling this preparation as a spherical volume of brain tissue surrounded by a spherical layer of culture medium, the current density induced in the brain tissue can be estimated to be approximately $1 \, nA/m^2$ (64). This is six orders of magnitude lower than Bernhardt's value.

There have been several attempts to develop models that can explain the interaction of biological tissues with extremely small ELF electric fields. These models have been reviewed recently (84,85). At present, none of these models has been developed to the point where it can make convincing quantitative predictions. In addition, none of them, without the addition of ad hoc assumptions, seem to offer much chance of explaining the intensity windows that have also been observed in the calcium-release experiments (79,81).

Protective measures for electric field coupling

Protection from electric field exposure can be relatively easily achieved using shielding. At ELF frequencies, virtually any conducting surface will provide substantial shielding. One practical approach for personnel working in high field strength areas is to provide them with clothes that are electrically conductive. This practice is commonly used in the electric utility industry by linemen who work on high-voltage transmission lines using "bare-hand" techniques.

The other method to obtain protection from electric field exposure is to limit the access of individuals to areas where electric field strengths are large.

Capacitive discharges and contact currents

Conducting objects placed in an electric field have potentials induced on them, currents induced inside them, and electrical energy stored in their capacitances. If two such objects, initially separated from one another, are brought into contact, two phenomena occur that can be of biological significance. At the instant of contact, and perhaps just before this instant, a capacitive discharge occurs in which single or multiple transient pulses of

195

current pass between the two bodies to reduce the potential difference between them. If the difference in potentials is sufficiently large, the discharge manifests itself as a spark at the point of contact between the two bodies. For this reason, the discharge is often called a spark discharge, but it also has a non-spark component that occurs when actual physical contact is made between the two bodies. This discharge is often perceptible, sometimes annoying, and occasionally painful. Among utility linemen and electricians who work in environments where strong power-frequency electric fields are present, the most common complaint is of the repeated spark discharges they experience when handling tools and other equipment.

Once contact is made between two conducting bodies, a steady-state ELF current generally passes between them. If sufficiently strong, this current can produce perception, muscular tetanus and/or cardiac fibrillation in humans and animals. The following two sections present more detailed discussion of capacitive discharges and contact currents.

Capacitive discharges
Work on capacitive discharges has been performed by several groups *(8,14,17,71,86–90)*. The most extensive study is that performed by Reilly & Larkin *(87–90)*. The following paragraphs summarize the results of these authors pertinent to questions of personnel protection.

Fig. 9 shows typical voltage and current waveforms where a grounded person has just approached and tapped with the finger an electrode connected to a 400 pF capacitor charged to *(a)* 405 V or *(b)* 990 V *(90)*. The observed waveforms can be explained as follows: the electric field between the subject's finger and the electrode increases as the distance between them decreases and, at some point prior to physical contact, an electric arc will be initiated if the voltage between the subject and the electrode is larger than about 450 V *(90)*. (Evidently this voltage is required to produce a strong enough electric field to exceed the dielectric strength of the insulating layers of the skin.) If the voltage does exceed 450 V (Fig. 9B) a spark discharge occurs and the voltage between the two bodies is quickly reduced to about 450 V where the discharge terminates because of the aforementioned insulating properties of the skin. A second discharge, of longer duration, occurs when the finger actually contacts the electrode. (This is the only discharge present in Fig. 9A because the voltage was too low to produce a spark discharge.) It is interesting to note that this second discharge, which we shall refer to as a contact discharge, appears not to reduce the voltage between the bodies to zero but to a plateau of about 100 V.

Capacitive discharges are characterized by two basic parameters: the capacitance being discharged and the initial voltage of the discharge. Another significant characteristic is whether the voltage being used to energize the capacitor is a.c. or d.c. The majority of the measurements of Reilly et al. were made with d.c. energization, coupled through a high source resistance, so that only one discharge occurred as the finger neared and contacted the energized electrode. However, they made sufficient a.c. measurements to be able to scale their d.c. data to a.c. cases.

196

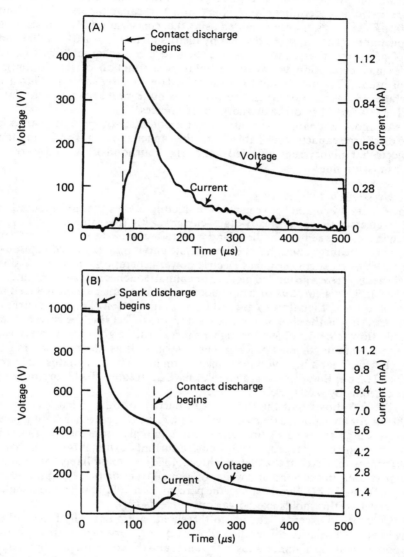

Fig. 9. Voltage and current waveforms for discharges
to a person tapping an energized capacitor

Note. In (A) the initial voltage of 405 V was too small for a spark discharge to occur. A
discharge did occur, however, when actual physical contact was made between the
capacitor and the subject's finger. In (B) the initial voltage of 990 V was sufficient for a
spark discharge to occur about 110 μsec prior to actual physical contact. A second
discharge occurred when contact was made.

Source: Reilly & Larkin *(90).*

197

Fig. 10 shows the proportions of a sample of 124 healthy adult humans (74 men, 50 women) who perceived capacitive discharges of various voltages from a 200 pF capacitance. Fig. 11 gives similar data for discharges that were rated as "definitely annoying" and for capacitances of 200 and 6400 pF. These data were obtained using d.c. energization of the discharge capacitance. They must be divided by factors of 1.75 and 2.04 for 200 and 6400 pF, respectively, to extrapolate them to a.c. discharges (90). Fig. 12 gives a.c. perception voltage thresholds as a function of the discharge capacitance. The data in this figure are based on only five subjects. They are included in this chapter to provide an approximate means of scaling the data in Fig. 10 and 11 to other discharge capacitances.

An important parameter in assessing the potential for spark discharges is the discharge capacitance. Table 5 lists the capacitances for some common objects that might be encountered in the neighbourhood of an electric power transmission line.

Steady-state contact currents
Once contact is made between two conducting bodies that are exposed to ELF electric fields, a steady-state current with the same frequency is exchanged between the bodies. Under certain circumstances, this current may become large enough to be perceptible and it may even be dangerous. The case of most concern near electric power generation and transmission facilities is where a grounded person (or animal) touches a large ungrounded object such as a car, bus or truck. Some fraction of the current induced in this object may then flow to ground through the person. The worst current that can flow is the short-circuit current of the object (i.e. the current that would flow from the object through a short circuit to ground). Table 6 lists the short-circuit currents of a number of large objects that a person might touch while near a high-voltage transmission line. The data in this table are for 60 Hz, but they may be scaled to other ELF frequencies by multiplying the currents by $f/60$ (50).

Table 6 shows that the short-circuit current of a large truck is about 500 mA for an exposure field strength of 1 kV/m. Under transmission lines, fields as high as 10–15 kV/m can be found, in which case the short-circuit current could be as large as 7.5 mA. Such a current flowing through the hand and arm would exceed the "let-go threshold" (i.e. would likely cause muscular tetanus in the hand and wrist) of a small fraction ($< 1\%$) of the adult population (74,77). Fig. 13 shows the percentiles of adult males and females whose let-go thresholds exceed various currents (74).

Currents in excess of the let-go thresholds are potentially fatal, either through the initiation of cardiac fibrillation or, very rarely, through the production of muscular tetanus in the chest and the consequent suppression of respiration (71,74,91). Threshold values for these phenomena are given in an IEC report (92).

Protective measures for capacitive discharges and contact currents
The potential for capacitive discharges and contact currents exists when conducting objects in an electric field have different electrical potentials

198

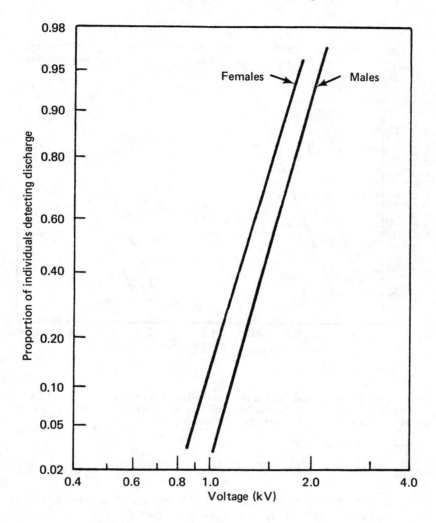

Fig. 10. Cumulative probability that an adult can perceive
the discharge that occurs when tapping with the fingertip
an electrode connected to a 200 pF capacitor
charged to a positive d.c. voltage

Note. The voltages shown on the abscissa must be divided by 1.75 (discharge
capacitance = 200 pF) to approximately scale the probability data from d.c. to 60 Hz
a.c. discharges.

Source: Reilly & Larkin *(90).*

Fig. 11. Cumulative probability that an adult will rate as "definitely annoying" the discharge that occurs when tapping with the fingertip an electrode connected to a 200 or 6400 pF capacitor charged to a positive d.c. voltage

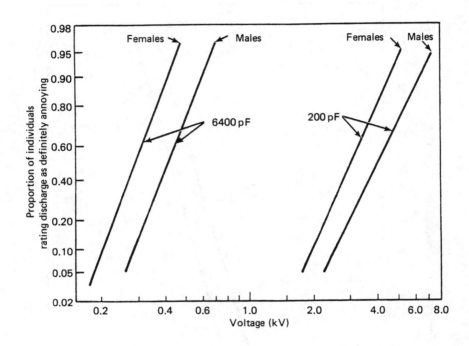

Note. The voltages shown on the abscissa must be divided by 1.75 and 2.04 for discharge capacitances of 200 and 6400 pF, respectively, to approximately scale the probability data from d.c. to 60 Hz a.c. discharges.

Source: Reilly & Larkin *(90)*.

induced on them. The basic means of protection is, therefore, to insure that all objects in an electric field have equal potentials. In theory, this can be accomplished by electrically grounding all conducting objects exposed to electric fields. For large, stationary objects this is a practical method, but for mobile objects (e.g. a worker) this technique may not be so useful.

Capacitive discharges and contact currents can also be prevented in particular situations by electrical shielding of the affected objects. Finally, protection can be obtained by limiting access of people and animals to areas where electric field strengths are large.

200

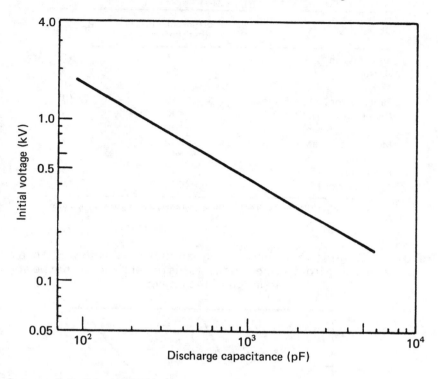

Fig. 12. Voltage threshold for the perception of 60 Hz
a.c. capacitance discharges that occur when
an energized electrode is tapped with a finger

Source: Reilly & Larkin *(90).*

Table 5. Capacitances of some objects that may
be encountered near high voltage transmission lines

Object	Capacitance (pF)
Person	100
Small car	1000
Medium-sized car	1600
Large station wagon	2000
Single-decker bus	2700
Large truck	3600

Table 6. Short-circuit currents for objects resting on ground and exposed to a 1 kV/m, 60 Hz electric field[a]

Object	Short-circuit current (mA/(kV·m))	Reference
Adult human	15	51
Horse	27	8
Cow	24	8
Farm tractor (length 3.7 m)	60	14
Car (length 4.6 m)	90	8
Car (length 5.5 m)	100	14
Metal wagon (length 5 m)	200	16
Bus (length 8.3 m)	300	16
Bus (length 10.4 m)	390	16
Truck (length 12 m)	500	16

[a] These data may be extrapolated to 50 Hz by multiplying the current values by 0.833.

Fig. 13. Cumulative probability that an individual with a 60 Hz a.c. current passing from an electrode into his or her hand will not be able to let go of the object

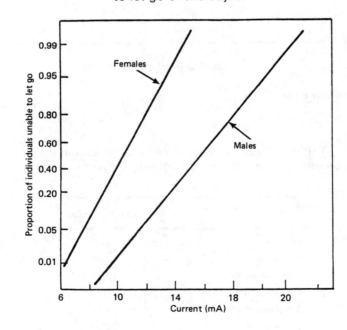

Source: Dalziel (74).

Theory of magnetic field coupling

In contrast to electric field exposure, the bodies of humans and other living organisms cause almost no perturbation in an ELF magnetic field to which they are exposed. This is true because: (a) excluding a few highly specialized tissues that contain magnetite, living tissues contain no magnetic materials and therefore have magnetic properties almost identical to those of air; and (b) the modification in the applied magnetic field, due to the secondary magnetic fields produced by currents induced in the body of the subject, is small (64,65).

Faraday's law of induction states that time-varying magnetic fields generate electric fields through induction. Therefore, a living organism exposed to ELF magnetic fields will also be exposed to an induced electric field from this source. The induced electric field causes current to flow in any conductive body. These currents, called eddy currents, circulate in closed loops that tend to lie in planes perpendicular to the direction of the magnetic field.

A fairly useful model for a human or animal exposed to a uniform ELF magnetic field is a homogeneous ellipsoid (57). An ellipsoid is defined by three parameters — the semi-major axes — which are the x, y and z coordinates where the surface of the body intersects the x, y and z axes, respectively. (Assume that the symmetry axes of the ellipsoid coincide with the coordinate axes.) The electric field induced by a magnetic field of frequency f parallel to, for example, the z axis is:

$$E_m = \frac{2\pi f B}{a^2 + b^2} \cdot (b^4 x^2 + a^4 y^2)^{1/2} \tag{7}$$

where a and b are the semi-major axes of the ellipsoid in the x and y directions, respectively.

Data on magnetic field coupling

Very little theoretical or experimental work has actually been carried out on the coupling of magnetic fields to living organisms. Spiegel (61) published a paper describing magnetic and electric ELF field coupling to spherical models. The magnetic field portion of this work essentially involved an application of equation (7), where a = b = radius of the sphere. Gandhi et al. (93) calculated induced current densities in the torso of a person exposed to an alternating magnetic field, using a technique that simulates an exposed object with a multidimensional lattice of impedance elements. They only considered a two-dimensional simulation, but there appears to be no fundamental problem in extending the method to three dimensions. The method offers a simple and apparently powerful way to analyse the coupling of people and animals to ELF magnetic fields.

Biophysical analysis of magnetic field coupling

As discussed above, alternating magnetic fields induce electric fields inside the bodies of exposed people and animals. External alternating electric fields also induce electric fields inside bodies. The distributions of the fields

203

induced by these two types of coupling are different, but at the level of the cell there would appear to be no fundamental difference. Thus, the biophysical analysis provided earlier in this chapter for electric field coupling can also be applied to the electric fields generated by magnetic field coupling.

One interesting question is what are the magnetic fields that will induce in humans ELF current densities of $1 A/m^2$ and $10^{-3} A/m^2$. These two current density levels correspond to the minimum values required, respectively, to stimulate excitable cells and to be smaller than endogenous levels.

Magnetic induction of currents can be modelled using a simple ellipsoidal approximation of a man. A typical man has a height of 1.7 m, a mass of 70 kg *(94)* and a ratio of body width to body thickness of about 2. An ellipsoid with semi-major axes of 0.85 m, 0.2 m and 0.1 m has the same body height, the same width-to-thickness ratio, and a body volume of $0.017 m^3$. Using equation (7), the maximum electric field E_m induced in this model can be shown to occur when the magnetic flux density vector B is horizontal and perpendicular to the front of the body. The actual value of E_m is $1.3fB$ (equation (7)), where f is the frequency of the field in Hz. Equation (3) can be used to estimate the induced current density from E_m. Using an average tissue conductivity of 0.2 S/m *(2–5)* the maximum induced current density is:

$$J_m \sim 0.25fB \qquad (8)$$

It is now possible to estimate magnetic flux densities, B_s, that will produce internal current densities exceeding the $1 A/m^2$ and $10^{-3} A/m^2$ thresholds discussed above. At 60 and 600 Hz, $B_s \sim 65$ and 6.5 mT (50 and 5 kA/m), respectively, for a current density of $1 A/m^2$, and $B_s \sim 65$ and 6.5 μT (50 and 5 A/m), respectively, for a current density of $10^{-3} A/m^2$. The former values are much larger than the flux densities produced by electric power transmission lines, but similar values may be found in certain specialized industrial environments. The latter values are, in contrast, comparable in size to levels produced by transmission lines.

In addition to the induction of currents in living bodies, ELF magnetic fields will act directly on certain parts of living tissues. The remainder of this section presents a discussion of the direct coupling of magnetic fields to biological tissues.

As noted earlier, ELF magnetic fields penetrate living organisms without significant perturbation. These fields will exert Lorentz forces on any charged particle that is in motion within the body. Equation (4) and the discussion following it describe the dependence of this force on the particle's velocity, its electrical charge, and the magnetic flux density.

The most prevalent types of motion in matter are the motion of electrons in atoms and the intrinsic spins of the electrons, protons and neutrons that make up matter. This motion leads to the existence of magnetic dipole moments that may be permanent or induced by the applied magnetic field. A magnetic dipole interacts with a uniform magnetic field in such a way that it experiences a torque but no net force. This torque is in a direction to align the dipole parallel to the magnetic field, but this alignment is resisted by random thermal motion.

204

At body temperature and at a flux density of about $20\mu T$ (16 A/m) that is characteristic of a heavily loaded transmission line (13) it can be shown (64,65) that the magnetically induced alignment of magnetic dipoles in living tissues is less than about one part in 10^8. The effect on the magnetic dipoles that are part of the body of a subject exposed to an ELF magnetic field of any realistic strength is very small. Of course, every single dipole is subject to this effect and, conceivably, some sort of cooperative physiological process that is sensitive to the average response of a great many of these dipoles might be affected by even a small ELF magnetic field.

Magnetic dipoles placed in a non-uniform magnetic field experience a net force. This force can be estimated, for the maximum flux density produced by a transmission line, to be extremely small in comparison to typical chemical forces (65).

Charged particles are also carried by the bulk motion of various parts of the body. For example, charged ions are carried by blood flow. These ions are both positively and negatively charged and will experience Lorentz forces in opposite directions, resulting in a separation of the two polarities of electric charge and, therefore, in the generation of electric potentials. It has been shown that these potentials can produce electrical signals that can be recorded in the electrocardiograms of rats (95) and monkeys (96) exposed to static magnetic fields, although at magnetic field strengths much larger than those generated by electric power generation, transmission and distribution facilities. (This result, although obtained with a static magnetic field, is applicable to situations involving exposure to very strong power frequency magnetic fields.)

Protective measures for magnetic field coupling

In virtually all cases there is no practical way of shielding against exposure to ELF magnetic fields. Thus, the only practical protective method is to limit exposure, either by limiting access of personnel to areas where magnetic fields are strong or by limiting to safe levels all magnetic fields to which people could be exposed.

REVIEW OF CELLULAR STUDIES

The effects of electric fields on systems *in vitro* have been studied in a few laboratories. A great advantage of such *in vitro* studies is the potential for large sample size and a high degree of control over experimental variables. These studies also provide a more direct investigation of possible mechanisms of interaction between a biological system and an ELF field. The most serious difficulties with *in vitro* experiments are those of dosimetry and extrapolation. The dosimetric relationship between exposure in cellular systems and in whole animals is unclear, and extrapolation of results from less complicated systems to human beings is extremely uncertain.

205

Growth and metabolism

Experiments using cultured Chinese hamster ovary cells exposed to 3.7 V/m (measured in the culture medium) showed no effects on cell survival, growth or mutation rate (97). Cell-plating efficiency, however, reflecting a possible alteration in the cell membrane, was reduced in cells exposed to 60 Hz fields at strengths greater than 0.7 V/m. At the same field strength, slime mould showed frequency-dependent effects on mitotic rate, cell respiration and protoplasmic streaming after several months of exposure (98,99). Moreover, these effects, observed using electric or magnetic fields, were enhanced when electric and magnetic fields were simultaneously applied.

Studies examining a variety of models (100,101) have produced some-what confusing results. Effects on cell division, growth and metabolism may appear at field strengths of the order of tenths of a V/m or tenths of a mT (0.1 mT = 80 A/m) in the medium. On the other hand, electrical cell rotation and fusion (102) appear at much higher field strengths, in the range of 10–100 kV/m.

Membrane effects and activity

Experimental findings suggest that a principal site of interaction of an ELF field in a living system is likely to be at the cell membrane (83,103–106). These findings include a 10–20% alteration in calcium exchange from chick or cat brain tissues exposed either to radiofrequency (RF) carrier waves amplitude modulated at ELF frequencies or directly to ELF electric fields (82,105,107). The calcium effect is windowed in frequency, with maximum effects occurring at 16 Hz. In the case of direct ELF exposure, Blackman et al. (80,81) reported several frequency windows, for field strengths in air less than 100 V/m (tissue field strengths $< 1\mu$V/m), centred at 15 Hz and at harmonic multiples of 15 Hz up to 105 Hz. Amplitude (i.e. field strength) windows have also been observed in these experiments (80–82,105). Bawin et al. (105) found a relationship between calcium efflux from brain tissue exposed to ELF fields and the ionic composition of the bathing medium. A calcium-efflux effect is also reported for *in vivo* studies on cats (104).

In the case of the ELF modulation of an RF field, the magnitude of the effective ELF field that acts on calcium-binding sites is presumably determined by a demodulation process at some unidentified site. Assuming complete demodulation, the effective ELF field in tissue would be similar to that induced by an external ELF field of about 100 kV/m in air (83). (Use of an RF carrier results in no significant heating of tissue (108) and no known artifacts, such as spark discharges.)

Possible underlying biophysical mechanisms and a relationship to the electric properties of the brain (EEG waves) are discussed by Grodsky (109). However, the physiological implications and functional significance of the calcium-efflux phenomenon are unknown.

An investigation (110) in which invertebrate neurons from the sea hare *Aplysia* were exposed *in vitro* to a low frequency electric field indicated a strong frequency dependence of cellular activity in response to extracellular currents that included synchronization with the applied field. The neurons were most sensitive at frequencies below 1 Hz, close to the natural neuronal

206

firing, and for a particular orientation of the cell with respect to the field. Other data reported by Sheppard et al. *(111)* showed transient changes in firing rate and increased variability during exposure to an electric field of 0.25 V/m. Episodic synchronization between the neuron and the applied field was also reported at the same exposure level.

In summary, the results of *in vitro* studies suggest that time-varying ELF electric fields may change the properties of cell membranes and modify cell function. Several theoretical explanations have been proposed, and it seems conceivable that several parallel mechanisms exist. However, no comprehensive and experimentally confirmed theory has yet been found.

REVIEW OF ANIMAL STUDIES

Although the interaction of people with electric and magnetic fields is of prime importance and concern, many areas of biological investigation are more efficiently and appropriately conducted using laboratory animals. Most experiments have been performed with rats and mice, but other species, including insects, birds, cats, dogs, pigs and non-human primates, have also been used. A broad range of exposure levels has been employed and an equally large number of biological endpoints have been examined for evidence of possible electric and/or magnetic field effects. These studies have been covered by several excellent reviews *(24,112–120)* and only summaries of important findings will be presented here. The summaries are arranged according to the biological systems that appear to be principally involved.

By far the largest body of biological data comes from experimental work conducted at power frequency fields of 50 and 60 Hz. Hence, frequency will be explicitly stated in the remainder of this chapter only when it differs from a power frequency. More limited work, including many cellular studies, has been conducted at lower frequencies (15–35 Hz). Relatively few studies have been performed at frequencies between 100 and 300 Hz.

Neural and neuroendocrine systems
Many of the biological effects observed in animals exposed to ELF fields appear to be associated, either directly or indirectly, with the nervous system. This apparent relationship is not altogether surprising, since electrochemical processes are involved in nervous system function and since the nervous system is fundamentally involved in the interaction of animals with their environment. The major segments of such interactions — transmission of sensory input from external stimuli, central processing of such information, and subsequent efferent innervation of tissues and organs — may provide both the mechanisms and the explanations for possible links between ELF exposure and observed biological consequences. Additionally, interactions between the nervous system and other biological systems are often mediated indirectly through neuroendocrine or endocrine responses.

In early experimental studies, nervous system parameters were measured only occasionally, even though many of the observed effects, primarily behavioural, were related to nervous system function. Prior to 1977, studies on ELF exposure relating to nervous system function generally could be divided into three categories: (*a*) assessments of activity or startle–response behaviour, (*b*) an evaluation of stress-related hormones such as corticosteroids, and (*c*) general measurements of central nervous system response such as electroencephalograms and inter-response times. The results were often contradictory, with claims of both effects and non-effects resulting from ELF exposure. However, owing to the apparent sensitivity of the nervous system to ELF fields, studies were subsequently expanded to include a wider range of neurological assessments. Concomitant with this increased emphasis, specific nervous system responses, in addition to behaviour, began to be used as experimental endpoints. This effort was mounted, first, to determine the extent and nature of ELF–tissue interaction and, second, in an attempt to understand the mechanisms underlying observed biological effects.

Behaviour
Among the most sensitive measures of perturbation in a biological system are tests that determine modifications in the behavioural patterns of animals. This sensitivity is especially valuable in studying environmental agents of relatively low toxicity.

Studies in several species provide evidence of field perception and imply that electric fields may directly alter behaviour. The threshold for the detection of 60 Hz electric fields by rats is reported by Stern et al. *(121)* to be in the range 4–10 kV/m. Thresholds for electric field perception in mice *(122)*, pigs *(67)*, pigeons *(123)* and chickens *(123, 124)* have been measured to be in the 25–35 kV/m range.

Preference/avoidance behaviour in animals has been studied in experiments where the subject had the option of being or not being exposed to an ELF electric field. At 100 V/m, no effect of exposure was evident in monkeys, either in preference behaviour or temporal discrimination *(125)*. At 25 kV/m, rats preferred to spend their inactive period in the field, while at 75–100 kV/m they avoided exposure *(126)*. Pigs tended to remain out of the field (30 kV/m) at night but demonstrated few other observable behavioural changes *(127)*.

Alterations in activity have been reported in animals exposed to ELF fields, including a transitory increased response to initial exposure of rats or mice at 25–35 kV/m *(122,126)*.

Most of the behavioural work with nonhuman primates has been performed at very low field strengths (7–100 V/m) *(64)*. Gavalas *(128)* and Gavalas-Medici *(129)* observed changes in the inter-response time of monkeys during exposure, but no other effects were seen.

In recent studies examining the effect of ELF magnetic fields on behaviour, many of the investigations performed at low flux densities showed behavioural alterations, primarily activity changes *(130–133)*. In contrast, studies conducted at higher flux densities have shown no evidence of a

208

magnetic-field-associated effect on animal behaviour *(134,135)*. It is unknown whether the discrepancy between these two sets of experiments is due to differences in experimental methodologies or is, instead, due to an unusual "window-type" of dose–response relationship.

Biological rhythms

A number of investigations have been conducted to examine the effects of ELF fields on natural biological rhythms. Following Wever's significant findings *(136)* on the influence of 10 Hz electromagnetic fields on internal timing mechanisms in humans (reviewed later in this chapter), several studies have been reported. Dowse *(137)* claimed that a 10 Hz, 150 V/m field affected the locomotor rhythm of individual fruit flies. Duffy & Ehret *(138)* and Ehret et al. *(139)* used metabolic indicators to examine both circadian and ultradian rhythms in rats and mice exposed to 60 Hz electric fields. These researchers observed no effects in rats but they did report that activity and rhythms of oxidative metabolism could be phase-shifted in male mice by exposure.

Wilson and co-workers *(140,141)* examined another aspect of circadian activity in rats by measuring, at multiple points in time, the production of indolamines and enzymes in the pineal gland. Significant reductions in the normal nocturnal rise of melatonin and one of its biosynthetic enzymes (*N*-acetyl transferase) were observed in rats exposed to either 1.5 or 40 kV/m, but this change in pineal response occurred only after 3 weeks of chronic exposure *(142)*. In three recent studies, nocturnal pineal components in mice and rats were shown to be sensitive to circularly polarized magnetic fields *(143–145)*.

Recently Sulzman *(146)* has begun to investigate the effects of ELF fields on circadian function in squirrel monkeys. Preliminary data from this study suggest that electric and magnetic fields at 39 kV/m, 100μT (80 A/m) may alter biological rhythms in this system.

It is evident that ELF fields alter circadian timing mechanisms in mammals. However, it is unknown if there are any health consequences related to these effects. It is possible that effects mediated by such a system could have significance in other areas where effects have been observed, such as behaviour, reproduction and development.

Neurophysiology

The relationship between the neurotransmitters norepinephrine/epinephrine and the physiological responses of stress/arousal is well established. Measurements of the concentrations of these transmitters in serum, urine and brain tissue have been used to determine whether ELF fields act as mild stressors *(147,148)*.

Groza et al. *(149)* measured catecholamines in both urine and blood following exposure of rats to 100 kV/m, 60 Hz fields. They reported significant increases in epinephrine levels in both blood and urine following acute (6-hour to 3-day) exposure, but no changes in norepinephrine or epinephrine with longer-term (12-day) exposure.

A report of increased norepinephrine in the serum of rats exposed to 50 Hz (50 V/m and 5.3 kV/m) comes from the work of Mose (150). A companion paper by Fischer et al. (151) examined norepinephrine content in brain tissue from rats exposed to 5.3 kV/m for 21 days. After 15 minutes of exposure, the levels increased rapidly. However, after 10 days of exposure a significant decrease in levels was observed when compared to a control group. Portet and co-workers (152,153) reported no changes in adrenal epinephrine or norepinephrine in 2-month-old rats exposed for 8 hours per day to a 50 kV/m, 50 Hz field.

Examining another neurochemical parameter, Kozyarin and co-workers (154,155) measured the levels of the enzyme acetylcholinesterase (AChE) in rats exposed to 50 Hz electric fields. They reported that serum AChE activity was elevated by approximately 25% in both young and old animals exposed to 15 kV/m for 30 minutes per day for 60 days. Brain levels of AChE decreased in exposed animals, although not by such large percentages. Further important measurements were made to estimate the course of recovery from the observed effects. All values had returned to normal one month following cessation of exposure. The authors concluded that electric fields can cause changes in the functional condition of the CNS, although these changes do not appear to be permanent.

In general, neurochemical data provide some putative evidence that exposure to electric fields in the power frequency range may cause slight changes in nervous system function. The number of experiments is not large and there are significant questions about the validity of several of the studies. Nevertheless, the findings are generally consistent and support the hypothesis that ELF exposure may increase arousal in the animal.

Measurements of corticosteroids in animals exposed to electric fields have resulted in a somewhat confusing picture, perhaps due to the quick stimulus/response nature of these adrenal steroids (115). In studies done at Pennsylvania State University (156), a transient increase in corticosterone levels was observed in the plasma of mice exposed to 25 or 50 kV/m. These changes returned to normal within a day after the initiation of exposure. Studies conducted in the USSR (147) showed increases in corticosteroids in rats exposed for 1, 3 or 4 months to 5 kV/m electric fields.

Marino et al. (157) reported a decrease in serum corticosteroid in animals exposed for 30 days to 15 kV/m. The statistical analysis used in this work may, however, have been incorrect, since the variance estimates used by these authors were based on pooled serum samples and thus did not include the effects of inter-subject differences. Another criticism of these studies was that the exposed rats were individually housed while the control animals were housed in groups.

Results that contradict those of Marino et al. (157) and Dumansky et al. (147) have been reported by Free et al. (158), Portet (159) and Portet et al. (153). In these studies, no effects on corticosterone levels were observed in animals exposed for 30 and 120 days to 100 or 50 kV/m electric fields. Additional support for these results comes from experiments in which dogs (160) and rabbits (159) were exposed to 15 and 50 kV/m, respectively. In neither case were there indications of any effects on steroid secretion resulting from electric field exposure. Quinlan et al. *(161)* collected blood samples via carotid artery cannulas during exposure or sham exposure to an 80 kV/m, 60 Hz electric field. No differences in corticosterone levels were noted.

Several laboratories have examined brain tissue morphology in animals exposed to ELF electric fields. Carter & Graves *(162)* and Bankoske et al. *(124)* exposed chicks to 40 kV/m electric fields and saw no effects on CNS morphology. This was supported by the findings of Phillips et al. *(163)* who examined rats exposed to 100 kV/m for 30 days and found no morphological evidence of any electric field effects. Subsequently, however, in a study in Sweden *(164,165)*, dramatic changes resulting from 14 kV/m electric field exposure were reported in the structure of cells from the cerebellum of rabbits. Exposed animals showed disintegration of Nissl bodies and the three-dimensional endoplasmic reticulum structure, as well as the abnormal presence of many lamellar bodies, particularly in Purkinje cells. Reduced numbers of mitochondria, reduced arborization of the dendritic branches, and an absence of hypolemmal cisterns were also evident in these cells. These data must be interpreted with caution, however. The animals were exposed outdoors and showed evidence of significant deterioration of health, though whether this was due to the electric field, to other environmental conditions, or to some combination of these factors is unclear. Furthermore, results from these studies conflict with experiments conducted by Portet et al. *(153)* in which no ultrastructural changes occurred in the cerebella of a few young rabbits exposed to 50 kV/m. These questions concerning neuroanatomical changes have yet to be resolved. The lack of obvious significant CNS functional deficits in the thousands of rats exposed to date, however, suggests that the reported dramatic morphological alterations in the CNS of exposed rabbits may result from conditions unrelated to electric field exposure. The possibility of synergistic effects from the electric field and a stressful environment cannot be ruled out.

It has been assumed that the nervous system would be particularly sensitive to influence by external ELF fields. There are, at present, a confusing array of studies reporting effects and non-effects of exposure to ELF fields. For example, consider those experiments that have studied a commonly used measure of general CNS activity, the electroencephalogram (EEG). In chicks exposed over 3 weeks to 60 Hz electric fields of up to 80 kV/m, Graves et al. *(123)* noted no changes in EEGs recorded via electrodes implanted after exposure. Similarly, no effects were observed in the EEGs of cats exposed to 80 kV/m at 50 Hz *(166)*. Blanchi et al. *(167)*, however, reported significant changes in EEG activity when guinea pigs were exposed for 30 minutes to a 100 kV/m, 50 Hz field. Takashima et al. *(168)* examined EEGs in rabbits exposed to 1 to 10 MHz fields amplitude modulated at a frequency of 15 Hz. Silver electrodes, placed on the animals' skulls for recording purposes, were present during exposure. Abnormal EEGs developed after 2–3 weeks of exposure. Gavalas et al. *(128)* noted that 7 and 10 Hz electric fields of only 7 V/m affected EEGs recorded from monkeys via implanted electrodes. The significance of these results is unclear; they may conceivably be due to an artifact caused by the presence of the electrodes.

Electroencephalographic results from cats exposed to 50 Hz magnetic fields for 8 hours per day at 20 mT (16 kA/m) showed short-term alterations *(169)*. This response was observed only during the period shortly after the magnetic field was energized.

Jaffe et al. *(170)* performed visual-evoked-response measurements on 114 rats exposed *in utero* and for 20 days postpartum to assess for a possible effect of exposure on the development of the visual system. The exposure of the dams, fetuses and subsequent pups was to a 65 kV/m, 60 Hz electric field. No consistent, statistically significant effects of exposure were observed.

Three other excellent neurophysiological studies have identified clear, replicable results. In the first, Jaffe et al. *(171)* examined synaptic junctions from chronically exposed rats (60 Hz, 100 kV/m, 30 days). In these studies, presynaptic fibres were removed from the body and stimulated with a pair of supra-threshold pulses. The height ratio of the resultant compound action potentials, observed as a function of the interspike interval, demonstrated an enhanced neuronal excitability in nerves from exposed animals. Many other parameters of nerve function were tested without effects being observed. In a second experiment *(172)* in rats exposed to 100 kV/m electric fields for 30 days, a wide range of peripheral nervous system and neuromuscular junction parameters were tested. The only effect observed was slightly faster recovery from fatigue, after chronic stimulation, in the soleus (a slow-twitch) muscle. A third study *(173)*, in which brain tissue slices were exposed *in vitro* to a 1 V/m electric field (calculated in the tissue itself), demonstrated long-lasting changes in tissue excitability. Although these studies were conducted at field strengths (and hence current densities) much greater than would be found in the tissues of animals exposed to environmental fields, they are of the same approximate strength as EEG fields.

In summary, numerous studies have been conducted to determine to what extent an environment containing electric or magnetic fields of 1–300 Hz affects the nervous system. The neurological effects reported in many of the experiments have not yet confirmed any pathological effects, even after prolonged exposures to high-strength (100 kV/m) electric fields *(174)* and high-intensity magnetic (5 mT, 4 kA/m) fields *(37)*. As demonstrated above, areas in which effects have been observed in association with the nervous system include: altered neuronal excitability, altered circadian levels of pineal hormones, behavioural aversion to or preference for the field, and changes in the EEG waveform. In addition, in several instances where unconfirmed or controversial data exist, observed effects (e.g. changes in serum catecholamines, corticosteroids and brain morphology) may or may not be caused by electric field exposure. It is not yet known whether these and other reported effects are due to a direct interaction of the electric field with tissue or to an indirect interaction, such as a physiological response due to detection and/or sensory stimulation by the field. The nature of the physical mechanisms involved in field-induced effects is obscure, and such knowledge is one of the major goals of current research.

212

The behavioural tests that most frequently showed an effect of exposure were those relating to detection of the field or to activity. Most other types of behaviour did not change with exposure to ELF fields.

Reproduction and development

It is generally assumed that developing organisms, including pre- and post-natal mammals, are more sensitive to physical or chemical agents than are adult organisms (175). This greater sensitivity, when it occurs, is thought to originate in the processes and controls that guide the developing cellular interactions. A number of studies have been conducted to examine the effects of ELF exposure on reproduction and development of both mammalian and non-mammalian species. These studies have been reviewed (163,176,177) and will only be briefly summarized here.

Most of the non-mammalian studies have been performed in birds, either chickens or pigeons. Electric field exposure of chicks at several field strength levels, both before and after hatching, did not produce any significant effects on viability, morphology, behaviour or growth (178–180). In one series of experiments, however, chicks exposed to 40 or 80 kV/m on days 1–22 following hatching showed a significant reduction in motor activity during the week following removal from the field (181).

Very few studies have been performed to examine the effects of ELF magnetic fields on growth and development. Krueger (178) exposed chicks of 0–28 days of age to a non-uniform 45 Hz, 14 mT (110 A/m) field. Growth rates were depressed but no other parameters were affected. A great deal of interest has been shown in the recent reports by Delgado et al. (182). A marked increase in malformation rate was observed in chicken eggs exposed to low levels of pulsed magnetic fields (0.12 or 12 μT, 0.1 or 10 A/m). It was subsequently reported that an important determinant of the results was the shape of the pulse (183).

Numerous studies have been reported in which rats, rabbits and mice were exposed to 20, 50, 100, 200 or 240 kV/m without observed effects on reproduction, survival, growth or development (152,184–190). There have been relatively few studies in which exposures of prenatal mammals to electric fields have been reported to produce deleterious effects on postnatal growth and survival (164,191–194).

Developmental effects have not generally been observed except in one teratological study of pigs exposed for long durations to 30 kV/m electric fields (177).

A circularly polarized magnetic field (50 μT to 1.5 mT, 40 A/m to 1.2 kA/m) was used to expose pregnant rats during various stages of gestation. Differences were noted between exposed and sham-exposed offspring, including increased thyroid and testis weights in exposed pups and increased responsiveness in exposed offspring when tested under suppressed-response conditions (195,196).

Although the majority of studies indicate no deleterious effects on mammalian reproduction and development of exposure to ELF electric or magnetic fields, the few studies that do report such effects point out the necessity for further investigation.

Other biological functions

Bone growth and repair
In animals exposed to 60 Hz electric fields, one report *(197)* indicated that bone growth *per se* was not affected by exposure to 100 kV/m. However this study, as well as another report *(198)*, suggested that repair of bone fracture was retarded in rats and mice exposed to 5 or 100 kV/m, 60 Hz fields but not to very low (1 kV/m) field strengths. McClanahan & Phillips *(197)* suggest that exposure affected the rate of healing but not the strength of the healed bone.

Cardiovascular system
Cardiovascular function has been assessed by measuring blood pressure and heart rate and by performing electrocardiograms (ECGs). Early studies reported, as possible effects of exposure to electric fields, a decrease in heart rate and cardiac output in dogs exposed to 15 kV/m *(160)* and an increase in the heart rates of chickens exposed to 80 kV/m *(162)*. A more recent and comprehensive study in rats exposed to 100 kV/m showed no effects of exposure, even when the animals were subjected to cold stress following exposure *(199)*. Cerretelli & Malaguti *(200)* reported transient increases in blood pressure in dogs exposed to electric field strengths greater than 10 kV/m. Hilton & Phillips *(199)* were unable to confirm a report by Blanchi et al. *(167)* of changes in ECGs of animals exposed to 100 kV/m.

Magnetic field exposure of dogs (50 Hz, 2.4 T, 1.9 MA/m) caused stimulation of the heart in the diastolic phase, with salvos of ectopic beats appearing in the ECGs *(169)*.

Serum chemistry appears to be relatively unaffected by exposure to either ELF electric or magnetic fields *(148,201–203)*. Haematological data, however, present a more confusing picture. With electric fields, white cell count was often elevated in exposed populations of mice and rats *(202,204)*. It is interesting to observe that, with the exception of one report *(205)*, published studies on haematological effects of magnetic field exposure have shown no consistent field-associated effects *(206–209)*. The occasional reports of positive or negative changes in haematological parameters must be carefully evaluated. Often, apparent sporadic effects are not statistically significant when appropriate multivariate techniques are used to analyse the data *(202)*.

Immunology
In most studies, exposure of animals to electric fields does not appear to affect the immune system. In a comprehensive investigation of the humoral and cellular aspects of the immune system, Morris *(210–212)* observed no effects of exposure at very low field strengths (150–250 V/m) in mice or rats. LeBars et al. *(213)* also found no significant effects of extended exposure (1–6 months) to 50 kV/m electric fields on mice, rats or guinea pigs. Using radiofrequency fields amplitude modulated at 60 Hz, Lyle et al. *(214)* observed significant decrements in the cytolytic capacity of lymphocytes

214

exposed *in vitro*. Dosimetric relationships between ELF electric field exposure and modulated radiofrequency field exposure have not yet been determined.

ELF magnetic fields are reported to strongly influence immune system response to mitogens and antigens *(215–217)*.

Carcinogenesis and mutagenesis

In cellular studies, no effects have been observed that might suggest an effect of electric field exposure on mutagenesis or carcinogenesis *(218–220)*. There is, however, considerable research interest and activity concerning possible connections between magnetic fields and human cancer risk (see below). As yet, there are no published results of laboratory studies that bear directly on this question.

REVIEW OF HUMAN STUDIES

Concern about possible deleterious effects on human health resulting from exposure to ELF electric and magnetic fields grew out of observations in the USSR during the late 1960s and early 1970s. These reports claimed a variety of exposure-related symptoms, including headache, poor digestion, cardio-vascular changes, decreased libido, loss of sleep and increased irritability, in switchyard workers after prolonged exposure to 50 Hz electric fields of up to 26 kV/m *(221,222)*. Because of the somewhat subjective nature of these studies, however, it was not possible to conclude that these functional changes were the exclusive result of exposure to electric and magnetic fields, especially in view of the presence of spark discharges at the substations. However, the findings led to increased research in the USSR and throughout the rest of the world. Numerous studies were initiated in a wide range of investigations to assess the potential biological consequences of the exposure of people to ELF electric and/or magnetic fields.

In studies conducted to date, the principal sources of information on the effects of ELF fields on man are surveys of workers and of people living in the vicinity of high-voltage lines, a few laboratory and clinical studies, and several epidemiological studies. The value of many of these studies has been compromised by one or more of three serious problems: (*a*) small sample sizes with extremely limited statistical power; (*b*) failure to obtain quantitative data on levels and durations of exposure; and (*c*) uncertainty about what constitutes an appropriate control group. While these difficulties do not necessarily invalidate the results of such studies, it is important to recognize that such problems exist when evaluating the results.

Laboratory investigations

Effects of ELF exposure on circadian activity have been investigated by Wever *(136)*. He examined interactive effects in people between naturally occurring electric and magnetic fields and artificially generated electric fields (2 V/m, 10 Hz). Repeated experiments demonstrated that these fields could alter human circadian periodicity *(136,223)*.

The most extensive laboratory experiments on human physiology were conducted by Hauf and co-workers. Beginning in 1974, they conducted a comprehensive clinical evaluation of over 100 volunteers exposed for relatively brief periods to 50 Hz electric fields of 1, 15 or 20 kV/m *(224,225)*. Among the many parameters tested, the few field-related effects observed included small decreases in reaction time, small increases in white blood cell counts, and a slight elevation of norepinephrine levels. The authors proposed that the latter change might reflect minor stress. In a further attempt to define accurately the cause of the slight variations indicated above, Hauf and co-workers conducted a study on the effects of $200\mu A$ currents applied through surface electrodes. These currents, calculated to equal the displacement currents present in their previous electric field experiments, caused no alterations in reaction time or EEGs. Hauf thus concluded that the slight effects previously observed in subjects exposed to an electric field were probably due to "nonspecific excitation" effects *(226)*. The results of Hauf et al. were confirmed and extended by Kuhne et al. *(227)*. Data from these studies are summarized in the chapter by Hauf in the first edition of this book *(24)*.

In a similar study conducted in Poland *(228)*, reactions to sound and light stimuli were studied in 35 volunteers exposed to 50 Hz electric fields. At field strengths greater than 10 kV/m, the reaction was prolonged for both types of stimuli.

All human studies in laboratories have involved short-term exposures of less than 5 hours. They therefore provide little guidance for the evaluation of possible biological effects arising from long-term exposure. For this reason, Hauf and co-workers *(229,230)* examined, in a combined epidemiological/medical study, 32 people occupationally exposed to electric and magnetic fields produced by 380 kV systems for periods exceeding 20 years. These individuals showed no differences from controls in multiple physical, haematological, biochemical, hormonal and behavioural tests. These subjects were also exposed in the laboratory to a 50 Hz, 20 kV/m electric field to test for the possibility that their previous exposure history has sensitized them to electric fields. No signs of such sensitization were detected.

Another study involving laboratory exposure conducted by Cabanes & Gary *(72)* evaluated the threshold for perception of a 50 Hz electric field. Seventy-five people were exposed to various electric fields in three body positions. Threshold levels for field perception ranged from 0.35 kV/m in 4% of the subjects to greater than 27 kV/m for 40% of the subjects when their arms were held against their bodies. Similar data for people exposed to 60 Hz electric fields were collected by Deno & Zaffanella *(25)*. They reported that 5% of exposed subjects were able to detect a 1 kV/m field and that the median value for perception was 7 kV/m, a value substantially less than that obtained by Cabanes & Gary *(72)*. If such a large perceptual threshold variation holds true in other species, it might provide some explanation for the great variability in biological data obtained to date.

Haematology and serum chemistry were the biological endpoints of interest in three laboratory studies that involved short-term exposures of people to ELF magnetic fields *(231–233)*. With the exception of one report of increased serum triglycerides in exposed subjects (231) no effects of ELF magnetic field exposure were observed. Additional observations were made in more than 100 people exposed to time-varying magnetic fields (5 Hz–1 kHz, B less than 100 mT (80 kA/m)). No effects of exposure were found in measurements of EEG, ECG, blood pressure and body temperature (234).

The most widely studied biological effect of ELF magnetic field exposure in man is a visual phenomenon known as "phosphenes" (235). This phenomenon appears to arise from the induction of electric currents in the eye that stimulate the retina. It is highly frequency-dependent, with maximum effects occurring at 20 Hz where the threshold magnetic flux density is about 5 mT (4 kA/m) (236).

Since the early 1970s, pulsed magnetic fields with repetition rates in the ELF range have been used in clinical applications as a noninvasive means of stimulating bone union and treating pseudoarthroses (45,237–239). Pulsed field applicators presently in use for fracture therapy give localized exposure levels of approximately 2 mT (1.6 kA/m) (240). Initial clinical trials with such magnetic stimulation have achieved a high rate of success (up to 85% in bone fracture healing). The biophysical mechanisms of the interaction between bone tissue and pulsed magnetic fields has been investigated in a number of studies. These investigations have suggested that bidirectional pulsed fields may enhance the synthesis of collagen and alter the synthesis of cell surface glycoproteins in cultures of bone-forming cells and fibroblasts (241–243). Two excellent recent reviews present detailed information on the interaction of time-varying ELF magnetic fields with living tissue (96,240).

The most recent laboratory studies on people were conducted at the Midwest Research Institute and at Manchester University. In the former study, male volunteers were exposed for 8-hour periods to both electric and magnetic fields (60 Hz) at levels below the perception threshold (244,245). Although some differences appear to exist between exposed and sham-exposed subjects, particularly in some endocrine parameters and heart rate, the effects are not consistent across time and lie within the range of normal biological variation. The Manchester work (246,247) involved male volunteers subjected for 5½ hours to 50 Hz currents introduced via surface electrodes in such a manner as to simulate exposure to a 36 kV/m electric field. Tests of mood, memory, attention and verbal reasoning skills have revealed no clear effects of this exposure.

Epidemiological assessments
In addition to the reports of epidemiological studies from the USSR, other investigations of ELF exposures were reported in the late 1960s and early 1970s. Kouwenhoven (248) and Singewald (249) studied 10 linemen exposed over a four-year period to unperturbed fields of up to 25 kV/m. In contrast to the USSR studies on substation personnel noted above, they observed no correlation between exposure and ill health. This study, however, covered only a small number of people with uncertain exposure histories. In addition, descriptions of the experimental protocol and results were incomplete.

217

Another four-year study was performed by Filippov (250) in which 20 employees of high-voltage installations were given health examinations. Nonspecific functional disorders of the cardiovascular and central nervous systems, as well as changes in peripheral blood, were reported. Follow-up laboratory studies were performed, but only cursory information is reported. An additional four-year study was conducted by Malboysson (251) in which many aspects of the emotional and physical health of two groups of electrical workers were studied. No biochemical changes and, apparently, no pathological effects that could not otherwise be explained were evident when a low-exposure group of 94 men and a group of 84 linemen (lines of 138, 220 and 400 kV) were compared.

Studies of non-electrical workers exposed to electric fields have been conducted in several countries. Dumansky et al. (252) found no effects in workers exposed to ELF fields of 12 kV/m for $1\frac{1}{2}$ hours per day, but fatigue and mild physiological effects were reported by workers exposed in similar environments to 16 kV/m. Eighteen farmers in Ohio working near 765 kV transmission lines reported no adverse health effects (253). Similarly, a study of the use of doctors and pharmacies by 70 families living and working close to 200 and 400 kV lines in France revealed no problems attributable to proximity to the lines (254). In a 10-year medical study of 110 linemen, using protective clothing and working "bare hand" on transmission lines of 110–380 kV, Issel et al. (255) observed no adverse health effects when comparing the exposed individuals to a control group.

Health surveys of occupationally exposed male workers include two particularly thorough studies by Stopps & Janischewski (34) in Canada and Knave et al. (256) in Sweden. Both groups examined a wide range of biological variables and reported no significant effects on nervous system function, blood chemistry, cardiovascular function or general physical condition. The Stopps study is one of the few epidemiological studies in which field characteristics and length of exposure were estimated. Knave et al. reported differences between exposed and unexposed people in scores obtained on psychological questionnaires (the exposed group scored better) and in fertility (fewer offspring and a lower percentage of male infants fathered by exposed workers). These differences, however, could not be ascribed to electric field exposure, and they may have been due to variations in the level of education or other differences between the two populations. Unfortunately, both studies are based on small numbers of subjects (30 and 53 men, respectively). In a recently completed study in the United Kingdom (257) no significant correlation was found between the measured exposure of 287 power transmission and distribution workers to 50 Hz electric fields and indices of their general health.

A study in Sweden (258) showed reproductive disturbances in families where the father had been working in energized high-voltage substations. There was an increased rate of congenital malformations among the children of the exposed workers (12 out of 119 children compared to 9 out of 225 in the reference group). Although showing statistical significance, this result must be interpreted with caution; the total number of children with malformations was small and, since each individual served as his or her own control (i.e. with data taken before exposure), the "exposed" and "control" groups were not matched in age. In addition, the frequency of incidents varied strongly with time.

Another finding of the Swedish study was an increased difficulty in bearing children in couples where the male worked in a 400 kV substation. In connection with this study, chromosomal aberrations in lymphocytes of substation workers were studied (259). The rate of chromatid and chromosome breaks in a group of 20 substation workers was found to be significantly higher than that in 17 controls. This finding has been followed up in a recent study of 19 workers from 400 kV substations and the results again showed a significant increase in chromosome breaks (260). The authors suggest that the observed chromosomal aberrations may be caused by spark discharges rather than as a direct effect of electric or magnetic field exposure.

In contrast to the Swedish findings, a study in the Federal Republic of Germany of 32 workers occupationally exposed for more than 20 years in 380 kV switchyards revealed no increase in structural chromosome changes or sister-chromatid-exchange frequencies (229).

Recently, there has been a great deal of concern that exposure to ELF magnetic fields may be associated with an increased incidence of cancer. Wertheimer & Leeper (261) reported an increased incidence of leukaemia among children whose houses were judged to have electrical wiring configurations suggestive of higher current flow. Presumably, these high-current configurations resulted in increased magnetic field levels in the house. In a later study, the same investigators extended their research and found in adults an increase in some kinds of cancer, but not in leukaemias (262). A comparable study by Fulton et al. (263) found no correlation between childhood leukaemia and the wiring configuration in the house. From a recent study in Sweden, Tomenius (264) reports increased cancer rates in children whose homes had magnetic fields (50 Hz) greater than $0.3 \mu T$ (0.24 A/m). (It should be noted that only one summertime field measurement per residence was made.) Several aspects of these studies limit the strength of their conclusions and may, perhaps, explain the discrepancies between them. These aspects include lack of validation of the exposure measure, residential mobility restrictions, and failure to factor in possible confounding variables. The Tomenius report lacks sufficient detail for a critical evaluation of the methodology or the results to be made.

There have been other recent reports of an association between various cancers and exposure to electric and/or magnetic fields. Lin et al. (265) reported an association between brain tumour mortality and estimated occupational exposures to ELF fields. In another epidemiological study, conducted in the United States, it was reported that children of fathers occupationally exposed to electric and magnetic fields had a significantly increased risk of cancer of the central nervous system (266). Recent examinations of deaths in Washington State (267) and in England & Wales (268), as well as cancer incidence in Los Angeles (269) and London (270), show an association between cancer and occupations loosely categorized as "electrical". Because of limitations in such studies, particularly those based on an analysis of cancer registries or death certificates, one cannot be sure of the significance of these results. Nevertheless, the appearance of such data from multiple studies points out the need for additional research on potential relationships between weak electromagnetic fields and cancer.

Cardiac pacemakers
When examining the interaction between electromagnetic fields and man, the cardiac pacemaker deserves special attention because its operation depends on the sensing of small voltages resulting from the electrical activity of the heart.

Three major studies have been conducted to investigate potential problems caused by the effects of ELF fields on cardiac pacemakers *(271–273)*. These studies included both laboratory testing and *in vivo* testing of pacemakers during actual exposure of patients to ELF fields. Griffin *(274)* concludes, in a comprehensive review of the effects of electromagnetic interference on pacemakers, that the potential for an interaction between implanted pacemakers and ELF electric fields exists. With most pacemakers, ELF interference has no effect on operation. However, in some unipolar models operation may revert to the asynchronous or noise mode, while in still others aberrant signal activity may occur. The levels of exposure necessary to affect a susceptible pacemaker system vary considerably. Abnormal pacing characteristics were exhibited in 20 of 26 pacemakers exposed to 60 Hz magnetic fields with amplitudes ranging from 0.1 to 0.41 mT (80–320 A/m) (induced body currents of 26 to $>200\mu$A are necessary to affect the unit). The response varies greatly with the brand of pacemaker, some showing much greater immunity to ELF interference than others.

It should be noted that clinically documented cases where ELF interference has caused functional problems in pacemakers are extremely rare. Also, since the interference, if it does occur, usually causes the pacemaker to revert to the asynchronous mode, the patient is presented with only slightly increased risk. Obviously, for the minority of patients who depend on the pacemaker for a cardiac rhythm, the unit with a potential for aberrant behaviour (e.g. cessation or pronounced slowing of pacing) would present a definite hazard in strong ELF electric or magnetic field environments.

CONCLUSIONS AND RECOMMENDATIONS[a]

Conclusions
Extremely low frequency (ELF) fields are quantified in terms of the electric field strength E (V/m) and either the magnetic field strength H (A/m) or the magnetic flux density B (T). Natural environmental field strengths are

---

[a] These conclusions and recommendations are those pertaining to ELF fields made by the WHO Working Group on Health Implications of the Increased Use of NIR Technologies and Devices, Ann Arbor, USA, October 1985.

normally very low but, with the widespread use of electrical energy, field strengths have increased considerably. In non-occupational settings, people are seldom exposed to ELF electric and magnetic fields with strengths above about 100 V/m and $30\mu T$ (24 A/m) respectively. The highest occupational exposures may be of the order of tens of kV/m and tens of mT (tens of kA/m) but typical exposure levels are 10–100 times lower.

ELF magnetic fields are used in a variety of therapeutic and diagnostic applications, including healing of bone fractures, promotion of nerve regeneration, and acceleration of wound healing. These applications involve partial body exposures in the range 1–30 mT (0.8–24 kA/m). Exposures to time-varying ELF fields also result from the medical use of magnetic resonance imaging devices.

Instrumentation to measure ELF fields is commercially available, but some skill is required to achieve reliable results. Separate measurements of the electric and magnetic fields are necessary.

Whole-body exposure to ELF electric fields may involve effects related to stimulation of the sensory apparatus at the body surface (hair vibration, possible direct neural stimulation) and effects within the body caused by the flow of current. Magnetic fields may interact predominantly by the induction of internal current flow.

Internal current flow is described in terms of current density, J (A/m^2), in tissue. Ohm's law permits an equivalent expression of current density in terms of internal electric field strength E (V/m). It is not known whether J or E is the more useful and relevant physical quantity in understanding mechanisms of biological effects. Internal current densities produced by exposure to external electric or magnetic fields at practical levels (up to approximately 100 kV/m and 1 mT (800 A/m)) are far lower than current density levels that produce electric shocks.

Data on neuromuscular stimulation (including respiratory tetanus and cardiac fibrillation) indicate that current densities above about 1 A/m can be dangerous. At lower levels, biological effects may occur; however, the implications for health of exposure at such levels are not clear.

For sinusoidal fields, the magnitude of the internal current density induced by external ELF electric and/or magnetic fields is directly proportional to frequency. For other waveforms, the rate of change of the field may be relevant. The duration of current flow is also important.

Biological effects observed in a living organism may depend on the electric fields induced inside the body, possibly on the magnetic fields penetrating into the body, and on the fields acting at the surface of the body. Accurate relationships can be determined between field and current levels inside and at the surface of a body and the external unperturbed fields to which it is exposed. These relationships depend on frequency, the orientation and shape of the body, in some cases its size, and tissue composition. Because these relationships can be determined, exposure limits can be defined in terms of external unperturbed electric and magnetic field strengths, frequencies and exposure durations.

Most biological data come from experimental work conducted with electric fields of 50 and 60 Hz. Fewer studies have been conducted at lower frequencies and at frequencies between 100 and 300 Hz. An overview of available literature suggests that ELF electric and magnetic fields are environmental agents of relatively low potential hazard to biological systems. Many of the biological effects that have been reported are quite subtle. Furthermore, many indices of general physiological status appear relatively unaffected by exposure to ELF electric and/or magnetic fields.

Areas in which effects have been observed often appear to be associated with the nervous system, including altered neuronal excitability and neurochemical changes, altered hormone levels, changes in behavioural responses, and changes in biological rhythms. No studies unequivocally demonstrate deleterious effects of ELF electric or magnetic field exposure on mammalian reproduction and development, but several suggest such effects.

Exposure to ELF electric and magnetic fields does produce biological effects. However, except for fields strong enough to induce current densities above the threshold for the stimulation of nerve tissues, there is no consensus as to whether these effects constitute a hazard to human health. Human data from epidemiological studies, including reported effects on cancer promotion, congenital malformations, reproductive performance and general health, though somewhat suggestive of adverse health effects, are not conclusive.

Recommendations
1.   The recommendations in Environmental Health Criteria No. 35 *(1)* are endorsed, and can be summarized as follows.

   — Occupational exposure to strong electric fields is generally intermittent, and field strengths where spark discharges are prevalent should be avoided. Available scientific knowledge about the effects of exposure to electric fields indicate no present need to limit public access to regions where field strengths are below about 10 kV/m. However, it is prudent to limit long-term exposure to such fields to levels as low as can reasonably be achieved.

   — Although guidelines on limiting exposure to electric fields have been proposed by several countries, guidance on ELF magnetic fields has been provided only by the Federal Republic of Germany. It is desirable to develop a basis for uniform international standards for both electric and magnetic ELF fields.

2.   Exposure to electric and magnetic fields should be considered in the design, siting and shielding of ELF sources.

3.   Several protective measures are possible, and consideration should be given to the following.

   — Shielding from electric field exposure can be accomplished relatively easily by placing conducting surfaces around the personnel to be protected. No practical means of shielding personnel from ELF magnetic fields exists.

222

— Field strengths decrease rapidly with distance from the source.

— Protective clothing is available to prevent spark discharges to workers and to shield them from ELF electric fields.

— Large metallic structures in the close vicinity of high-voltage transmission lines should be well grounded to eliminate the possibility of electric shock to individuals who may touch them.

— Conventional electrical safety procedures should be followed.

— Appropriate education and training of workers is essential.

4. Measurements should be performed by trained individuals and standard reporting procedures should be adopted.

5. There is a need for more information on the levels and conditions of exposure in occupational situations and for the general population. Magnetic field strengths, frequency content, and the spatial and temporal characteristics of the fields require special attention.

6. Personal meters to assess the level and duration of individual exposure to ELF electric and magnetic fields are needed, as well as instrumentation for the measurement of induced electric fields and current densities within living systems.

7. Dosimetric data are needed to take account of the often complex characteristics of exposure, such as temporal variations and spatial non-uniformity of induced currents and fields. It is important to determine which physical quantity best represents the interaction of ELF fields with biological systems. Dosimetric considerations are essential to extrapolate data from animal and *in vitro* studies to human exposure conditions.

8. Further research is needed to investigate the mechanisms of those effects that display non-linear responses in field strength and frequency.

9. Because of emerging information on immunity, cancer, cell–cell communication, growth, differentiation and cell repair, there is a need to understand the electrochemical, cellular and biochemical changes that may occur as a result of exposure to ELF electric and magnetic fields. Further study is needed on the influence of electric and magnetic fields on cellular and animal systems, particularly in the areas of the nervous system and the reproductive system. Possible combined effects of field exposure and chemicals need further investigation.

10. Epidemiological studies should be continued and developed to include both people with known occupational exposure to ELF electric and magnetic fields and those exposed in residential settings. Emphasis should be

given to validating recent findings that suggest an association between cancer and exposure to ELF fields. Consideration should also be given to epidemiological studies of people exposed to ELF fields for medical purposes. Because the lack of exposure data is the greatest source of uncertainty in investigations of human health effects, increased effort is needed to improve exposure assessment techniques for future human studies.

REFERENCES

1. **Suess, M.J., ed.** *Extremely low frequency (ELF) fields.* Geneva, World Health Organization, 1984 (Environmental Health Criteria 35).
2. **Geddes, L.A. & Baker, L.E.** The specific resistance of biological material — a compendium of data for the biomedical engineer and physiologist. *Medical and biological engineering,* **5**: 271–293 (1967).
3. **Schwan, H.P.** Electrical properties of tissue and cell suspensions. *Advances in biological and medical physics,* **5**: 147–209 (1957).
4. **Schwan, H.P. & Kay, C.F.** Specific resistance of body tissue. *Circulation research,* **4**: 664–670 (1956).
5. **Schwan, H.P. & Kay, C.F.** The conductivity of living tissues. *Annals of the New York Academy of Sciences,* **65**: 1007–1013 (1957).
6. **Misakian, M.** Measurement of electric and magnetic fields from alternating current power lines. *IEEE transactions on power apparatus and systems,* **97**: 1104–1114 (1978).
7. **Misakian, M. et al.** Miniature ELF electric field probe. *Review of scientific instruments,* **49**: 933–935 (1978).
8. *Transmission line reference book — 345 kV and above.* Palo Alto, CA, Electric Power Research Institute, 1975.
9. **Kaune, W.T.** A prototype system for exposing small laboratory animals to 60-Hz vertical electric fields. *In: Biological effects of extremely-low-frequency electromagnetic fields. 18th Annual Hanford Life Sciences Symposium, Richland, WA, 16–18 October 1979.* Springfield, VA, National Technical Information Service, 1979.
10. **Israel, H.** *Atmospheric electricity.* Springfield, VA, US Department of Commerce, National Technical Information Service, 1973.
11. **Toland, R.B. & Vonnegut, B.** Measurement of maximum electric field intensities over water during thunderstorms. *Journal of geophysical research,* **82**: 438–440 (1977).
12. **Bracken, T.D.** Field measurements and calculations of electrostatic effects of overhead transmission lines. *IEEE transactions on power apparatus and systems,* **PAS-95**: 494–504 (1976).
13. **Bridges, J.E. & Preache, M.** Biological influences of power frequency electric fields — a tutorial review from a physical and experimental viewpoint. *Proceedings of the Institute of Electrical and Electronics Engineers,* **69**: 1092–1120.
14. **Deno, D.W. & Zaffanella, L.E.** *Electrostatic and electromagnetic effects of ultrahigh-voltage transmission lines.* Palo Alto, CA, Electric Power Research Institute, 1978 (Final Report, Research Project 566-1).

224

15. **Deuse, J. & Pirotte, P.** [Calculation and measurement of electric field strength near H.V. structures]. *In: International Conference on Large High Voltage Electric Systems,* Paris, CIGRE, 1976 (CIGRE report No. 36-04).

16. **Dietrich, F.M. & Kolcio, N.** Corona and electric field effects at the Apple Grove project and an 800 kV line in the USA. *In: International Conference on Large High Voltage Electric Systems,* Paris, CIGRE, 1976.

17. **Maruvada, P.S. et al.** Electrostatic field effects from high voltage power lines and in substations. *In: International Conference on Large High Voltage Electric Systems,* Paris, CIGRE, 1976.

18. **Mihaileanu, C.** Electrical field measurement in the vicinity of HV equipment and assessment of its bio-physical perturbing effects. *In: International Conference on Large High Voltage Electric Systems,* Paris, CIGRE, 1976 (CIGRE Report No. 36-08).

19. **Schneider, K.H. et al.** [Displacement currents to the human body caused by the dielectric field under overhead lines]. *In: International Conference on Large High Voltage Electric Systems.* Paris, CIGRE, 1974 (CIGRE Report No. 36-04).

20. **Stringfellow, G.C.** *Electric fields near CEGB transmission plant.* Leatherhead, Central Electricity Research Laboratories, 1980 (Laboratory Note RD/L/N 174/80).

21. **Utmischi, D.** *Das elektrische Feld unter Hochspannungsleitungen* [The electric field under high-tension lines]. Munich, Technical University, 1976.

22. *The electrostatic and electromagnetic effects of AC transmission lines.* Piscataway, NJ, Institute of Electrical and Electronics Engineers, 1979 (IEEE Course Text 79 EH0145-3-PWR).

23. **Poznaniak, D.T. et al.** *Simulation of transmission line ground-level gradient for biological studies on small plants and animals.* Piscataway, NJ, Institute of Electrical and Electronics Engineers, 1977 (IEEE Paper A 77 718-0, 1977 Summer Meeting of the Power Engineering Society).

24. **Hauf, R.** Electric and magnetic fields at power frequencies, with particular reference to 50 and 60 Hz. *In:* Suess, M.J., ed. *Nonionizing radiation protection.* Copenhagen, WHO Regional Office for Europe, 1982, pp. 175–198 (WHO Regional Publications, European Series, No. 10).

25. **Deno, D.W. & Zaffanella, L.** Electrostatic effects of overhead transmission lines and stations. *In: Transmission line reference book, 345 kV and above,* 2nd ed. Palo Alto, CA, Electric Power Research Institute, 1982.

26. **Bracken, T.D.** *Comparison of electric field exposure monitoring instrumentation.* Palo Alto, CA, Electric Power Research Institute, 1985 (Final Report, Research Project 799-19).

27. **Chartier, V.L. et al.** BPA study of occupational exposure to 60-Hz electric fields. *IEEE transactions on power apparatus and systems,* **PAS-104**: 733–744 (1985).

28. **Deno, D.W.** Monitoring of personnel exposed to a 60-Hz electric field. *In: Biological effects of extremely-low-frequency electromagnetic fields. 18th Annual Hanford Life Sciences Symposium, Richland, WA, 16–18 October 1979.* Springfield, VA, National Technical Information Service, 1979, pp. 93–108.

29. **Deno, D.W. & Silva, M.** Method for evaluating human exposure to 60 Hz electric fields. *IEEE transactions on power apparatus and systems,* **PAS-103**: 1699–1706 (1984).

30. **Lattarulo, F. & Mastronardi, G.** Microprogrammed meter for recording human exposure times under varying electric fields. *Microcomputer applications,* **1**: 16–18 (1982).

31. **Looms, J.S.T.** Power frequency electric fields: dosimetry. *In: Biological effects and dosimetry of non-ionizing radiation: radiofrequency and microwave energies.* New York, Plenum, 1983 (NATO Advanced Study Institute Series, Series A: Life Sciences).

32. **Lovstrand, K.G. et al.** Exposure of personnel to electric fields in Swedish extra-high-voltage substations: field strength and dose measurements. *In: Biological effects of extremely-low-frequency electromagnetic fields. 18th Annual Hanford Life Sciences Symposium, Richland, WA, 16–18 October 1979.* Springfield, VA, National Technical Information Service, 1979.

33. **Silva, M. et al.** An activity systems model for estimating human exposure to 60 Hz electric fields. *IEEE transactions on power apparatus and systems,* **104**: 1923–1929 (1985).

34. **Stopps, G.J. & Janischewsky, W.** *Epidemiological study of workers maintaining HV equipment and transmission lines in Ontario. Research report.* Montreal, Canadian Electrical Association, 1979.

35. *Magnetic fields.* Geneva, World Health Organization, 1987 (Environmental Health Criteria 69).

36. **Grandolfo, M. & Vecchia, P.** Natural and man-made environmental exposures to static and ELF electromagnetic fields. *In:* Grandolfo, M. et al., ed. *Biological effects and dosimetry of non-ionizing radiation: static and ELF electromagnetic fields.* New York, Plenum, 1985.

37. **Tenforde, T.S.** Biological effects of ELF magnetic fields. *In: Biological and human health effects of extremely low frequency electromagnetic fields.* Arlington, VA, American Institute of Biological Sciences, 1985, pp. 79–128.

38. **Caola, R.J., Jr. et al.** Measurements of electric and magnetic fields in and around homes near a 500 kV transmission line. *IEEE transactions on power applications and systems,* **PAS-102**: 3338–3347 (1983).

39. **Male, J.C. et al.** Human exposure to power-frequency electric and magnetic fields. *In:* Anderson, L.E. et al., ed. *Interaction of biological systems with static and ELF electric and magnetic fields.* Proceedings of the 23rd Annual Hanford Life Sciences Symposium. Springfield, VA, National Technical Information Service, 1987.

40. **Stuchly, M.A. et al.** Extremely low frequency electromagnetic emissions from video display terminals and other devices. *Health physics,* **45**: 713–722 (1983).

41. **Haubrich, H.J.** Das Magnetfeld im Nahbereich von Drehstrom-freileitungen [The magnetic field in the proximity of polyphase alternating current overhead transmission lines]. *Elektrizitätswirtschaft,* **73**: 511–517 (1974).

42. **Naval Electronic Systems Command.** Extremely low frequency (ELF) communications system and its electromagnetic fields. *In: Biological and human health effects of extremely low frequency electromagnetic fields.* Arlington, VA, American Institute of Biological Sciences, 1985.

43. **Scott-Walton, B. et al.** *Potential environmental effects of 765 kV transmission lines: views before the New York State Public Service Commission, cases 26529 and 26559, 1976–1978.* Springfield, VA, National Technical Information Service, 1979 (Report No. DOE/EV-0056).

44. **Lovsund, P. et al.** ELF magnetic fields in electrosteel and welding industries. *Radio science,* **17**: 35S–38S (1982).

45. **Bassett, C.A.L. et al.** Pulsing electromagnetic field treatment in ununited fractures and failed arthrodeses. *Journal of the American Medical Association,* **247**: 623–628 (1982).

46. **Budinger, T.F. & Lauterbur, P.C.** Nuclear magnetic resonance technology for medical studies. *Science,* **226**: 288–298 (1984).

47. **Margulis, A.R. et al., ed.** *Clinical magnetic resonance imaging.* San Francisco, CA, University of California, 1983.

48. **Mills, C.J.** The electromagnetic flowmeter. *Medical instrumentation,* **11**: 136–138 (1977).

49. **Reitz, J.R. & Milford, F.J.** *Foundations of electromagnetic theory.* Reading, PA, Addison-Wesley Publishing Company, 1960.

50. **Kaune, W.T. & Gillis, M.F.** General properties of the interaction between animals and ELF electric fields. *Bioelectromagnetics,* **2**: 1–11 (1981).

51. **Deno, D.W.** Currents induced in the human body by high voltage transmission line electric field — measurement and calculation of distribution and dose. *IEEE transactions on power apparatus and systems,* **PAS-96**: 1517–1527 (1977).

52. **Kaune, W.T.** Power-frequency electric fields averaged over the body surfaces of grounded humans and animals. *Bioelectromagnetics,* **2**: 403–406 (1981).

53. **Kaune, W.T. & Miller, M.C.** Short-circuit currents, surface electric fields, and axial current densities for guinea pigs exposed to ELF electric fields. *Bioelectromagnetics,* **5**: 361–364 (1984).

54. **Kaune, W.T. & Phillips, R.D.** Comparison of the coupling of grounded humans, swine and rats to vertical, 60-Hz electric fields. *Bioelectromagnetics,* **l**: 117–129 (1980).

55. **Barnes, H.C. et al.** Rational analysis of electric fields in live line working. *IEEE transactions on power apparatus and systems,* **PAS-86**: 482–492 (1967).

56. **Bayer, A. et al.** [Experimental research on rats for determining the effect of electrical a.c. fields on living beings]. *Elektrizitätswirtschaft,* **76**: 77–8l (1977) (in German).

227

57. **Hart, F.X. & Marino, A.A.** ELF dosage in ellipsoidal models of man due to high voltage transmission lines. *Journal of bioelectricity,* l: 129–154 (1982).

58. **Kolesnikov, S.V. & Chuckhlovin, B.A.** Interaction of a power-line electric field with human and animal organisms. *Soviet technical physics letters,* **4**: 377–378 (1978).

59. **Lattarulo, F. & Mastronardi, G.** Equivalence criteria among man and animals in experimental investigations of high voltage power frequency exposure hazards. *Applied mathematical modelling,* **5**: 92–96 (1981).

60. **Shiau, Y. & Valentino, A.R.** ELF electric field coupling to dielectric spheroidal models of biological objects. *IEEE transactions on biomedical engineering,* **BME-28**: 429–437 (1981).

61. **Spiegel, R.J.** ELF coupling to spherical models of man and animals. *IEEE transactions on biomedical engineering,* **BME-23**: 387–391 (1976).

62. **Spiegel, R.J.** High-voltage electric field coupling to humans using moment method techniques. *IEEE transactions on biomedical engineering,* **BME-24**: 466–472 (1977).

63. **Spiegel, R.J.** Numerical determination of induced currents in humans and baboons exposed to 60-Hz electric fields. *IEEE transactions on electromagnetic compatability,* **EMC-23**: 382–390 (1981).

64. **Kaune, W.T.** Coupling of living organisms to ELF electric and magnetic fields. *In: Biological and human health effects of extremely low frequency electromagnetic fields.* Arlington, VA, American Institute of Biological Sciences, 1985, pp. 25–60.

65. **Kaune, W.T.** *Physical interaction of humans and animals with power-frequency electric and magnetic fields.* Piscataway, NJ, Institute of Electrical and Electronics Engineers, 1986 (Special paper, winter meeting of the IEEE Power Engineering Society).

66. **Kaune, W.T. & Phillips, R.D.** Dosimetry for extremely low-frequency electric fields. *In:* Grandolfo, M. et al., ed. *Biological effects and dosimetry of non-ionizing radiation: static and ELF electromagnetic fields.* New York, Plenum, 1985.

67. **Kaune, W.T. et al.** A method for the exposure of miniature swine to vertical 60-Hz electric fields. *IEEE transactions on biomedical engineering,* **BME-25**: 276–283 (1978).

68. **Guy, A.W. et al.** Determination of electric current distributions in animals and humans exposed to a uniform 60-Hz high-intensity electric field. *Bioelectromagnetics,* **3**: 47–71 (1982).

69. **Kaune, W.T. & Forsythe, W.C.** Current densities measured in human models exposed to 60-Hz electric fields. *Bioelectromagnetics,* **6**: 13–32 (1985).

70. **Kaune, W.T. et al.** Comparison of the coupling of grounded and ungrounded humans to vertical 60-Hz electric fields. *In:* Anderson, L.E. et al., ed. *Interaction of biological systems with static and ELF electric and magnetic fields. Proceedings of the 23rd Annual Hanford Life Sciences Symposium.* Springfield, VA, National Technical Information Service, 1987.

71. **Reilly, J.P.** Electric and magnetic field coupling from high voltage ac power transmission lines — classification of short-term effects on people. *IEEE transactions on power apparatus and systems,* **PAS-97**: 2243–2252 (1978).

72. **Cabanes, J. & Gary, C.** La perception directe du champ electrique. [Direct perception of the electric field]. *In: International Conference on Large High Tension Electric Systems,* Paris, CIGRE, 1981 (CIGRE Report No. 233-08).

73. **Gillis, M.F. & Kaune, W.T.** Hair vibration in ELF electric fields. *In: Biological effects of electromagnetic waves.* Brussels, International Union of Radio Science, 1978.

74. **Dalziel, C.F.** Electric shock hazard. *IEEE spectrum,***9**: 41–50 (1972).

75. **Bernhardt, J.** The direct influence of electromagnetic fields on nerve and muscle cells of man within the frequency range of 1 Hz to 30 MHz. *Radiation and environmental biophysics,* **16**: 309–323 (1979).

76. **Bernhardt, J.H.** *On the rating of human exposition to electric and magnetic fields with frequencies below 100 kHz.* Ispra, Italy, Commission of the European Communities, Joint Research Centre, 1983.

77. **Bernhardt, J.H.** Assessment of experimentally observed bioeffects in view of their clinical relevance and the exposures at work places. *In:* Bernhardt, J.H., ed. *Proceedings of the Symposium on Biological Effects of Static and ELF-Magnetic Fields.* Munich, MMV Medizin Verlag, 1985.

78. **Schwan, H.P.** Field interaction with biological matter. *Annals of the New York Academy of Sciences,* **103**: 198–213 (1977).

79. **Bawin, S.M. & Adey, W.R.** Sensitivity of calcium binding in cerebral tissue to weak environmental electric fields oscillating at low frequency. *Proceedings of the National Academy of Sciences,* **73**: 1999–2003 (1976).

80. **Blackman, C.F. et al.** Effects of ELF (1–120 Hz) and modulated (50 Hz) RF fields on the efflux of calcium ions from brain tissue *in vitro. Bioelectromagnetics,* **6**: 1–11 (1985).

81. **Blackman, C.F. et al.** Effects of ELF fields on calcium ion efflux from brain tissue *in vitro. Radiation research,* **92**: 510–520 (1982).

82. **Blackman, C.F. et al.** Induction of calcium-ion efflux from brain tissue by radio-frequency radiation: effects of modulation frequency and field strength. *Radio science,* **14**: 93–98 (1979).

83. **Adey, W.R.** Tissue interactions with non-ionizing electromagnetic fields. *Physiological reviews,* **61**: 435–514 (1981).

84. **Postow, E. & Swicord, M.L.** Modulated field and "window" effects. *In:* Polk, C., ed. *Handbook of biological effects of electromagnetic radiation.* Boca Raton, FL, CRC Press, 1985.

85. **Swicord, M.L.** Possible biophysical mechanisms of electromagnetic interactions with biological systems. *In: Biological and human health effects of extremely low frequency electromagnetic fields.* Arlington, VA, American Institute of Biological Sciences, 1985.

86. **Jaczewski, M. & Pilatowicz, A.** Transient touch current near EHV line. *In: International Conference on Large High Voltage Electric Systems,* Paris, CIGRE, 1976.

87. **Larkin, W.D. & Reilly, J.P.** Strength/duration relationships for electrocutaneous sensitivity: stimulation by capacitive discharges. *Perception & psychophysics,* **36**: 68–78 (1984).

88. **Reilly, J.P. & Larkin, W.D.** Electrocutaneous stimulation with high voltage capacitive discharges. *IEEE transactions on biomedical engineering,* **BME-30**: 631–641 (1983).

89. **Reilly, J.P. & Larkin, W.D.** Mechanisms for human sensitivity to transient electric currents. *In:* Bridges, J.E., ed. *Electrical shock safety criteria.* New York, Pergamon Press, 1983.

90. **Reilly, J.P. & Larkin, W.D.** *Human reactions to transient electric currents — summary report.* Laurel, MD, Applied Physics Laboratory, Johns Hopkins University, 1985.

91. **Kupfer, J.** Untersuchungen zur Herzkammerflimmerschwell bei 50-Hz Wechselstrom [Investigation of the heart fibrillation threshold at 50 Hz]. *Der Elektro-Praktiker,* **36**: 387–390 (1982).

92. *Effect of current passing through the human body.* Geneva, International Electrotechnical Commission, 1984 (IEC Publication No. 479, part 2).

93. **Gandhi, O.P. et al.** Impedance method for calculation of power deposition patterns in magnetically induced hyperthermia. *IEEE transactions on biomedical engineering,* **BME-31**: 644–651 (1984).

94. **International Commission on Radiation Protection.** *Report of the Task Group on Reference Man.* Oxford, Pergamon Press, 1975 (ICRP Publication No. 23).

95. **Gaffey, C.T. & Tenforde, T.S.** Alterations in the rat electrocardiogram induced by stationary magnetic fields. *Bioelectromagnetics,* **2**: 357–370 (1981).

96. **Tenforde, T.S. & Budinger, T.F.** Biological effects and physical safety aspects of NMR imaging and in vivo spectroscopy. *In:* Thomas, S.R. & Dixon, R.L., ed. *NMR in medicine: instrumentation and clinical applications.* New York, American Association of Physicists in Medicine, 1986 (AAPM Medical Monograph 14).

97. **Frazier, M.E. et al.** Viabilities and mutation frequencies of CHO-K1 cells following exposure to 60-Hz electric fields. *In:* Anderson, L.E. et al., ed. *Interaction of biological systems with static and ELF electric and magnetic fields. Proceedings of the 23rd Annual Hanford Life Sciences Symposium.* Springfield, VA, National Technical Information Service, 1987.

98. **Goodman, E.M. et al.** Bioeffects of extremely low frequency electromagnetic fields: variation with intensity, waveform, and individual or combined electric and magnetic fields. *Radiation research,* **78**: 485–501 (1979).

99. **Marron, M.T. et al.** Mitotic delay in the slime mould *Physarum polycephalum* induced by low intensity 60- and 75-Hz electromagnetic fields. *Nature,* **254**: 66–67 (1975).

100. **Greenebaum, G. et al.** Effects of extremely low frequency fields on slime mold: studies of electric, magnetic, and combined fields, chromosome numbers, and other tests. *In: Biological effects of extremely-low-frequency electromagnetic fields. 18th Annual Hanford Life Sciences*

Symposium. Richland, WA, 16–18 October 1979. Springfield, VA, National Technical Information Service, 1979, pp. 117–131.

101. **Inoue, M. et al.** Growth rate and mitotic index analysis of *Vicia faba* L. roots exposed to 60-Hz electric fields. *Bioelectromagnetics,* **6**: 293–304 (1985).

102. **Pohl, H.A.** *Dielectrophoresis: the behaviour of matter in non-uniform electric fields.* London, Cambridge University Press, 1978.

103. **Adey, W.R.** Evidence for cooperative mechanisms in the susceptibility of cerebral tissue to environmental and intrinsic electric fields. *In:* Schmitt, F.O. et al., ed. *Functional linkage in biomolecular systems.* New York, Raven Press, 1975.

104. **Adey, W.R.** Frequency and power windowing in tissue interactions with weak electromagnetic fields. *Proceedings of the Institute of Electrical and Electronics Engineers,* **68**: 119–125 (1980).

105. **Bawin, S.M. et al.** Possible mechanisms of weak electromagnetic field coupling in brain tissue. *Bioelectrochemistry and bio-energetics,* **5**: 67–76 (1978).

106. **Sheppard, A.R. & Adey, W.R.** The role of cell surface polarization in biological effects of extremely low frequency fields. *In: Biological effects of extremely low frequency electromagnetic fields. 18th Annual Hanford Life Sciences Symposium, Richland, WA, 16–18 October 1979.* Springfield, VA, National Technical Information Service, 1979.

107. **Bawin, S.M. et al.** Effects of modulated VHF fields on the central nervous system. *Annals of the New York Academy of Sciences,* **247**: 74–80 (1975).

108. **Tenforde, T.S.** Thermal aspects of electromagnetic field interactions with bound calcium ions at the nerve cell surface. *Journal of theoretical biology,* **83**: 517–521 (1980).

109. **Grodsky, I.T.** Neuronal membrane: a physical synthesis. *Mathematical biosciences,* **28**: 191–219 (1976).

110. **Wachtel, H.** Firing pattern changes and transmembrane currents produced by extremely low frequency fields in pacemaker neurons. *In: Biological effects of extremely low frequency electromagnetic fields. 18th Annual Hanford Life Sciences Symposium, Richland, WA, 16–18 October 1979.* Springfield, VA, National Technical Information Service, 1979.

111. **Sheppard, A.R. et al.** ELF electric fields alter neuronal excitability in *Aplysia* neurons. *Bioelectromagnetics,* **1**: 227 (1980).

112. **Conti, R. et al.** Possible biological effects of 50 Hz electric fields: a progress report. *Journal of bioelectricity,* **4**: 177–193 (1985).

113. **Graves, H.B., ed.** *Biological and human health effects of extremely low frequency electromagnetic fields.* Arlington, VA, American Institute of Biological Sciences, 1985.

114. **Male, J.C. & Norris, W.T.** *Are the electric fields near power-transmission plants harmful to health? A brief review of present knowledge and proposed action.* Leatherhead, Central Electricity Research Laboratories, 1980 (Note No. RD/L/N 2/80).

115. **Michaelson, S.M.** Analysis of studies related to biologic effects and health implications of exposure to power frequencies. *Environmental professional,* **1**: 217–232 (1979).
116. **National Academy of Sciences.** *Biologic effects of electric and magnetic fields associated with proposed Project Seafarer. Report of the Committee on Biosphere Effects of Extremely-Low-Frequency Radiation.* Washington, DC, National Research Council, 1977.
117. **Phillips, R.D. & Kaune, W.T.** *Biological effects of static and low-frequency electromagnetic fields: an overview of United States literature.* Palo Alto, CA, Electric Power Research Institute, 1977 (EPRI Special Report EA-490-SR).
118. **Schaefer, H.** *Über die Wirkung elektrischer Felder auf den Menschen.* Berlin/Heidelberg/New York/Tokyo, Springer-Verlag, 1983.
119. **Shandala, M.G. & Dumansky, Y.D.** [Biological effects of low-frequency (50 Hz) electric fields]. *In: Proceedings of the Third Soviet-American Working Conference on the Problem of Biological Effects of Physical Environmental Factors, Kiev, 1982* (in Russian).
120. **Sheppard, A.R. & Eisenbud, M.** *Biologic effects of electric and magnetic fields of extremely low frequency.* New York, New York University Press, 1977.
121. **Stern, S. et al.** Behavioral detection of 60-Hz electric fields by rats. *Bioelectromagnetics,* **4**: 215–247 (1983).
122. **Rosenberg, R.S. et al.** Relationship between field strength and arousal response in mice exposed to 60-Hz electric fields. *Bioelectromagnetics,* **4**: 181–191 (1983).
123. **Graves, H.B. et al.** Perceptibility and electrophysiological response of small birds to intense 60-Hz electric fields. *IEEE transactions on power apparatus and systems,* **PAS-97**: 1070–1073 (1978).
124. **Bankoske, J.W. et al.** *Ecological influence of electric fields.* Palo Alto, CA, Electric Power Research Institute, 1976 (EPRI Report No. EX-178).
125. **de Lorge, J.** *A psychobiological study of rhesus monkeys exposed to extremely low-frequency low-intensity magnetic fields.* Springfield, VA, National Technical Information Service, 1974 (USN Report NAMRL-1203).
126. **Hjeresen, D.L. et al.** Effects of 60-Hz electric fields on avoidance behavior and activity of rats. *Bioelectromagnetics,* **1**: 299–312 (1980).
127. **Hjeresen, D.L. et al.** A behavioral response of swine to 60-Hz electric field. *Bioelectromagnetics,* **2**: 443–451 (1982).
128. **Gavalas, R.J. et al.** Effect of low-level, low-frequency electric fields on EEG and behavior in *Macaca nemestrina. Brain research,* **18**: 491–501 (1970).
129. **Gavalas-Medici, R. & Day-Magdaleno, S.R.** Extremely low frequency, weak electric fields affect schedule-controlled behavior of monkeys. *Nature,* **261**: 256–259 (1976).
130. **Andrianova, L.A. & Smirnova, N.P.** [Motor activity of muscles in a magnetic field of varying intensity]. *Kosmičeskaja biologija i aviakosmičeskaja medicina,* **11**: 54–58 (1977) (in Russian).

131. **Persinger, M.A.** Open-field behavior in rats exposed prenatally to a low intensity-low frequency, rotating magnetic field. *Developmental psychobiology,* **2**: 168–171 (1969).

132. **Persinger, M.A. & Foster, W.S., IV.** ELF rotating magnetic fields: prenatal exposures and adult behavior. *Archiv für Meteorologie, Geophysik und Bioklimatologie, Ser. B,* **18**: 363–369 (1970).

133. **Smith, R.F. & Justesen, D.R.** Effects of a 60-Hz magnetic field on activity levels of mice. *Radio science,* **12**: 279–285 (1977).

134. **Creim, J.A. et al.** Exposure to 30-gauss magnetic fields does not cause avoidance behavior in rats. *In:* Anderson, L.E., ed. *Interaction of biological systems with static and ELF electric and magnetic fields. Proceedings of the 23rd Annual Hanford Life Sciences Symposium.* Springfield, VA, National Technical Information Service, 1987.

135. **Davis, H.P. et al.** Behavioral studies with mice exposed to DC and 60-Hz magnetic fields. *Bioelectromagnetics,* **5**: 147–164 (1984).

136. **Wever, R.** Influence of electric fields on some parameters of circadian rhythms in man. *In:* Menaber, M., ed. *Biochronometry.* Washington, DC, National Academy of Sciences, 1971, pp. 117–132.

137. **Dowse, H.B.** The effects of phase shifts in a 10 Hz electric field cycle on locomotor activity rhythm of *Drosophila melanogaster. Journal of interdisciplinary cycle research,* **13**: 257–264 (1982).

138. **Duffy, P.M. & Ehret, C.F.** Effects of intermittent 60-Hz electric field exposure: circadian phase shifts, splitting, torpor, and arousal responses in mice. *In: Abstracts, 4th Annual Scientific Session, Bioelectromagnetics Society, Los Angeles, CA, 28 June–2 July 1982.* Gaithersburg, MD, Bioelectromagnetics Society, 1982.

139. **Ehret, C.F. et al.** Biomedical effects associated with energy transmission systems: effects of 60-Hz electric fields on circadian and ultradian physiological and behavioral functions in small rodents. *In: Annual Report, US Department of Energy, Division of Electric Energy Systems, Washington, DC, 1980.*

140. **Wilson, B.W. et al.** Chronic exposure to 60-Hz electric fields: effects on pineal function in the rat. *Bioelectromagnetics,* **2**: 371–380 (1981).

141. **Wilson, B.W. et al.** (Erratum) Chronic exposure to 60-Hz electric fields: effects on pineal function in the rat. *Bioelectromagnetics,* **4**: 293 (1983).

142. **Anderson, L.E. et al.** Pineal gland response in animals exposed to 60-Hz electric fields. *In: Abstracts, 4th Annual Scientific Session, Bioelectromagnetics Society, Los Angeles, CA, 28 June–2 July 1982.* Gaithersburg, MD, Bioelectromagnetics Society, 1982.

143. **Kavaliers, M. et al.** Magnetic fields abolish the enhanced nocturnal analgesic response to morphine in mice. *Physiology and behavior,* **32**: 261–264 (1984).

144. **Semm, P.** Neurobiological investigations on the magnetic sensitivity of the pineal gland in rodents and pigeons. *Comparative biochemistry and physiology,* **76A**: 683–689 (1983).

145. **Welker, H.A. et al.** Effects of an artificial magnetic field on serotonin N-acetyltransferase activity and melatonin content of the rat pineal. *Experimental brain research,* **50**: 426–432 (1983).

146. **Sulzman, F.M.** Effects of electromagnetic fields on circadian rhythms. *In: Assessments and viewpoints on the biological and human health effects of extremely low frequency electromagnetic fields.* Arlington, VA, American Institute of Biological Sciences, 1985, pp. 337–350.

147. **Dumansky, Y.D. et al.** [Hygiene assessment of an electromagnetic field generated by high-voltage power transmission lines]. *Gigiena i sanitarija,* No. 8, pp. 19–23 (1976) (in Russian).

148. **Marino, A.A. & Becker, R.O.** Biological effects of extremely low-frequency electric and magnetic fields: a review. *Physiological chemistry and physics,* **9**: 131–147 (1977).

149. **Groza, P. et al.** Blood and urinary catecholamine variations under the action of a high voltage electric field. *Physiologie,* **15**: 139–144 (1978).

150. **Mose, J.R.** Problems of housing quality. *Zentralblatt für Bakteriologie, Parasitenkunde, Infektionskrankheiten und Hygiene. I. Abt. Orig. B,* **166**: 292–304 (1978).

151. **Fischer, G. et al.** Übt das netzfrequente Wechselfeld zentrale Wirkungen aus? [Does a 50-cycle alternating field cause central nervous system effects?] *Zentralblatt für Bakteriologie, Parasitenkunde, Infektionskrankheiten und Hygiene, I. Abt. Orig. B,* **166**: 381–385 (1978).

152. **Portet, R.T. & Cabanes, J.** Development of young rats and rabbits exposed to a strong electric field. *Bioelectromagnetics,* **9**: 95–104 (1988).

153. **Portet, R.T. et al.** Développement du jeune lapin soumis à un champ électrique intense. *Comptes rendus des séances de la Société de biologie,* **178**: 142–152 (1984).

154. **Babovich, R.D. & Kozyarin, I.P.** [Effects of low frequency electrical fields (50 Hz) on the body]. *Gigiena i sanitarija,* No. 1, pp. 1–15 (1979) (in Russian).

155. **Kozyarin, I.P.** [Effects of low frequency (50-Hz) electric fields on animals of different ages]. *Gigiena i sanitarija,* No. 8, pp. 18–19 (1981) (in Russian).

156. **Hackman, R.M. & Graves, H.B.** Corticosterone levels in mice exposed to high intensity electric fields. *Behavioral and neural biology,* **32**: 201–213 (1981).

157. **Marino, A.A. et al.** *In vivo* bioelectrochemical changes associated with exposure to extremely low frequency electric fields. *Physiological chemistry and physics,* **9**: 433–441 (1977).

158. **Free, M.J. et al.** Endocrinological effects of strong 60-Hz electric fields on rats. *Bioelectromagnetics,* **2**: 105–121 (1981).

159. **Portet, R.T.** Etude de la thyroide et des surrénales de rats exposés chroniquement à un champ électrique intense. *Comptes rendus des séances de la Société de biologie,* **177**: 290–295 (1983).

160. **Gann, D.W.** *Biological effects of exposure to high voltage electric fields.* Palo Alto, CA, Electric Power Research Institute, 1976 (Report RP98-02).

161. **Quinlan, W.J. et al.** Neuroendocrine parameters in the rat exposed to 60-Hz electric fields. *Bioelectromagnetics,* **6**: 381–389 (1985).
162. **Carter, J.H. & Graves, H.B.** *Effects of high intensity AC electric fields on the electroencephalogram and electrocardiogram of domestic chicks: literature review and experimental results.* University Park, PA, Pennsylvania State University, 1975.
163. **Phillips, R.D. et al.** *Biological effects of 60-Hz electric fields on small laboratory animals. Annual report.* Washington, DC, Department of Energy, 1978 (DOE Report No. HCP/T1830-3).
164. **Hansson, H.-A.** Purkinje nerve cell changes caused by electric fields: ultrastructural studies on long term effects on rabbits. *Medical biology,* **59**: 103–110 (1981).
165. **Hansson, H.-A.** Lamellar bodies in Purkinje nerve cells experimentally induced by electric field. *Experimental brain research,* **216**: 187–191 (1981).
166. **Silny, J.** *Effects of electric fields on the human organism.* Cologne, Institut zur Erforschung elektrischer Unfälle, Medizinisch — Technischer Bericht, 1979.
167. **Blanchi, C. et al.** Exposure of mammals to strong 50-Hz electric fields: effect on the electrical activity of the heart and brain. *Archivio di fisiologia,* **70**: 33–34 (1973).
168. **Takashima, S. et al.** Effects of modulated RF energy on the EEG of mammalian brains: effects of acute and chronic irradiations. *Radiation and environmental biophysics,* **16**: 15–27 (1979).
169. **Silny, J.** The influence thresholds of the time-varying magnetic field in the human organism. *In:* Bernhardt, J., ed. *Proceedings of the Symposium on Biological Effects of Static and ELF Magnetic Fields. Neuherberg, May 13–15 1985.* Munich, MMV Medizin Verlag, 1985 (BGA-Schriftenreihe).
170. **Jaffe, R.A. et al.** Perinatal exposure to 60-Hz electric fields: effects on the development of the visual-evoked response in the rat. *Bioelectromagnetics,* **4**: 327–339 (1983).
171. **Jaffe, R.A. et al.** Chronic exposure to a 60-Hz electric field: effects on synaptic transmission and the peripheral nerve function in the rat. *Bioelectromagnetics,* **1**: 131–147 (1980).
172. **Jaffe, R.A. et al.** Chronic exposure to a 60-Hz electric field: effects on neuromuscular function in the rat. *Bioelectromagnetics,* **2**: 277–239 (1981).
173. **Bawin, S.M. et al.** Influences of sinusoidal electric fields on excitability in the rat hippocampal slice. *Experimental brain research,* **323**: 227–237 (1984).
174. **Anderson, L.E.** Interaction of ELF electric and magnetic fields with neural and neuroendocrine systems. *In: Biological and human health effects of extremely low frequency electromagnetic fields.* Arlington, VA, American Institute of Biological Sciences, 1985.
175. **Mahlum, D.D. et al.** *Developmental toxicology of energy-related pollutants.* Springfield, VA, National Technical Information Service, 1978 (NTIS Conf. No. 771017).

176. **Chernoff, N.** Reproductive and developmental effects in mammalian and avian species from exposure to ELF fields. *In: Biological and human health effects of extremely low frequency electromagnetic fields.* Arlington, VA, American Institute of Biological Sciences, 1985, pp. 227–240.

177. **Sikov, M.R. et al.** Evaluation of reproduction and development in Hanford Miniature Swine exposed to 60-Hz electric fields. *In:* Anderson, L.E., ed. *Interaction of biological systems with static and ELF electric and magnetic fields. Proceedings of the 23rd Annual Hanford Life Sciences Symposium.* Springfield, VA, National Technical Information Service, 1987.

178. **Krueger, W.F.** Influence of low-level electric and magnetic fields on the growth of young chickens. *Biomedical scientific instruments,* **9**: 183–186 (1972).

179. **Reed, T.J. & Graves, H.B.** *Effects of 60-Hz electric fields on embryo and chick development, growth, and behavior. Vol. l. Final report.* Palo Alto, CA, Electric Power Research Institute, 1984 (EPRI Project 1064-1).

180. **Veicsteinas, A. et al.** Effect of 50-Hz electric field exposure on growth rate of chicks. *In:* Anderson, L.E. et al., ed. *Interaction of biological systems with static and ELF electric and magnetic fields. Proceedings of the 23rd Annual Hanford Life Sciences Symposium.* Springfield, VA, National Technical Information Service, 1987.

181. **Graves, H.B. et al.** Responses of domestic chicks to 60-Hz electrostatic fields. *In:* Mahlum, D.D., ed. *Developmental toxicology of energy-related pollutants.* Springfield, VA, National Technical Information Service, 1978, pp. 317–329.

182. **Delgado, J.M.R. et al.** Embryological changes induced by weak, extremely low frequency electromagnetic fields. *Journal of anatomy,* **134**: 533–551 (1982).

183. **Ubeda, A. et al.** Pulse shape of magnetic field influences chick embryogenesis. *Journal of anatomy,* **137**: 513–536 (1983).

184. **Cerretelli, P. et al.** 1000-kV project: research on the biological effects of 50-Hz electric fields in Italy. *In: Biological effects of extremely-low-frequency electromagnetic fields. 18th Annual Hanford Life Sciences Symposium, Richland, WA, 16–18 October, 1979.* Springfield, VA, National Technical Information Service, 1979, pp. 241–257.

185. **Fam, W.Z.** Long-term biological effects of very intense 60-Hz electric fields on mice. *IEEE transactions on biomedical engineering,* **BME-27**: 376–381 (1980).

186. **Pafkova, H.** Possible embryotropic effect of the electric and magnetic component of the field of industrial frequency. *Pracovni lekarstvi,* **37**: 153–158 (1985).

187. **Rommereim, D.N. et al.** Reproduction and development in rats chronically exposed to 60-Hz fields. *In:* Anderson, L.E. et al., ed. *Interaction of biological systems with static and ELF electric and magnetic fields. Proceedings of the 23rd Annual Hanford Life Sciences Symposium.* Springfield, VA, National Technical Information Service, 1987.

188. **Savin, B.M. & Sokolova, I.P.** Effect of 50-Hz electric fields on embryogenesis and postnatal development of SHK white mice in a system of

236

generations. *In: Proceedings of US/USSR Workshop on Physical Factors — Microwaves and Low Frequency Fields, May 1985.* Research Triangle Park, NC, National Institute of Environmental Health Science, 1985.

189. **Seto, J.J.** *Pilot Study of 60-Hz electric fields induced biological subtle effects. Final report.* New Orleans, LA, Louisiana Power and Light Co, 1979 (EBRL Report TR- 27).

190. **Sikov, M.R. et al.** Studies on prenatal and postnatal development in rats exposed to 60-Hz electric fields. *Bioelectromagnetics,* **5**: 101–112 (1984).

191. **Andrienko, L.E.** [Effect of electromagnetic fields at industrial frequencies on reproductive function under experimental conditions]. *Gigiena i sanitarija,* No. 6, pp. 22–25 (1977) (in Russian).

192. **Knickerbocker, G.G. et al.** Exposure of mice to strong AC electric field — an experimental study. *IEEE transactions on power apparatus and systems,* **PAS-86**: 498–505 (1967).

193. **Marino, A.A. et al.** The effects of continuous exposure to low frequency electric fields on three generations of mice: a pilot study. *Experientia,* **32**: 565–566 (1976).

194. **Marino, A.A. et al.** Power frequency electric field induces biological changes in successive generations of mice. *Experientia,* **36**: 309–311 (1980).

195. **Ossenkopp, K.-P. et al.** Prenatal exposure to an extremely low frequency-low intensity rotating magnetic field and increases in thyroid and testicle weight in rat. *Developmental psychobiology,* **5**: 275–285 (1972).

196. **Persinger, M.A. & Pear, J.J.** Prenatal exposure to an ELF-rotating magnetic field and subsequent increase in conditioned suppression. *Developmental psychobiology,* **5**: 269–274 (1972).

197. **McClanahan, B.J. & Phillips, R.D.** The influence of electric field exposure on bone growth and fracture repair in rats. *Bioelectromagnetics,* **4**: 11–20 (1983).

198. **Marino, A.A. et al.** Fracture healing in rats exposed to extremely low-frequency electric fields. *Clinical orthopaedics,* **145**: 239–244 (1979).

199. **Hilton, D.I. & Phillips, R.D.** Cardiovascular response of rats exposed to 60-Hz electric fields. *Bioelectromagnetics,* **1**: 55–64 (1980).

200. **Cerretelli, P. & Malaguti, C.** Research carried on in Italy by ENEL on the effects of high voltage electric fields. *Revue générale de l'électricité,* Special Issue, pp. 65–74 (1976).

201. **Mathewson, N.S. et al.** *Extremely low frequency (ELF) vertical electric field exposure of rats: a search for growth, food consumption and blood metabolite alterations.* Springfield, VA, National Technical Information Service, 1977.

202. **Ragan, H.A. et al.** Hematologic and serum chemistry studies in rats and mice exposed to 60-Hz electric fields. *Bioelectromagnetics,* **4**: 79–90 (1983).

203. **Ragan, H.A. et al.** Clinical pathologic evaluations in rats and mice chronically exposed to 60-Hz electric fields. *In: Biological effects of extremely-low-frequency electromagnetic fields. 18th Annual Hanford Life Sciences Symposium, Richland, WA, 16–18 October 1979.* Springfield, VA, National Technical Information Service, 1979, pp. 297–325.

204. **Graves, H.B. et al.** Biological effects of 60-Hz alternating-current fields: a Cheshire cat phenomenon? *In: Biological effects of extremely-low-frequency electromagnetic fields. 18th Annual Hanford Life Sciences Symposium, Richland, WA, 16–18 October 1979.* Springfield, VA, National Technical Information Service, 1979, pp. 184–197.

205. **Tarakhovsky, M.L. et al.** Effects of constant and variable magnetic fields on some indices of physiological function and metabolic processes in albino rats. *Fiziologičnyj žurnal,* **17**: 452–459 (English translation by Joint Publications Research Service, Arlington, VA, USA, 1974).

206. **de Lorge, J.** *Operant behavior of rhesus monkeys in the presence of low-frequency low-intensity magnetic and electric fields: experiment 1.* Springfield, VA, National Technical Information Service, 1972 (USN Report NAMRL-1155).

207. **de Lorge, J.** *Operant behavior of rhesus monkeys in the presence of low-frequency low-intensity magnetic and electric fields: experiment 2.* Springfield, VA, National Technical Information Service, 1973 (USN Report NAMRL-1179).

208. **de Lorge, J.** *Operant behavior of rhesus monkeys in the presence of low-frequency low-intensity magnetic and electric fields: experiment 3.* Springfield, VA, National Technical Information Service, 1973 (USN Report NAMRL-1196).

209. **Fam, W.A.** Biological effects of 60-Hz magnetic field on mice. *IEEE transactions on magnetics,* **MAG-17**: 1510–1513 (1981).

210. **Morris, J.E. & Phillips, R.D.** Effects of 60-Hz electric fields on specific humoral and cellular components of the immune system. *Bioelectromagnetics,* **3**: 341–348 (1982).

211. **Morris, J.E. & Phillips, R.D.** (Erratum) Effects of 60-Hz electric fields on specific humoral and cellular components of the immune system. *Bioelectromagnetics,* **4**: 294 (1983).

212. **Morris, J.E. & Ragan, H.A.** Immunological studies with 60-Hz electric fields. *In: Biological effects of extremely-low-frequency electromagnetic fields. 18th Annual Hanford Life Sciences Symposium, Richland, WA, 16–18 October 1979.* Springfield, VA, National Technical Information Service, 1979, pp. 326–334.

213. **LeBars, H. et al.** Les effets biologiques des champs electriques — effets sur le rat, la souris, le cobaye. *Recueil de médecine vétérinaire,* **159**: 823–837 (1983).

214. **Lyle, D.B. et al.** Suppression of T-lymphocyte cytotoxicity following exposure to sinusoidally amplitude-modulated fields. *Bioelectromagnetics,* **4**: 281–292 (1983).

215. **Conti, P. et al.** Reduced mitogenic stimulation of human lymphocytes by extremely low frequency electromagnetic fields. *FEBS letters,* **162**: 156–160 (1983).

238

216. **Mizushima, Y. et al.** Effects of magnetic field on inflammation. *Experientia,* **21**: 1411–1412 (1975).
217. **Odintsov, Y.N.** [The effect of a magnetic field on the natural resistance of white mice to *Listeria* infection]. *Tomsk voprosy epidemiologii, mikrobiologii i immunologii,* **16**: 234–238 (1965) (in Russian).
218. **Frazier, M.E. et al.** Effects of 60-Hz electric fields on CHO-K1 cells. *In: Abstracts, 4th Annual Scientific Session, Bioelectromagnetics Society, Los Angeles, CA, 28 June–2 July 1982.* Gaithersburg, MD, Bioelectromagnetics Society, 1982.
219. **Mittler, S.** *Low frequency electromagnetic radiation and genetic aberrations.* Springfield, VA, National Technical Information Service, 1972.
220. **Phillips, R.D. et al.** *Biological effects of high strength electric fields on small laboratory animals.* Washington, DC, US Department of Energy, Division of Electric Energy Systems, 1979 (DOE/TIC-10094).
221. **Asanova, T.P. & Rakov, A.I.** [The state of health of persons working in electric fields of outdoor 400- and 500-kV switchyards]. *In: [Hygiene of labour and professional diseases],* No. 5, 1966. English translation issued as Special Publication 10 of the IEEE Power Engineering Society, Piscataway, NJ, 1975.
222. **Korobkova, V.A. et al.** Influence du champ électrique dans les postes à 500 kV et 750 kV sur les équipes d'entretien et les moyens de leur protection [Influence of the electric field in 500- and 750-kV switchyards on maintenance staff and means for its protection]. *In: International Conference on Large High Tension Electric Systems,* Paris, CIGRE, 1972 (CIGRE Report No. 23-06).
223. **Wever, R.** ELF effects on human circadian rhythms. *In:* Persinger, M.A., ed. *ELF and VLF electromagnetic field effects.* New York, Plenum, 1974.
224. **Hauf, R.** Wirkung von 50-Hz-Wechselfeldern auf den Menschen [Effect of 50 Hz alternating fields on man]. *Elektrotechnische Zeitschrift,* **B26**: 318–320 (1974).
225. **Hauf, R.** Einfluss elektromagnetisches Felder auf den Menschen [Effect of electromagnetic fields on humans]. *Elektrotechnische Zeitschrift,* **B28**: 181–183 (1976).
226. **Hauf, R.** Influence of 50 Hz alternating electric and magnetic fields on human beings. *Revue générale de l'électricité,* Special issue, pp. 31–49 (1976).
227. **Kuhne, B.** Einfluß elektrischer Felder auf lebende Organismen [The effect of electric fields on living organisms]. *Der praktische Arzt,* **15**: (1978).
228. **Szuba, M. & Nosol, B.** [Duration of conscious reaction in those exposed to electric field of 50 Hz frequency]. *Medycyna pracy,* **36**: 21–26 (1985) (in Polish).
229. **Bauchinger, M. et al.** Analysis of structural chromosome changes and SCE after occupational long-term exposure to electric and magnetic fields from 380-kV systems. *Radiation and environmental biophysics,* **19**: 235–238 (1981).

239

230. **Hauf, R.** Untersuchungen zur Wirkung energietechnischer Felder auf den Menschen [Investigations on the effect of electromagnetic fields at power frequencies on man]. *Beitrage zur Ersten Hilfe und Behandlung von Unfallen durch electrischen Strom,* Heft 9/1981, Berichte der Forschungsstelle fur Elektropathologie, Freiburg, Federal Republic of Germany.

231. **Beischer, D.E. et al.** *Exposure of man to magnetic fields alternating at extremely low frequency.* Pensacola, FL, Naval Aerospace Medical Research Laboratory, 1973 (USN Report NAMRL-1180).

232. **Mantell, B.** *Untersuchungen über die Wirkung eines magnetischen Wechselfeldes 50 Hz auf den Menschen* [Investigations into the effects on man of an alternating magnetic field at 50 Hz]. Thesis, Freiburg University, 1975.

233. **Sander, R. et al.** Laboratory studies on animals and human beings exposed to 50-Hz electric and magnetic fields. *In: International Conference on Large High Voltage Electrical Systems (abstracts).* Paris, Number 36-01, 1–9 September 1982.

234. **Silny, J.** Influence of low-frequency magnetic field on the organism. *Proceedings of the 4th Symposium on Electromagnetic Compatibility, Zurich, 10–12 March, 1981,* pp. 175–180.

235. **Lovsund, P. et al.** Influence on vision of extremely low frequency electromagnetic fields. *Acta ophthalmologica,* **57**: 812–821 (1979).

236. **Lovsund, P. et al.** Magnetophosphenes: a quantitative analysis of thresholds. *Medical & biological engineering & computing,* **18**: 326–334 (1980).

237. **Bassett, C.A.L.** Pulsing electromagnetic fields: a new approach to surgical problems. *In:* Buchwald, H. & Varco, R.L., ed. *Metabolic surgery.* New York, Grune & Stratton, 1978.

238. **Bassett, C.A.L.** Acceleration of fracture repair by electromagnetic fields: a surgically noninvasive method. *Annals of the New York Academy of Sciences,* **238**: 242–262 (1974).

239. **Sedel, L. et al.** Résultats de la stimulation par champ électromagnétique de la consolidation des pseudarthroses. *Revue de chirurgie orthopédique et réparatrice de l'appareil moteur,* **67**: 11–23 (1981).

240. **Tenforde, T.S.** Interaction of time-varying ELF magnetic fields with living matter. *In:* Polk, C. & Postow, E., ed. *Handbook of biological effects of electromagnetic fields.* Boca Raton, FL, CRC Press, 1986.

241. **Cain, C.S. et al.** Pulsed electromagnetic field effects on PTH stimulated cAMP accumulation and bone resorption in mouse calvariae. *In:* Anderson, L.E., ed. *Interaction of biological systems with static and ELF electric and magnetic fields. Proceedings of the 23rd Annual Hanford Life Sciences Symposium.* Springfield, VA, National Technical Information Service, 1987.

242. **Fitton-Jackson, S. & Bassett, C.A.L.** The response of skeletal tissues to pulsed magnetic fields. *In:* Richards, R.J. & Rajan, K.T. *Tissue culture in medical research.* New York, Pergamon, 1980.

243. **Norton, L. A.** Effects of pulsed electromagnetic field on a mixed chondroblastic tissue culture. *Clinical orthopaedics and related research,* **167**: 280–290 (1982).

244. **Fotopoulos, S.S. et al.** Effects of 60-Hz fields on human neuroregulatory, immunologic, hematologic, and target organ activity. *In:* Anderson, L.E. et al., ed. *Interaction of biological systems with static and ELF electric and magnetic fields. Proceedings of the 23rd Annual Hanford Life Sciences Symposium.* Springfield, VA, National Technical Information Service, 1987.

245. **Graham, C.** A double-blind evaluation of 60-Hz field effects on human performance, physiology, and subjective state. *In:* Anderson, L.E. et al., ed. *Interaction of biological systems with static and ELF electric and magnetic fields. Proceedings of the 23rd Annual Hanford Life Sciences Symposium.* Springfield, VA, National Technical Information Service, 1987.

246. **Bonnell, J.A. et al.** Can induced 50 Hz body currents affect mental functions? *In: International Conference on Electric and Magnetic Fields in Medicine and Biology, London, 4–5 December 1985.*

247. **Stollery, B.T.** Human exposure to 50 Hz electric currents. *In:* Anderson, L.E. et al., ed. *Interaction of biological systems with static and ELF electric and magnetic fields. Proceedings of the 23rd Annual Hanford Life Sciences Symposium.* Springfield, VA, National Technical Information Service, 1987.

248. **Kouwenhoven, W.B. et al.** Medical evaluation of man working in ac electric fields. *IEEE transactions on power apparatus and systems,* **PAS-86**: 506–511 (1967).

249. **Singewald, M.L. et al.** Medical follow-up study on high voltage linemen working in ac electric fields. *IEEE transactions on power apparatus and systems,* **PAS-92**: 1307–1309 (1973).

250. **Filippov, V.** Der Einfluss von elektrischen Wechselfeldern auf den Menschen [The effect of alternating electric fields on man]. *In: Second International Colloquium für die Verhutung von Arbeitsunfällen und Berufskrankheiten durch Elektrizität.* Cologne, Berufsgenössenschaft der Feinmechanik und Elektrotechnik, 1972, pp. 170–177.

251. **Malboysson, E.** Medical control of men working within electromagnetic fields. *Révue générale de l'électricité,* Special issue, pp. 75–80 (1976).

252. **Dumansky, Y.D. et al.** [Effects of low-frequency (50 Hz) electromagnetic field on functional state of the human body]. *Gigiena i sanitarija,* No. 12, pp. 32–35 (1977) (in Russian).

253. **Busby, K. et al.** *A field survey of farmer experience with 765 kV transmission lines.* Albany, NY, Agricultural Resources Commission, 1974.

254. **Strumza, M.V.** Influence sur la santé humaine de la proximité des conducteurs de l'electricité à haute tension. *Archives des maladies professionnelles, de médecine du travail et de sécurité sociale,* **31**: 269–276 (1970).

255. **Issel, I. et al.** Tauglichkeits- und Eignungsuntersuchungen an Elektromonteuren — Erläuterung an einer neuen Untersuchungsanweisung

[Occupational, medical and psychological examinations on electrical fitters — comment on a new regulation]. *Deutsche Gesundheitswesen,* **32**: 1526–1531 (1977).

256. **Knave, B. et al.** Long-term exposure to electric fields: a cross-sectional epidemiologic investigation of occupationally exposed industrial workers in high-voltage substations. *Electra,* **65**: 41–54 (1979).

257. **Broadbent, D.E. et al.** Health of workers exposed to electric fields. *British journal of industrial medicine,* **42**: 75–84 (1985).

258. **Nordstrom, S. et al.** Reproductive hazards among workers at high voltage substations. *Bioelectromagnetics,* **4**: 91–101 (1983).

259. **Nordenson, I. et al.** Clastogenin effects in human lymphocytes of power frequency electric fields: *in vivo* and *in vitro* studies. *Radiation and environmental biophysics,* **23**: 191–201 (1984).

260. **Nordstrom, S. et al.** Genetic and reproductive hazards in high voltage substations. *In:* Anderson, L.E. et al., ed. *Interaction of biological systems with static and ELF electric and magnetic fields. Proceedings of the 23rd Annual Hanford Life Sciences Symposium.* Springfield, VA, National Technical Information Service, 1987.

261. **Wertheimer, N. & Leeper, E.** Electrical wiring configuration and childhood cancer. *American journal of epidemiology,* **109**: 273–284 (1979).

262. **Wertheimer, N. & Leeper, E.** Adult cancer related to electrical wires near the home. *International journal of epidemiology,* **11**: 345–355 (1982).

263. **Fulton, J.P. et al.** Electrical wiring configurations and childhood leukemia in Rhode Island. *American journal of epidemiology,* **111**: 292–296 (1980).

264. **Tomenius, L. et al.** 50-Hz Electromagnetic environment and the incidence of childhood tumors in Stockholm county. *Bioelectromagnetics,* **7**: 191–207 (1986).

265. **Lin, R.S. et al.** Occupational exposure to electromagnetic fields and the occurrence of brain tumors. *Journal of occupational medicine,* **27**: 413–419 (1984).

266. **Spitz, M. & Johnson, C.** Neuroblastoma and paternal occupation: a case-control analysis. *American journal of epidemiology,* **121**: 924–929 (1985).

267. **Milham, S., Jr.** Mortality from leukemia in workers exposed to electrical and magnetic fields. *New England journal of medicine,* **307**: 249 (1982).

268. **McDowall, M.E.** Leukemia mortality in electrical workers in England and Wales. *Lancet,* **1**: 246 (1983).

269. **Wright, W.E. et al.** Leukemia in workers exposed to electrical and magnetic fields. *Lancet,* **2**: 1160–1161 (1982).

270. **Coleman, M. et al.** Leukemia incidence in electrical workers. *Lancet,* **1**: 982–983 (1983).

271. **Bridges, J.E. & Frazier, M.J.** *The effects of 60 Hz electric and magnetic fields on implanted cardiac pacemakers.* Palo Alto, CA, Electric Power Research Institute, 1979 (EPRI Final Report EA-1174).

272. **Butrous, G.S. et al.** The effect of power-frequency high-intensity electric fields on implanted cardiac pacemakers. *Pacing and clinical electrophysiology,* **6**: 1282–1292 (1983).

273. **Jenkins, B.M. & Woody, J.A.** Cardiac pacemaker responses to power frequency signals. *In: IEEE International Symposium on Electric and Magnetic Compatibility.* New York, Institute of Electrical and Electronics Engineers, 1978, pp. 273–277 (IEEE Report No. chl304-5/78-0000-0273).

274. **Griffen, J.** The effects of ELF electric and magnetic fields on artifical cardiac pacemakers. *In: Assessments and viewpoints on the biological and human health effects of extremely low frequency electromagnetic fields.* Arlington, VA, American Institute of Biological Sciences, 1985, pp. 173–184.

243

6

Ultrasound

C.R. Hill & G. ter Haar

Revised by G. ter Haar & C.R. Hill

CONTENTS

	Page
Introduction	246
Physical properties	246
Sound speed	247
Transmission through interfaces	248
Attenuation	248
Absorption and heating	250
Structure of sound fields	251
Measurement of sound fields	254
Cavitation	258
Applications	260
Processing applications	260
Underwater target location	261
Testing and measurement applications	261
Medical diagnostic methods	261
Ultrasound physiotherapy	262
Ultrasound hyperthermia	263
Surgical techniques	263
Applications in dentistry	264
Biological action	265
Characterization of ultrasound exposures used in the bioeffects literature	265
Isolated biomolecules	265
Isolated cells and cell cultures	266
Multicellular organisms	270
Human epidemiology	272
Effects of industrial and airborne ultrasound	273
Criteria for appropriate use	274
Surgery and cancer therapy	275
Applications in physical medicine	275
Medical diagnosis and investigation	276
Airborne ultrasound	278

Protection measures 279
Conclusions and recommendations 280
    Conclusions 280
    Recommendations 281
References 281

INTRODUCTION

Ultrasound is a form of mechanical energy that has found increasingly widespread application over the past 50 years. Many of its uses entail exposure of human beings, either incidentally or, as in the case of medical applications of ultrasound, as an essential part of the procedure. The fact of such exposures inevitably raises the question of the possible existence of any corresponding hazard to the individual exposed.

This is a question that has been the subject of investigation for a long time (1,2). Since the publication of the first edition of this book several detailed reviews of some aspects of the subject have appeared (3-7) but it remains the case that there is still no thoroughly satisfactory answer. This chapter is designed to provide a concise review of the principal lines of evidence that constitute our knowledge in this field, and to outline some of the scientific approaches that may be helpful in achieving refinement of this knowledge.

The physical nature of ultrasound, the physical laws that determine its behaviour, and the principal methods that are available for measuring the characteristics of ultrasound fields are first described in outline, after which the main practical applications of ultrasound are considered from the point of view of their potential for exposure of humans to ultrasound fields. The evidence that exists for the induction of biological change is considered, with emphasis on quantitative relationships between induced changes of a potentially damaging nature and the physical exposures by which these changes may be induced. Finally, guidelines are put forward for the appropriate use of ultrasound in various situations involving human exposure.

In choosing the references to be included in this chapter, those results that have been available for critical comment in refereed journals have been preferred. However, some reports are necessarily too recent to have been refereed, or are less easily available.

The original specifications of this review required that it should discuss both airborne and liquid-borne ultrasound phenomena. It will be seen, however, that both the scientific and practical protection aspects of these two categories of phenomena are considerably different and essentially separable.

PHYSICAL PROPERTIES

The term ultrasound refers to the set of mechanical vibration phenomena that occur in the frequency range above the upper frequency limit for human hearing, which may be taken as lying at about 16 kHz. The upper frequency limit is set by practical considerations in the generation of mechanical vibration and, with present technology, lies at about 100 GHz. Frequencies above 1 GHz are sometimes referred to as hypersound. However, as will be shown later, the practical upper limit for concern as to human exposure is about 20 MHz.

In contrast to electromagnetic radiation, ultrasound is fundamentally related to the existence of a material medium and cannot be sustained *in vacuo*. The physical nature of the vibration phenomena that may occur is strongly dependent on the phase of the medium concerned — solid, liquid or gas — and will also be greatly influenced by vibration frequency over the very wide spectrum referred to above (16 kHz–100 GHz). It is not appropriate here to describe these phenomena in detail; for this purpose, reference should be made to the specialized texts available *(8,9)*. It will be useful, however, to discuss in simplified outline the main factors that determine the physical nature of relevant phenomena, and thus determine their application and the consequent possibilities for exposure of humans.

The generation of ultrasound vibrations in a medium entails the existence of fluctuations in the values, at different points in the medium, of certain of its physical parameters. In particular, in considering any very small "particle" of the medium,[a] it will generally be the case that fluctuations occur in the displacement s (a vector quantity) of the particle from its resting position. Similar fluctuations will occur in the instantaneous values of the particle velocity v, the particle acceleration a, and the instantaneous pressure p. In the particular case that the fluctuation is sinusoidal in form, the field parameters s, v, a, p will exhibit maxima whose magnitudes, S, V, A, P, are referred to as displacement, velocity, acceleration and pressure "amplitudes". For certain simplified situations, such as that of a linear plane progressive wave disturbance, these four parameters are simply related and a measurement of the spatial distribution of any one of them may serve to characterize completely, or to map, the sound field.

Sound speed

An ultrasound disturbance that is induced in any medium will travel away from its source with a certain speed[b] that is characteristic of the medium. It is this speed c which, taken together with the frequency f of the vibration

---

[a] In this context a "particle" can be considered as a volume of the medium with dimensions small in relation to an ultrasound wavelength (see p. 248) but large in relation to molecular dimensions.

[b] In some texts sound speed is incorrectly referred to as sound "velocity". In physics, however, the term velocity is conventionally used to mean a component of movement in a specified direction (i.e. a vector) whereas sound speed is a property of a medium that is generally independent of direction.

induced, determines the wavelength λ ($= c/f$) of the waves that are propagated. For example, the propagation speed of ultrasound in most human soft tissues is approximately 1500 m/s, so that an induced frequency of 1 MHz leads to a wavelength of 1.5 mm. It follows from the fundamental physical principles of diffraction that ultrasound diagnostic procedures carried out, for example, at frequencies around 3 MHz have the potential for providing anatomical detail of the order of 0.5 mm. Values of the sound speed for some other media are given in Table 1.[a]

Transmission through interfaces

A second property of a medium, which is of fundamental importance in determining the extent to which ultrasound energy is transmitted through an interface separating two continuous isotropic media, is its characteristic acoustic impedance Z, which is defined as the product ρc, where ρ is the density of the medium and c is the speed of sound in the medium. The quantitative role of characteristic acoustic impedance can be seen from the expressions (for normal incidence of an ultrasound wave on a plane interface) for the fractions P_t/P_i and P_r/P_i respectively, of the incident sound pressure amplitude P_i transmitted through P_t and reflected back from the interface:

$$P_t/P_i = 2Z_2/(Z_1 + Z_2) \tag{1}$$

$$P_r/P_i = (Z_2-Z_1)/(Z_2 + Z_1) \tag{2}$$

where P_i is the acoustic pressure amplitude in the wave incident on the interface, P_r is that in the reflected wave, P_t is that in the transmitted wave, and Z_1 and Z_2 are the characteristic acoustic impedances of the media on either side of the interface.

Representative values of the characteristic acoustic impedance are listed in Table 1. From inspection of these values and the above relationships it will be seen, for example, that an interface between air and human tissue permits transmission of only some 3% of the incident pressure (or 0.1% of incident intensity, which varies as the square of pressure), the remainder being reflected. This is an extremely important fact because of its bearing on the possibilities for exposure of humans by airborne ultrasound or, more generally, by sources shielded by an air gap. Strong reflections also occur at bone/soft tissue interfaces, and this constitutes another of the important limitations on the accessibility of human anatomical regions to ultrasound investigation and exposure.

Attenuation

A third property of a medium that needs to be considered is its ability to attenuate a propagating wave. The pressure p_x in a plane progressive sound

---

[a] It may be noted that in unbounded solids, ultrasound may be propagated in two different modes: as compressional (longitudinal) waves, as in the case of liquids and gases, or in the form of shear (transverse) waves. The propagation speed for the two modes is generally different.

Table 1. Ultrasound properties of various media[a]

Medium	Sound speed, c (compression wave) (m/s)	Characteristic acoustic impedance, z (kg/(s·m²))	Attenuation coefficient α at 1 MHz (neper/m)[b]	Frequency dependence of attenuation coefficient, n
Air	330	4.50×10^2	15	2
Water	1500	1.50×10^6	0.02	2
Liver	1540	1.54×10^6	8	1.2
Fat	1440	1.40×10^6	5–25	1.4
Lens of eye	—	1.84×10^6	—	—
Blood	1520	1.61×10^6	2.3	1.3
Skeletal muscle	1520	1.70×10^6	10–35	1.1
Skull bone	3360	6.99×10^6	115	1.7
Perspex (Lucite)	2680	3.16×10^6	23	2

[a] Approximate room temperature values for illustrative purposes only. Accurate measurement of ultrasound attenuation in human tissues is difficult, and published data are not always reliable. Furthermore, actual values may show quite marked variability with such factors as temperature.

[b] To convert to "engineering" units of dB/cm, the values in this column should be multiplied by 0.087.

wave of initial pressure amplitude P_0, after travelling a distance x in any attenuating medium, is described by the relationship:

$$P_x = P_0 \cdot e^{-\alpha x} \tag{3}$$

where α is termed the attenuation coefficient of the medium (in nepers per unit propagation distance)[a] and e is the base of natural logarithms. If the decibel notation is used, the corresponding attenuation coefficient in dB per unit distance is equal to 8.686α (Table 1).

The attenuation phenomenon is important in the present context from two points of view. In the first place, at least part of the attenuation will in general be due to a process of absorption, in which propagating energy is transferred to the medium in the form of heat, with the possibility of causing permanent modification as a result of the corresponding rise in temperature. Second, the attenuating property of one region of a structure will evidently

[a] In WHO publications it is normal to use SI units, in which the unit of length is the metre (see p. xiv). In ultrasound practice and its associated literature, characteristic dimensions of interest are expressed in centimetres. For the convenience of the reader not yet accustomed to SI units, the following conversion factor applies: $10\,kW/m^2 = 1\,W/cm^2$. Similarly $10\,kJ/m^2 = 1\,J/cm^2$.

affect the extent to which propagating energy penetrates to deeper regions. An important point to note in both these connections is that ultrasound attenuation by a medium generally increases with increasing frequency in a manner that can be expressed (as an approximation over limited frequency ranges) in the form:

$$\alpha_f = \alpha_{f_0}(f/f_0)^n \tag{4}$$

where α_f is the attenuation coefficient of the medium at frequency f (MHz) and α_{f_0} is the attenuation coefficient of the medium at the reference frequency f_0. For many human soft tissues n has a value close to 1 in the low megahertz frequency region, so that attenuation is approximately proportional to frequency.

It should be noted that attenuation in human soft tissues is fairly high. It is this that corresponds, on the one hand, to the effective interactions that are exploited in ultrasound therapy (referred to below) and, on the other, to the existence of a practical upper limit of about 20 MHz (beyond which most ultrasound diagnostic techniques are invalidated over appreciable path lengths).

Absorption and heating

In most materials, including human tissues, a substantial part of the process of attenuation is that corresponding to the direct conversion of acoustic energy to heat, and generally referred to as absorption. Quantitatively it can be written that:

$$\alpha_f = \alpha_{af} + \alpha_{sf} \tag{5}$$

where α_{af} and α_{sf} are the absorption and scattering coefficients of the medium respectively, at frequency f, and where the process of scattering is one in which incident energy is deflected out of the volume of interest without conversion to heat.

The corresponding generation of heat within a small volume (the absorbed power) is given, in a plane travelling wave (and similar fields), by:

$$<\dot{h}> = 2I \cdot \alpha_{af} \text{ (W/m}^3) \tag{6}$$

where I is the ultrasound intensity incident on the volume. Thus, in the absence of any process of removal of heat, the resulting rate of rise of temperature in the volume will be:

$$\dot{T} = 2I \cdot \alpha_{af}/C_v \text{ (K/s)} \tag{7}$$

where C_v is the heat capacity of the medium per unit volume: approximately $4.2 \times 10^6 \text{ J} \cdot \text{m}^{-3} \cdot \text{K}^{-1}$ for human soft tissues.

Using the data in Table 1, and making the conservative assumption that most attenuation is due to absorption, it follows that the maximum initial rates of temperature rise in tissue, consequent on exposure to intensity I (W/m^2) at frequency f (in MHz) will be of the order of $3.8 \times 10^{-6} If^{1.2}$ K/s for soft tissues and approximately $55 \times 10^{-6} If^{1.7}$ K/s for bone (assuming heat capacity similar to that for soft tissue).

In practice, there will almost always be substantial heat removal from the exposed volume, which will set a limit to the maximum temperature

250

achieved in prolonged exposure. The quantitative factors affecting this dynamic equilibrium can, however, be quite variable. In an exposure of a relatively small fraction of the vascularized soft tissue volume in the human (e.g. using a 20 mm diameter beam, a typical situation in physical therapy) it is necessary to use spatial average, temporal average values of intensity of about 20 kW/m^2 at 0.75 MHz to achieve an equilibrium temperature of 43 °C (the approximate threshold for thermal tissue damage) anywhere in the tissue (10). Use of the same beam experimentally to heat a small animal such as a mouse may lead to much higher equilibrium temperatures, since a large fraction of the animal is exposed, leaving no effective heat sink. Absorption coefficients in bone and cartilage are substantially higher than those for most "soft" tissues and this, combined with their generally lower capacity for vascular removal of heat, is a reason for expecting them to constitute critical tissues for some forms of ultrasound exposure.[a]

The above discussion has assumed that the tissue volumes in question are essentially homogenous. Even if this is not the case, however, and specifically if the tissue contains inclusions of material of high attenuation but with dimensions small relative to typical beam diameters, it can be shown by calculation that the resulting absorbed heat will be conducted away sufficiently rapidly that no substantial hot spot will occur. A good systematic treatment of this and other aspects of the thermal action of ultrasound is given in NCRP Report No. 74 (5).

Structure of sound fields

Later sections of this chapter will demonstrate the importance of being able, in a logical and quantitative manner, to relate human ultrasound exposures to the available experimental evidence on induction of biological change. An important aspect of this is being able to carry out valid and relevant dosimetry. To lay the groundwork for discussion of such dosimetric principles it is necessary here to examine in some detail the nature and structure of sound fields.

Many of the sound fields encountered in relation to the actual exposure of humans, or in related biological experimentation, may be quite complex. They will often have strongly converging (i.e. focused) rather than plane wave fronts and propagation may be markedly nonlinear (11,12). It is nonetheless instructive to ignore these complications temporarily and to consider two idealized types: the plane progressive or travelling wave linear disturbance and the stationary wave field.

In the first of these, each of the four parameters listed above (s, v, a and p) fluctuates with its own amplitude which (for a lossless plane wave only) is the same at all points in the field. In the more complex fields (such as, for

---

[a] There is another way, however, in which bone can play a critical role in setting exposure limits. This arises from the fact that, due to an effect known as "mode conversion", an ultrasound compression wave propagating across a soft tissue/bone interface may give rise there to a component vibrating as a shear wave. The attenuation coefficient for shear wave propagation in liquids and soft tissues (not shown in Table 1) is extremely high, with the consequence that all the energy propagated in this mode will be absorbed in the layer very close to the bone surface.

example, occur very close to an extended sound source, the so-called "near field") the four parameters may vary in a complex relationship to each other. However, for many practical purposes measurement of any one of them, as a function of time and of three spatial coordinates, may adequately characterize a field; as will be seen below, the development of internationally agreed practice is moving towards measurement of instantaneous pressure, p, for this purpose.

The structure of an actual travelling wave field due to a finite "piston" source, as typically used in many medical applications, is illustrated in Fig. 1. This demonstrates the following features that are important in the present context: the complex spatial structure of the "near field" close to the transducer (Fig. 1B and 1C); the existence of a single axial spatial peak in instantaneous pressure (Fig. 1B and 1C); the complex pattern of time variation of pressure within a pulse period (Fig. 1A, part of which is due to effects of nonlinear propagation); and the asymmetry between positive and negative pressure excursions (Fig. 1A), which is again related to the non-linearity in propagation that arises in practice in high amplitude pulses of the kind used in many medical diagnostic machines. The spatial peak illustrated here is a common feature of both focused and unfocused fields, and its existence is an important factor in the interpretation of some of the biological data presented below and in the formulation of resulting recommendations. A further quantity that has been very commonly used hitherto in describing the magnitude of ultrasound fields (and that has been introduced above in the context of absorption and heating) is intensity: the flux of energy, per unit area, along the direction of propagation. In conditions of a linear, sinusoidal continuous travelling plane wave or similar field, such a description can be made in a manner that is entirely consistent with a description in terms of fundamental field parameters. Quantitatively, under these conditions, time average intensity is related to pressure amplitude by the relationship (cf. Fig. 1B):

$$I = P^2/2\rho c \tag{8}$$

However, under other conditions, and particularly those that may arise in pulse–echo diagnostic beams, as illustrated in Fig. 1, the relationship may become quite complex, particularly in the near field. It may in particular be noted here that the values of "derived intensity" indicated in Fig. 1B are calculated from measurements of pressure, using equation 8, and their validity in the near field is therefore limited.

Normal practice hitherto has nonetheless been to make frequent use of direct measurements of intensity or power, and emissions and exposures are almost invariably reported in these terms in the current literature, including that reviewed later in this chapter. Referring again to Fig. 1, there will in principle be some value of intensity attributable to the main axial peak of the field (i.e. the maximum intensity obtained across the beam profile); this is normally referred to as the spatial peak intensity, $I(sp)$. Equally there will be some value of spatial average intensity, $I(sa)$, which is the power, per unit cross-sectional area, that propagates through a given cross-section that is appropriately defined (e.g. arbitrarily as the surface area of the transmitting

252

Fig. 1. Representations of measured sound fields

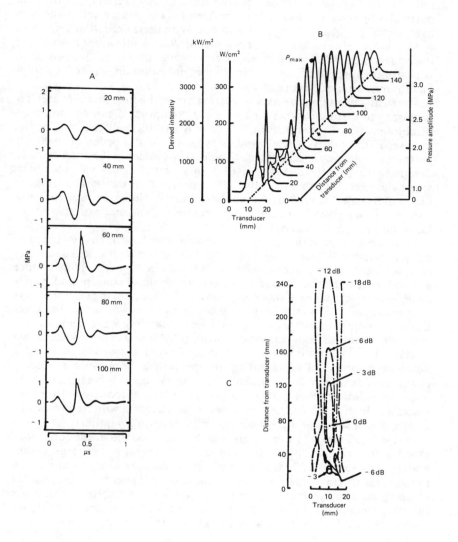

A. Temporal pressure waveforms measured at a succession of points along the axis of a pulse–echo diagnostic beam (3.5 MHz centre frequency, 13 mm diameter transducer nominally focused between 40 and 100 mm). Development of a nonlinear waveform is clearly apparent.

B. Lateral pressure profiles (with intensities derived from equation 8 also indicated) along a similar beam (1.5 MHz centre frequency, 20 mm diameter circular transducer).

C. Iso-intensity representation of the data from Fig. 1B.

transducer element). Fig. 1 demonstrates that the concepts of spatial peak intensity and spatial average intensity need to be carefully qualified, in particular by reference to the distance from the transducer and to the area over which spatial averaging is carried out. With these definitions, the ratio $I(sp):I(sa)$ is always greater than unity; if $I(sa)$ is an average over the transducer face, then for unfocused beams it generally has a value close to 3, while for the focused beams used in some high-resolution diagnostic equipment it may be of the order of 100.

Intensity averaging can also be carried out in the time domain, and it is necessary to distinguish between time (or "temporal") average and time peak intensities, $I(ta)$ and $I(tp)$. The ratio $I(tp):I(ta)$ will generally be the inverse of the "duty factor" and is typically of the order of 1000 for pulse–echo diagnostic equipment (Table 2). Commonly it is necessary to designate whether the quantity cited is an average or peak value for both space and time; for example, "$I(sp,ta)$" represents the temporal average at the spatial peak. Definition of a number of specific intensity parameters in current use are given in a standard prepared by the American Institute of Ultrasound in Medicine (AIUM) *(22)*.

The second type of field, the pure standing wave field, is characterized by a spatial pattern of "nodes" and "antinodes" for each of the field quantities (s, v, a, p) and by zero net time-averaged flux of energy across any surface within the field. Such a stationary field is an idealized concept and, partly because of the existence of finite medium attenuation, will in practice always be accompanied by a travelling wave component, in which case the composite is referred to as a standing wave field. Standing wave fields tend to arise when continuous ultrasound is propagated into a confined space that contains materials with low absorption at the frequency concerned and is bounded by material representing an appreciable impedance mismatch (leading to strong reflection). This may be the case, for example, with airborne ultrasound in a small container in the absence of absorptive materials. In general, due to the increase of attenuation with frequency, the occurrence of standing wave fields becomes increasingly unlikely (except in very small volumes) at increasingly high frequencies. Under the conditions existing, for example, in medical diagnostic and therapy applications (generally in the range 1–20 MHz, as outlined below) the usual situation is a predominantly travelling wave field, although there may be an appreciable standing wave component if, for example, there is a bone/tissue or tissue/gas interface close to the transducer. The possibility of the occurrence of standing waves usually does not arise with pulsed ultrasound.

Measurement of sound fields
It will be evident from the foregoing that sound fields may vary considerably in their spatial and temporal patterns, and it is also to be expected that any hazardous or other biological effects that they produce may depend quantitatively on aspects of this pattern. It is therefore necessary to consider the methods available for making measurements of physical quantities that will be of relevance and practical value in the interrelation of particular human exposures with the results of biological experimentation.

254

Table 2. Reported values of instrumental and field parameters for medical ultrasound equipment[a]

Equipment type	Frequency range (MHz)	Area of typical source (mm²)	Typical duty factor	Average power (mW)	Spatial peak temporal average intensity (kW/m²)[b]	Temporal-spatial peak positive and negative pressure excursions (MPa)[c]
Diagnostic						
Pulse-echo (including pulsed Doppler)	1–20	100–3 000	0.001	0.01–80	1–10 000	0.1–7.4 (+) 0.1–3.9 (–)
Continuous-wave Doppler:						
abdominal/fetal heart	2–4	100	1	5.0–240	200–18 000	0.01–0.11 (±)
vascular	5–10	100	0.01–1	0.01–40	30–19 500	0.003–0.12 (±)
Therapeutic						
Continuous	0.75–3	500	1	0–15 000	0–100 000	0–0.25 (±)
Pulsed	0.75–3	500	0.2	0–3 000	0–20 000	0–0.25 (±)

[a] Based on reported measurements *(12–21)*, all of which were made in non-attenuating media.

[b] Spatial peak pulse average values in the case of pulsed beams. To obtain values in W/cm² the column figures should be divided by 10.

[c] The tabulated values are peak pressure excursions — positive (+) and negative (–) from the ambient value.

255

This subject falls, in practice, mainly into two parts: measurements on liquid-borne and airborne ultrasound, respectively. The phenomenon of solid-borne ultrasound (e.g. in bone) is also sometimes of interest, but will not be dealt with here.

In each case, measurement procedures will generally entail two inter-related operations: (a) detailed measurement of a particular field parameter of interest, using a device that is appropriately precise and convenient for the purpose but which will normally require calibration; and (b) calibration of the device in a manner that will be traceable to some generally agreed standard.

As will be seen from the discussion presented below on the biological effects of ultrasound, there is at present no agreement as to the complete set of mechanisms whereby ultrasound may induce biological effects in man, and consequently no certainty as to the identity of all the physical parameters that may be responsible. It thus appears prudent, at the present time, to employ types of measurement for this purpose that are capable, as far as practically possible, of completely characterizing the sound field of interest. As has already been indicated, this will entail measurement of at least one (and possibly more) of the four field parameters (s, v, a, p) over all relevant conditions of space and time. As far as liquid-borne (and tissue-borne) ultrasound is concerned, this has been the subject of extensive work and discussion over a period of years within the framework of the International Electrotechnical Commission (IEC). This has resulted in general agreement that acoustic fields of interest in the medical and biological context should be measured in terms of instantaneous pressure (p), detailed measurements being made by means of a piezoelectric hydrophone. Thus IEC has a programme for publishing a series of recommendations both on the desirable performance specifications of such hydrophones and on appropriate methods for their calibration (23). Examples of particular hydrophone techniques and calibration methods that may be useful in this context have been described by a number of authors (24–32).

As previously discussed, a set of acoustic field parameters that may sometimes be quantitatively related to the biological consequences of ultrasound exposure are those based on intensity. For situations where such parameters are of interest, good future practice is likely to be to derive values for them from the above pressure measurements. For the derivation of total acoustic power, however, common practice at present is to perform measurements by a method that relies on the existence of a fundamental relationship between the total acoustic power (W) in a travelling wave beam incident on a surface and the radiation force experienced by that surface when the radiation beam is reflected by or absorbed within it (33). A number of practical devices employing this principle have been described (15,34,35).

Dividing the power (W) by the area of a (hypothetical) surface through which the beam passes leads to the spatially averaged intensity over this area. The intensity so computed is also temporally averaged since the measured power is based on an average over the measurement time, typically a few seconds. The instrumental simplicity of this method for determining average intensity has led to its widespread application and may be responsible for what is sometimes a rather uncritical use of the parameter "time-averaged, space-averaged intensity" (I(sa,ta)) in reporting sound fields encountered in medical and biological applications.

Confusion arises in the use of "space-averaged" intensity when the area over which the average is taken is not specified. For example, in reports of investigations into biological effects, the average is sometimes over the aperture of a treatment vessel, or over part of the body surface of an animal. Probably the most common choice of averaging domain in the general literature is the area (or "effective area") of the transducer itself; the NCRP *(5)* has given a special symbol (I_t) to the average over this area. On the other hand, the AIUM standard *(22)* refers the spatial average temporal average (sa,ta) intensity to the beam cross-sectional area (carefully defined) in any designated plane normal to the beam axis; for a focused device a plane through the focal region is often selected.

There are a considerable number of parameters that can be used to describe some quantitative aspect of the acoustic field due to an ultrasound device (the "emission") and/or the modified ultrasound field at a region of interest, for example within the human body (the "exposure"). It will be seen from the section of this chapter dealing with the biological action of ultrasound that it is not yet possible to select any one, or even any unique combination, of these parameters as a universal quantitative predictor of biological change following exposure to ultrasound (in the manner that the gray or the sievert can be so used in ionizing radiation dosimetry). Two particular indications are nevertheless apparent. First, damage from excessive heating could be a problem under conditions that might arise during improperly controlled medical therapy and, possibly, in some diagnostic uses of the Doppler effect. Second, while it seems safe to assume that pulse–echo diagnostic use of ultrasound is unlikely to lead to heat damage in this way, some exposures in this class are known to lead to high instantaneous acoustic stress (as indicated, for example, by high acoustic pressures or particle accelerations); the biological consequences of this, while not known to be hazardous, are as yet incompletely explored.

The existence of such uncertainties underlines the desirability, as discussed above, of employing dosimetric practice that, as far as is reasonably possible, completely characterizes the sound fields of interest. There are, however, some immediate practical problems, such as in relation to protection guidelines (see p. 279) for which some quantitative indication of emission or exposure in terms of a single number would be of value, at least in relation to medical diagnostic equipment.

In this limited context, and strictly as an interim measure until general consensus can be reached on the basis of improved scientific evidence, it is suggested that the following parameters will provide useful indicators of emission and exposure, at least in the context of the appropriate use of ultrasound in diagnostic medicine.

1. *For pulse–echo devices*: the spatial peak pressure amplitude (i.e. the sum of the moduli of the peak positive and negative pressure amplitudes — twice the "mean-peak-cycle acoustic pressure" as defined by Livett & Preston *(36)*)

in pascals, where this is at its maximum value in the beam (generally near the focus), unless otherwise specified.[a]

2. *For therapeutic, surgical and diagnostic Doppler (continuous wave and pulsed) devices*: the total acoustic (temporally averaged) power, in watts.

In reporting and making use of the above pressure quantities it will be necessary to distinguish between pressure values obtained from a specific transmitting transducer under two different conditions: (*a*) those obtained under well defined free field measurement conditions, usually in water; and (*b*) those determined at the point of interest, e.g. within an organ in a living body.

These two values of pressure may differ significantly, depending on the attenuation in the body and the degree of non-linearity in the beam. Values used to specify ultrasound fields (Fig. 1) or the emission from ultrasound generators (Table 2) should always be determined as described in (*a*). To aid in interpreting investigations of bioeffects of ultrasound, it is desirable that exposure values at the point of interest also be given, as in (*b*) above, together with the method used to estimate or measure this value (*37*); it will also be desirable to document the duration of the exposure (including, for pulsed fields, a measure of duty cycle). In both cases a record of the effective ultrasound frequency should also be provided.

A specific problem encountered in the measurement of ultrasound fields in liquids and soft tissues, particularly in the lower frequency range (e.g. 20 kHz–3 MHz), is the phenomenon of cavitation, which may lead to major perturbations in the field to be measured. An account of this phenomenon is given in the following section.

As far as airborne ultrasound is concerned, the most important form, from a health point of view, is generally that emitted from high-power industrial and laboratory ultrasound processing equipment in the frequency range up to about 60 kHz. In addition to the airborne ultrasound at the operating frequency, some types of equipment, as a result of cavitation, also generate a broad band of noise at high audio frequencies. Measurement of the sound pressure level of airborne ultrasound can be made with a suitable microphone.

An urgent need exists for the development of an appropriate system of traceable standards, and corresponding instrumentation, in the field of ultrasound measurement.

Cavitation
The existence of an ultrasound field in a liquid can give rise, under certain conditions, to a phenomenon known as "cavitation", which has important

---

[a] There are substantial national and local variations in existing measurement practice and in related regulatory guidelines (*36*) and, while adoption of the use of pressure amplitude values should be the long-term aim in this context, a satisfactory alternative in the interim will be to use the quantity I(sp,pa), the spatial peak, pulse average intensity (*5*), again at its maximum value in the beam.

258

implications as a possible cause of biological damage to structures suspended in, or adjacent to, the liquid volume. Cavitation *(38–41)* involves oscillation of small bodies of gas or vapour (cavities or bubbles) and is typically a complex phenomenon that may include growth from preexisting nuclei and/or eventual catastrophic collapse. Its significance in the present context arises largely from its role as a mechanism whereby ultrasound energy can be converted, in discrete regions of a field (which may often be related to an existing standing wave pattern), to mechanically more damaging forms. Such converted energy forms may entail increases of several orders of magnitude in local amplitudes of the fluctuating field parameters, and also conversion from fluctuating to unidirectional ("microstreaming") motion, with consequent creation of high local values of fluid shear stress. A further significant consequence of cavitation can be the localized release, within the liquid, of rather high concentrations of chemically reactive free radical species, which can be of biological significance. For medical ultrasound frequencies, relevant bubbles are a few micrometres in diameter or smaller.

The physical conditions under which cavitation may occur appear to be quite variable and are not known in detail, although relevant investigations have been reported *(42–44)*. In general it appears that, for continuously applied radiation in ordinary air-saturated water, there exists for the phenomenon a pressure amplitude threshold whose value rises with frequency, reaching about 5×10^4 Pa at 0.25 MHz and 5×10^5 Pa at 5 MHz. (These pressure amplitudes correspond to intensities of approximately 0.03 and 3.0 W/cm^2 in a travelling, simple harmonic plane wave.) These thresholds are strongly dependent on the purity of the water, which determines the number of available nuclei from which cavities may be created. Intermittent (pulsed) radiation is generally less conducive to cavitation induction *(42, 43)*. It has been calculated that for the conditions employed in medical pulse–echo diagnosis (typically, 1 μs pulses spaced at intervals of 1 ms) pressure thresholds in the range 0.1–1.0 MPa will exist *(45)*. These predictions have still to be confirmed experimentally. Existing knowledge about the occurrence of cavitation in intact mammalian tissues is sparse. It is important to clarify the situation in order to determine the relevance of the considerable body of evidence of cavitation damage provided by biological experiments involving liquid suspension systems.

Three questions must be asked in order to shed light on this problem. First, do potential cavitation nuclei exist in tissue? Second, if they do, where they are smaller than active size, can they be made to grow under the action of ultrasound fields? Third, what sort of dynamic behaviour will they have? Recent studies have indicated that 0.75 MHz ultrasound fields can create "microbubbles" in mammalian tissues at acoustic pressures at the spatial peak above 10^5 Pa *(46, 47)*. Detection and quantification of cavitation activity *in vivo* is difficult *(38, 41–43, 48)*. Most available methods can be applied only in simple liquids, although methods using detection of acoustic harmonic or subharmonic energy that is specific to bubble activity look promising for use on intact tissues.

In addition to cavitation, other physical effects such as heating (temperature rise), particle aggregation, and structural changes due to acceleration, pressure or shear stress have frequently been discussed as mechanisms likely to induce bioeffects of ultrasound.

APPLICATIONS

As a result of the variety of contemporary practical applications of ultrasound, man may be exposed to it in several ways. In broad terms, the applications of ultrasound fall into three groups: (*a*) those in which its vibratory nature is exploited for the purpose of modifying or processing materials or objects; (*b*) those in which its radiation-like characteristics are used for investigative purposes, i.e. for the acquisition of information; and (*c*) those using either its radiation-like or vibratory properties for a broad range of medical therapeutic or surgical purposes. In general, the first group of applications employs frequencies at the lower end of the ultrasound spectrum (20–100 kHz), while many (though not all) of the investigative and therapeutic/surgical methods operate in the frequency region of a few megahertz, where wavelengths become relatively small and radiation-like behaviour can be achieved with corresponding precision.

The main classes of application of ultrasound that are believed to be of importance in the present context are described below, though not necessarily in order of importance. Moreover, the coverage is not comprehensive, and some of the other applications that are known to exist might conceivably be found to entail significant exposure of humans.

Processing applications

Ultrasound has found application in a wide variety of industrial processing techniques *(49,50)*. From the point of view of possible human exposure, these may usefully be divided into two groups, namely those in which cavitation induction in liquids is or is not involved as an essential mechanism. The latter group will be considered first.

Processes not involving cavitation induction principally entail the application of ultrasound vibrations as a means of machining, cutting, welding or otherwise treating materials in certain circumstances (e.g. brittle materials, complex shapes) where conventional methods are unsuitable. With properly designed equipment and procedures, physical coupling of ultrasound energy directly from the machine to the operator via dense media should be avoidable, and any actual exposure is thus likely to be to airborne ultrasound.

The other major group of processing techniques, that has quite widespread application in industry and elsewhere, is that in which ultrasound is used to induce cavitation, with the object of accelerating processes such as cleaning, chemical reactions, mechanical disruption, or mixing and emulsification within the liquid volume treated. Here, there is not only again a possibility of exposure to airborne ultrasound (and often, importantly, associated high levels of audible sound) but also a real possibility of effective liquid coupling to the fingers and hands of an operator, for example in working with ultrasound cleaning tanks. The action of high intensity ultrasound energy can result in the production of aerosols or of free radical chemical species, which may lead to the formation of toxic chemicals (although this possibility does not arise with the liquids normally used for ultrasound processing applications).

There appears to be little systematic information available as to the actual levels of exposure experienced in the above-mentioned applications.

Underwater target location

A major use of ultrasound is in the detection, location and identification of underwater objects. The application covers a wide variety of objectives, from submarine detection, through depth sounding and fish finding, to imaging of structures in turbid water. Depending on the application and the distance entailed, the frequency used may be anywhere from less than 10 kHz to about 1 MHz. Most techniques involve the pulse–echo method, in which returned echoes are detected in the period following transmission of a short pulse of sound, in a directional beam. Irradiation fields are thus characterized by pulse conditions with low duty factor (0.1% or less), although peak powers may be high and are often limited by the onset of cavitation. It is perhaps more than of historical interest that the first successful generation of high power ultrasound for this purpose led to the killing of fish irradiated by the beam (51).

It seems probable that divers, in carrying out underwater inspection of ships and structures such as oil rigs, may be appreciably exposed to underwater acoustic beams, but information on the values of the field parameters that might be involved appears to be lacking (at least in the open literature). The likelihood of operator exposure is subject to considerations similar to those applying in industrial processing equipment: with efficient design, the majority of the ultrasound energy will be projected into the work medium and relatively little should reach an operator. However, documented evidence on this point appears again not to be generally available.

Testing and measurement applications

A wide variety of physical investigative techniques employing relatively low power ultrasound, many of them falling within the field of industrial non-destructive testing, is now in practical use (52). The ultrasound fields used in these applications appear to be broadly similar, in terms of the range of field parameter values employed, to those that apply for medical diagnostic instruments. It seems reasonable, therefore, to consider together the possibilities of operator exposure in the two fields.

Medical diagnostic methods

This is a large and rapidly growing area of application of ultrasound and one that has found a major and apparently permanent investigative role in a variety of medical specialties including cardiology, vascular disease, ophthalmology, internal medicine, gynaecology and obstetrics. In a considerable number of countries, for example, it is already normal practice for a high proportion of women to be examined with ultrasound several times during the course of a pregnancy, and it is the expectation of WHO (53) that this practice will soon become even more widespread. Seen in this light it appears that this application will entail the numerically greatest single source of human exposure to ultrasound.

At the present time (and it seems unlikely that there will be any change in the foreseeable future) the predominant medical diagnostic application of ultrasound is in the frequency range 2–10 MHz, with occasional extension to 20 MHz for some eye examinations. Ultrasound irradiation falls into two categories — pulsed fields and continuous wave — the latter being used for Doppler frequency-shift identification of moving structures and for blood flow velocity measurement. Almost all accessible anatomical sites are liable to examination; those for which the technique has so far proved to be of most clinical value (e.g. the abdomen, pregnant uterus, heart and major vessels) are the most commonly irradiated. Accessibility is impaired to some extent by bone and particularly by air spaces; thus the inflated lung and, in some patients, gassy bowel constitute impenetrable barriers. As already noted, however, the reflection occurring at such sites could induce the formation of standing waves.

In view of the widespread recent interest in questions of safety in this area, surprisingly few systematic data have been published as to the magnitude of the acoustic fields produced by the devices used. Data based largely on five known published studies, two of which were carried out in the United Kingdom (12,17) and three in the United States (15,19,21), are summarized in Table 2.

These data relate to the beam that is designed for projection into the patient (i.e. to "emission"). Once again, the requirements of efficient design dictate that the energy in the main beam should be very much greater than any component that could be radiated backwards and reach the hand of the operator (the only organ that could conceivably be appreciably exposed). Even though actual levels have not been systematically measured, it seems safe to conclude that exposure of operators of this type of equipment (and, as noted above, of analogous industrial testing and measurement equipment) will be insignificant in relation to patient exposure. This conclusion does not, of course, apply when the operator deliberately projects ultrasound into his/her own body, as is frequently done for training or testing purposes.

A notable feature of the data in Table 2 is the large range of field parameters that result from the use of different pulse–echo equipment. Whilst there will undoubtedly be some thresholds in acoustic output parameters below which signal-to-noise ratios, and thus diagnostic performance, will significantly deteriorate, it is not clear that some of the particularly high output levels that have been recorded are entirely necessary in this sense. Some systematic study of this situation appears to be desirable.

Ultrasound physiotherapy

There is little doubt that ultrasound used in physiotherapy constitutes the most widespread source of ultrasound irradiation of humans at power levels sufficient to cause biologically significant rises in tissue temperature *(54–56)*. This is an old established technique *(8)* in which ultrasound beams in the approximate frequency range 0.75–3.0 MHz are directed into a patient with objectives such as stimulation of capillary circulation, relief of pain and acceleration of tissue regeneration *(57–60)*. Irradiation is either continuous or interrupted (for example 2 ms pulses repeated every 10 ms) and is generally applied over a period of 5–15 minutes per treatment by means of a hand-held transducer that is continually moved over the affected region in order to achieve an even distribution of treatment. Another means of applying ultrasound that is sometimes used, is to leave the transducer in a fixed position over the region to be treated. This method increases the exposure to a given area but is of dubious advantage, and is discouraged because of the possibility of standing wave formation. Some therapeutic effects of ultrasound may well be due to biophysical mechanisms other than those related to temperature rise, for example the observed acceleration of healing in chronic varicose ulcers *(61)*. A number of other effects, including cavitation, may occur simultaneously with local tissue heating. Many of the treatment methods currently in use involve such low spatial average, temporal average intensities that they are unlikely to be accompanied by temperature rises above that occurring normally as a result of diurnal variations.

The nominal acoustic power levels delivered to a patient by machines of this type do not normally exceed 15 W distributed over a transducer face area of the order of 500 mm² (Table 2), although it may be that nominal levels may differ considerably from those actually measured *(20)*. The extent of use of this technique, although in general not accurately known, is illustrated by the results of a survey carried out in the United States in 1973 in which, for a sample population of 500 000, it was found that ultrasound treatments were being carried out at a rate of over 3000 per month *(20)*. A survey carried out in Canada *(62)* reported that during 1977 nearly 4 million ultrasound treatments were given in a total population of about 20 million; presumably this involved several treatments per patient.

Ultrasound hyperthermia

Prolonged exposure to temperatures in the range 42–45 °C is lethal to mammalian cells, and this is used in the cancer therapy known as hyperthermia. In this technique, the tumour is heated to these levels, while an attempt is made to leave the surrounding normal tissue undamaged. Local heating can be achieved with ultrasound, although microwave and radiofrequency methods are also used *(63)*. Plane transducers are used for the heating of superficial volumes, whereas focused transducers or multi-transducer arrays are used for deeper seated tumours *(64,65)*. It has been shown that some human tumours respond to treatment by heat alone, but most benefit is obtained when heat is combined with either radiotherapy or chemotherapy.

Surgical techniques

A considerable variety of ultrasound techniques is used in connection with surgical procedures *(66)*. These fall into two groups: (*a*) those in which ultrasound vibration is used to increase or modify the effectiveness of a surgical instrument (67); and (b) those in which an ultrasound beam is generated and itself used to effect some surgical action. Few of these techniques in either group seem yet to have achieved more than limited, experimental use; the one with the most widespread application is probably the use of miniature beams in the treatment of Ménière's disease of the inner ear (66,68). Focused ultrasound surgery has been used with some success to treat glaucoma (69,70). Devices using 40 kHz ultrasound are now employed frequently in procedures for cataract removal. A small tip, rather like that of a hypodermic needle, is introduced into the anterior chamber of the eye, placed in contact with the lens, then set into longitudinal oscillation. This causes emulsification of the cataract body which, especially for mature cataracts, facilitates its removal by aspiration (71).

Tools driven by ultrasound have been used for many years for the disintegration of renal calculi (72). An important new application is the fragmentation of kidney stones using shock waves (extracorporeal shock wave lithotripsy). In this procedure, patients are carefully positioned in a large water bath such that the kidney stone lies at one focus of an ellipsoidal mirror. A spark discharge is generated at the other focus. The pressure amplitude of the shock wave is at its maximum at the stone, and causes its disintegration. Pressure amplitudes of 14 MPa have been reported at the focus (73). Little is known about the effect of the transit of such shock waves through soft tissues.

From the point of view of the human exposures that they entail, each of the numerous different techniques needs to be considered separately and although for some (such as the Ménière's disease application) good data on dosimetry are available, this is not so in all cases. In relation to the question of hazard, however, this category of ultrasound application constitutes a special case, to the extent that exposures will normally be carried out under professional supervision and in the light of informed judgement as to the relative benefits and risks entailed.

Applications in dentistry
Ultrasound scaling instruments are widely used for removing calculus and other deposits from teeth. Suitably shaped tips are mounted on the end of a vibrating rod in the ultrasound dental handpiece. The vibrating tip is slid over the tooth surface, applying a shearing force to the calculus layer being removed (74). Dental handpieces operate at frequencies of about 18–40 kHz, with a vibration amplitude of about 0.01–0.025 mm (75). Water is sprayed against the vibrating tip for cooling purposes and the cavitational effects in the liquid being sprayed reinforce the mechanical effect of the vibrations in removing deposits. Estimated total power output levels of 10 W have been reported (76).

No epidemiological studies on the possible effects of ultrasound dental scalers are known. However, it has been proposed that studies should be performed to assess the hazards of ultrasound in the oral cavity (76).

BIOLOGICAL ACTION

Numerous experiments have been performed to look for effects of ultrasound that would constitute serious health hazards to man. These have included studies of genetic damage, teratogenesis and general somatic damage. Evidence for these effects is obtainable at several levels of biological organization: isolated biomolecules, cells, intact tissues, multicellular organisms and human populations. From the point of view of questions of hazard, it would be of greatest value to have clear evidence of effects, or absence of effects, in man. This is generally very difficult to obtain, and the small amount of information of this type that may be available will need to be supplemented by evidence relating to other mammals and to lower animals. Even here, evidence tends to be empirical and it is necessary to review work on single cells and biomolecules in order to try to understand the nature of the underlying mechanisms of action in tissue. Moreover, available studies deal primarily with short-term effects, and delayed effects such as carcinogenesis would not have been observed.

It seems important to point out that, although the physics involved in the characterization of ultrasound fields has reached a reasonably sophisticated level, studies on the biology of bioeffects that have been reported show much less sophistication. It is therefore important that there should be close collaboration between biological experts and physical scientists, to ensure that complex biological endpoints are not studied without being fully understood, that the most up-to-date techniques are employed, and that ultrasound exposure systems (by being poorly designed and understood) do not introduce experimental artefacts.

Characterization of ultrasound exposures used in the bioeffects literature
In a complete description of the biological action of ultrasound, it is necessary to provide the exposure parameters used in each study. Unfortunately, there has been no consistency in the literature as to the parameters quoted. This makes comparison between reports difficult and in some cases impossible. In this section, therefore, it has been decided, for ease of reading and for comparative purposes, to divide quoted ultrasound exposures into three classes: the pulse–echo band (PE), the Doppler and therapy band (D,T) and the surgery band (S). The physical characteristics of these bands are given in Table 3. These bands may serve as a broad classification, although some overlap may exist. Where possible, the relevant exposure band will be included in the description of each study described.

Isolated biomolecules
Extensive work has been carried out on the action of nonthermal ultrasound on chemical systems and, in particular, on large molecules of biological interest *(77)*. The effects at this level are broadly of two kinds: (*a*) truly chemical effects, apparently attributable to the action of cavitation in releasing chemically active free radical species in irradiated solutions; and (*b*) mechanical degradation of large molecules having a relatively fragile structure.

265

Table 3. Exposure level "bands" used in this chapter
in reporting bioeffects literature

Band	Pulse length	Space averaged,[a] temporal averaged intensity
PE: pulse-echo	3–4 cycles	$\leq 0.8\,kW/m^2$
D,T: Doppler, therapy	≥ 10 cycles including continuous wave	$> 0.8\,kW/m^2$ $\leq 30\,kW/m^2$
S: surgery	1–∞ cycles	$\geq 30\,kW/m^2$

[a] Exposures are assigned to each category using quoted literature values; the definition of quantities may not always be clear from the text.

This latter effect has been particularly studied in solutions of DNA (78), where it has been shown that rupture of molecules occurs preferentially around their centre points, an observation that is consistent with the implication of fluid shear stress as a causal mechanism. The phenomenon has been studied mainly at low ultrasound frequencies, but it has also been shown to occur in the medical frequency range (42,43). From this work it is clear that many of the observed changes are attributable to cavitation and apparently do not occur in its absence. In addition, it should be noted that DNA in solution is more fragile than when bound within cells. The relative importance of the chemical and mechanical consequences of any cavitation activity that might arise from *in vivo* irradiation has not been definitely established, but there is some evidence (79) that, in practice, free radicals are largely inactivated by the buffering action of abundant chemical species present in the normal physiological environment, which thus act to protect critically important molecules such as DNA from chemical modification.

Experiments have been carried out in which a search was made for genetic effects of ultrasound after irradiation of isolated DNA (80); the findings here were negative, and this result supports those on intact cells (where it could be argued that the DNA is shielded from some chemical and mechanical effects of irradiation). No evidence for double- or single-strand breaks has been found in DNA isolated from cells irradiated with ultrasound (D,T) whilst intact (81).

Isolated cells and cell cultures
Studies in this area are particularly informative in providing, at a fundamental biological level, evidence both on the nature of ultrasound-induced changes that can occur and also useful evidence on the non-occurrence of certain specific effects under particular conditions of irradiation. The available experimental data presented below, and their interpretation, are arranged according to the particular biological endpoints that have been

266

considered. Emphasis is placed here on work that relates to endpoints of particular interest in connection with safety.

Cell survival

The evidence that ultrasound exposure of cells and microorganisms in suspension can lead to cell death by disintegration (lysis) is unequivocal, and there is no doubt that cavitation is an important mechanism in the process. What remains in some doubt is whether ultrasound at low megahertz frequencies causes cell death, either by disintegration or otherwise, in the absence of cavitation at normal physiological temperatures. A number of reports in the literature indicate that this may be so but, in view of the already mentioned technical difficulty of monitoring cavitation activity, these must be treated with some reservation. Where such investigations have been carried out systematically (D,T; S) (42,43,82,83) a clear correlation between cell death and cavitation has generally been found. No evidence exists for a time lag, after irradiation, before cell death occurs.

In experiments designed to reveal possible changes in cell proliferation pattern, such as those seen following exposure to ionizing radiation, no changes could be found following ultrasound irradiation in which cavitation had been inhibited by the use of very short pulse durations (84,85). An interesting finding however was that, when cavitation was allowed to occur, cell disintegration occurred preferentially during the mitotic phase of the cell cycle, possibly because of increased fragility of the extracellular membrane. However, Fu et al. (86) found their cell line to be relatively resistant to ultrasound immediately after mitosis. This effect may thus depend on culture conditions and cell line.

Genetic damage in cell systems

From the hazard point of view, gene mutation may be the most important endpoint for consideration; it also seems to be one that abounds in technical pitfalls. The evidence in the literature for ultrasound induction of genetic changes has been reviewed (87,88) and the conclusion drawn that, with the possible exception of situations where appreciable heat shock might be involved, there is little likelihood that genetic effects will be found to result from exposures to ultrasound at the levels currently used in medical applications. Damage to DNA is important from a hazard point of view if it leads to changes that can be inherited. Some DNA changes are not inherited, often because they are lethal to the cell. In one experimental investigation, an attempt was made to test the effect of extremely severe exposure to ultrasound by looking for evidence of genetic change in the survivors of cell populations exposed to cavitating ultrasound fields (D,T) (89). Here, changes were found only where appreciable heating or free radical formation occurred.

Considerable effort has been devoted to investigating chromosomal aberrations (PE; D,T) (85,90–94) and the results have been reviewed by Rott (95). The current consensus here is that the evidence is again overwhelmingly against the existence of any effect, at least under medical diagnostic exposure conditions, although the question has been raised as to whether ultrasound may have some synergistic action in X-ray induction of aberrations (96).

267

One method that is often used to detect the DNA-damaging ability of potential mutagens is to study sister chromatid exchange (SCE). From S phase until metaphase, a chromosome contains two (sister) chromatids. Exchanges occur between these sisters in the same chromosome, and may be detected using appropriate staining techniques. It is not understood how SCEs arise, but they do not appear to have any hereditary consequences, and the technique is a poor indicator for mutagenicity in some cases (e.g. for X-rays and γ-rays) *(88, 97)*. A report of increased SCE frequencies caused by irradiation of human lymphocytes and lymphoblastoid cells with ultrasound (PE) *in vitro (98)* prompted a large number of investigations on this topic. In almost all cases it was not found possible to reproduce the initial finding (see for example *5,98–104* and reviews by Miller *(105)* and Goss *(106)*).

Structural changes

Evidence for structural changes in cells irradiated with ultrasound has been sought by a number of investigators. These changes are not necessarily lethal. The cell surface is one obvious site for change, and it has been shown, by means of electrophoretic measurements on ascites tumour cells, that changes occur in the electrical charge density, and thus presumably in the structure, on the surface of treated cells (D,T) *(107,108)*. A change in charge density will also be seen if the cell volume is altered. In this work there was some indication that cavitation was not the effective mechanism, but subsequent work using cultured mammalian cells *(109)* has shown that these changes are correlated with disintegration death of other cells in the exposed population, almost certainly occurring as a result of cavitation. The later work also showed that the changes involved are reversible and nonlethal.

Chapman *(110)* has found that ultrasound can induce sublethal changes in the plasma membrane of thymocytes. *In vitro* treatment with ultrasound at 1.8 MHz and at I(sa,ta) intensities above $10 \, kW/m^2$ (D,T) was followed by an immediate decrease in potassium content, suggesting an alteration in membrane permeability and thus presumably in membrane structure. Changes in membrane permeability to sodium and calcium ions have also been shown for similar ultrasound exposures (cited by Dyson *(58)*).

Electron microscopy studies on organelles of treated cells have led to reports of modifications, and particularly of mitochondrial swelling *(111)*. In work on intact tissues, there have been claims of disruption of lysosomes and consequent cell destruction by released lysomal enzymes (D,T) *(112, 113)*. Slits and vacuoles may be seen in the cytoplasm of cells irradiated (D,T) either *in vitro (114)* or *in vivo (115)*. Damage to the plasma membrane on the luminal aspect of endothelial cells of blood vessels treated in standing wave fields (D,T) has also been reported *(115, 116)*.

Modifications of function and response

This is a potentially very large field of investigation since, in general, the various functions and responses of cells may depend on many factors, any one of which may be open to influence by an agent such as ultrasound. One such response that has been investigated is that of X-ray sensitivity. Here, a number of factors in the response, and notably that of the partial repair phenomenon, would seem to be open to the influence of ultrasound. Early reports on animal work suggested that there might be a synergistic effect between X-rays and ultrasound in relation to tumour regression, but follow-up work (D,T) using cultured mammalian cells appears to exclude nonthermal mechanisms in this effect (117). It has been shown that non-thermal mechanisms may act to enhance ultrasound-induced (D,T) cell killing of heated cells (held at temperatures in the range 41.5–44 °C) *in vitro (118,119)*.

A positive finding in respect of functional change has been reported in the form of an action spectrum for the mechanical flagellation function of rotifers (polynucleated aquatic microorganisms) exposed to ultrasound over a wide range of frequencies (120). In two frequency bands only (around 270 and 510 MHz) movement was inhibited and this change was found to be reversible for pulse durations up to 30 seconds but irreversible for much longer irradiations, acoustic intensities being of the order of 10 W/m^2.

Ultrasound may act not only to inhibit cell function, but also to cause stimulation. For example, Harvey et al. (121) have reported stimulation of protein synthesis in human fibroblasts irradiated (D,T) *in vitro*. Stimulation of bone repair has also been demonstrated (D,T) (58,122,123). Angiogenesis can be stimulated in chronically ischaemic tissue, resulting in increased blood flow and improving the local environment in a manner likely to assist in repair (124).

General conclusions from isolated cell work

The extrapolation of data obtained on isolated cells with the aim of reaching conclusions valid for complex organisms is notoriously dangerous. The situation is further complicated in ultrasound biology by phenomena such as cavitation that can produce major biological effects and are strongly dependent on the physical characteristics of the irradiation mechanism, and on the biology of the chosen culture technique. It is important to note, however, that an *in vitro* effect is a real biological effect: it is its relevance *in vivo* that must be considered.

Nevertheless, with this qualification, cellular ultrasound biology can be highly informative. In particular, genetic techniques can often be applied with greatest precision and economy at the cellular level. Cavitation-induced cell disintegration is a well established and common consequence of ultrasound irradiation, but detailed knowledge is still lacking about the physical parameters involved in its control, and particularly on its significance in relation to human irradiation *in vivo*. Evidence for nonthermal, non-cavitational mechanisms of action is still sparse.

In vitro experiments are important to suggest new endpoints for study and to give a basis for the design of *in vivo* experiments, and may identify fundamental inter- or intracellular interactions. Caution should be used in attempting to extrapolate *in vitro* results to the context of *in vivo* ultrasound exposures.

Multicellular organisms
A large number of reports have been published related to both positive and negative observations of biological change in animals and plants exposed to ultrasound. Of these reports, many are not directly relevant to the present review for various reasons. It is impossible, in many cases, to draw valid quantitative conclusions because of deficiencies in experimental design, statistical significance of the results, or physical dosimetry. In other cases the biological endpoint considered appears to bear no clear relationship to the question of hazard.

A limited number of reports, which are representative of the published material as a whole, are presented below.

Mammals
The main body of mammalian experimental work related to hazard has been carried out on rats and mice, which have been irradiated at exposure levels either comparable to, or considerably in excess of, those commonly experienced in ultrasound diagnosis. In this work, a search has generally been made for evidence of changes of a genetic or teratogenic nature and for consequent effects in the fetus.

Exposure of pregnant rodents to ultrasound during the later stages of pregnancy does not in general appear to have any significant effect on the offspring until intensities high enough to produce heating are used (125, 126).

Lyon & Simpson (127) found no evidence for any genetic effect when mouse gonads were irradiated with ultrasound at exposure levels (D,T) up to those sufficient to cause a degree of heat-induced sterilization.

It seems sensible to divide the available data (which are sparse) into different groups according to gestational age at the time of ultrasound exposure. It seems that intensities up to 10 kW/m² do not produce effects on the pre-implantation embryo unless there are adverse effects in the mother (126, 128–131). Where embryotoxic effects have been seen at this early stage of pregnancy, they may be attributed to significant thermal effects.

A number of investigators have studied the effects of ultrasound exposure of rodents during early organogenesis (126, 128, 132–137). Effects seen include fetal weight reduction, death and malformations. The occurrence of fetal death or malformations appears to be correlated with the higher intensities used (> 7.5 kW/m² (D,T)) and thus probably with heating effects. Shoji et al. (136) indicated that there might be a strain-dependent increase in fetal death following 5 hours of continuous exposure at 0.4 kW/m². This may, however, be due to prolonged induction of moderate temperature rise (138). Fetal weight reduction is less easily analysed since it is closely related to factors such as maternal stress and nutrition, which are not easily controlled.

An important area of uncertainty in interpreting all the above data concerns the effect of spatial concentration (focusing) of ultrasound energy. As far as is known, all the biological data cited here were obtained under unfocused beam conditions, where the ratio $I(sp) : I(sa) \backsim 3$ (see p. 254). In relation to irradiation with more strongly focused beams, it is not known at present whether spatial peak or spatial average intensity, or peak pressure, is the physical parameter that would be most closely associated with any possible biological effect. For the time being it will be prudent, therefore, when using such data for risk evaluation purposes, to assume that acoustic peak pressure may be an important parameter, and preferably to measure and report exposures in the form recommended above (p. 256).

The search for chromosomal aberrations and sister chromatid exchanges resulting from *in vivo* irradiation (PE; D,T) has generally proved negative *(100, 139–143)*.

The results discussed here do not allow straightforward conclusions to be drawn, although it appears that teratogenesis is strongly correlated with temperature *(144)*.

Dyson et al. *(145)* have shown that the irradiation (D,T) of blood vessels by a standing wave field can cause a modification of blood flow. The red cells in embryonic chick blood vessels clumped together to form bands at half-wavelength intervals. The effect was, in most cases, reversible. Similar effects have been seen in mammalian vessels (D,T) *(115)*. Dyson et al. *(146)* have also reported stimulation (D,T) of tissue regeneration in experiments on rabbit ear. More recently, work has been reported in which ultrasound (D,T) stimulated healing of varicose ulcers *(61)* and bone fractures *(58)*. A survey of the literature covering ultrasound-induced biological effects in mammalian tissues allows one to draw an empirical threshold in the product of intensity and exposure time below which no positive effects have been reported. Broadly, no independently confirmed significant biological effects have been seen when mammalian tissues are exposed *in vivo* either to intensities below $1 \, kW/m^2$ (sp,ta) or when the product of intensity and total exposure time is less than $500 \, kJ/m^2$ *(147)*.

Insects

In addition to work on mammalian systems, further animal-based evidence comes from investigations on insects, and specifically on changes in *Drosophila*. This insect contains microscopic gas bodies in its respiratory system and so is a useful animal model for tissues containing gas spaces.

Early work on changes following ultrasound treatment (D,T) of eggs *(148)* showed evidence of a variety of developmental abnormalities, which may perhaps have been due to the mechanical effects of (undetected) cavitation taking place in the water used as a suspension medium.

Drosophila has been used to investigate possible genetic hazards in animals irradiated with ultrasound *(148)*. In large-scale breeding experiments with flies surviving ultrasound irradiation, careful investigations indicated no significant increase in the frequency of recessive lethal mutations and chromosomal non-disjunction, even under exposure conditions (D,T) sufficient to kill a substantial proportion of the flies.

It has been shown that *Drosophila* larvae are particularly sensitive to pulsed ultrasound *(149)*. Short (1 ms) pulses were used, and the survival found to depend on the maximum intensity during the pulse rather than on the time-averaged intensity. Maximum intensities over $100 \, kW/m^2$ were found to increase mortality. It has been found that the eggs are most

271

sensitive to ultrasound just before they hatch, a stage at which they contain stabilized gas bodies *(150)*.

Plants
The effects of ultrasound on plants have been studied for many years and there are a number of reports of damage to chromosomes, cells and tissues resulting from continuous-wave exposures (D,T) at intensities above 3 kW/m^2 and at frequencies between 0.5 MHz and 10 MHz *(103)*.

Work in the 1950s (S; D,T) suggested that this damage can result from intratissue cavitation, which leads rapidly to the death of many of the cells involved and leaves surviving cells with no chromosome abnormalities *(151,152)*. More recent quantitative studies tend to support these findings. For example, effects on the growth of plant roots after exposure to ultra-sound (D,T) are very different from those found after irradiation with X-rays, with little or no delay in the depression of growth rate by ultrasound and no effect of fractionating the exposure *(153,154)*. None of the typical forms of chromosomal aberration, induced by such agents as ionizing radiation and various chemicals, were observed after exposure to ultra-sound, and other forms of chromosome disruption were not found to account for the rapid depression of growth rate in roots *(155)*. Recent evidence also supports the view that plant tissue is particularly vulnerable to the cavitational effects of ultrasound because of the presence of gas-filled spaces between cells *(156,157)*.

Human epidemiology
In any discussion of the safety of humans, regardless of the nature of the particular agent that may be under suspicion, no completely satisfactory conclusion can be drawn that does not rely on evidence from humans. Epidemiology is a demanding science, and production of fully satisfactory statistical evidence on the safety of ultrasound exposures would necessitate a study involving very large numbers of subjects *(4,5)*. The design and sample size of such a study must enable subtle effects to be detected with sufficient statistical power. The cost-effectiveness of such inevitably cumbersome epidemiological studies must be considered in the light of the validity and relevance of the end results.

It is convenient to consider the various studies that have been reported under the headings of prospective and retrospective, in which the former involves prior planning of the manner in which subjects are allocated between exposed and unexposed categories. An important subgroup of the retrospective studies is that employing the case-control method, in which a particular endpoint (e.g. incidence of childhood cancer) is in-vestigated and comparison is made between sets of cases and corre-sponding controls individually matched to the cases in all other important respects.

Two small-scale prospective studies have been reported from Norway. Bakketeig et al. *(158)* screened 510 out of 1009 pregnant women at 19 and 32 weeks of pregnancy. This study revealed no adverse short-term bio-logical effects from the ultrasound. Eik-Nes et al. *(159)* randomly allocated

272

1628 women into one of two groups. In weeks 18 and 32 of pregnancy, 809 women underwent ultrasound scans. Again, no adverse effects were detected.

Some retrospective surveys can also be found in the literature. In the largest of this nature, 1114 apparently normal pregnant women were examined by ultrasound in three different centres and at various stages of pregnancy (160). A 2.7% incidence of congenital abnormalities was found by physical examination of newborns in this group, as compared with a figure of 4.8% reported in a separate unmatched survey of women who had not had ultrasound diagnosis. Neither the time of gestation at which the first ultrasound examination was made, nor the number of examinations, seemed to increase the risk of fetal abnormality. Similar smaller studies, such as that by Koranyii et al. (161), have come to the same conclusion. In a more recent and rather thorough study (162), 425 exposed children and 381 matched controls were followed until 12 years of age, with a number of potentially adverse effects being investigated both at birth and also at a special examination between 7 and 12 years of age. Although some slight differences were seen (and particularly in dyslexia in the groups from all three participating hospitals) the study found no statistically significant differences, in any of the various measures of adverse effects, between exposed and unexposed children.

Two of the more quantitatively rigorous studies to have been published so far employ the case-control technique to search for the possibility of increased risk of cancer in children who had exposed to diagnostic ultrasound *in utero*. Both studies drew on the United Kingdom population: one dealt with 1731 children who died of cancer between 1972 and 1981 (163); and the second with 555 children having malignancies that were diagnosed in the period 1980–1983 (164). Neither study showed any excess of cancer or leukaemia in children who had been exposed to ultrasound *in utero*, when all age groups were taken together. When the data in the first study were subdivided by age at death, the risk of malignancy was found to be higher than expected for children aged 6 years and over, i.e. those who had been exposed during the early 1970s. There is strong evidence, however, that this outcome may have been an artefact arising from the fact that ultrasound at that time was being used selectively for only seriously abnormal pregnancies.

Although not scientifically objective, it should be noted that many patients subjected to diagnostic ultrasound undergo subsequent clinical examinations, and there have been no reports of any suspicion of ill effects due to ultrasound.

No comparable studies, even at the rather low level of epidemiological rigour of some of the above-mentioned investigations, appear to have been made in relation to either therapeutic or occupational exposure to ultrasound.

Effects of industrial and airborne ultrasound

Industrial occupational exposures to ultrasound will generally occur in one of two rather different forms: by direct coupling of ultrasound, via a liquid couplant, through the skin; and as an auditory effect resulting from the good acoustic impedance-matching properties of the ear. The physical and biological considerations that apply to the former type of exposure are essentially similar to those applying to the medical exposures previously described. Discussion in the present section will therefore concentrate on auditory effects. A comprehensive review dealing with both of the above classes of exposure has been published by Acton *(165)*.

The most common complaints made by people exposed to high levels of high-audible-frequency noise, such as that generated by processes involving cavitation, are of an unpleasant "fullness" or pressure in the ears; headaches, which do not persist after the exposure has ceased; tinnitus; and perhaps also nausea and mild vertigo *(166,167)*. These effects may also be caused by subharmonic distortion products of ultrasound frequencies at higher levels. The effects do not appear to be closely dependent on duration of exposure.

Various biological changes have been reported in industrial workers, but many of these effects have been similar to the changes brought about by exposure to other physical conditions or toxic agents, or to stress, and conclusions should be viewed cautiously in the absence of comparisons of results with those for control groups *(168)*. Grigor'eva *(169)* failed to demonstrate any significant physiological changes as a result of an hour's exposure to 110–115 dB at 20 kHz in a comparison with control subjects. Knight *(170)*, in a retrospective survey, did not find any permanent physiological or psychological changes in users of industrial ultrasound equipment, compared to a matched control group.

Neither temporary *(166)* nor permanent *(170)* losses of hearing were found in industrial workers exposed to levels of up to about 110–120 dB at low ultrasound frequencies. However, temporary shifts in the threshold of hearing were noted at subharmonic frequencies following exposure to discrete frequencies in the range 17–37 kHz at levels of 148–154 dB *(171)*. These effects probably result from nonlinear distortion of the eardrum, giving rise to subharmonics at levels of the same order as that of the fundamentals. Similar effects have been observed in the cochlear-microphonic potentials of guinea-pigs *(172)* and subharmonic frequencies have also been monitored in the sound field in front of the eardrum, using a probe-tube microphone *(173)*.

CRITERIA FOR APPROPRIATE USE

From the evidence presented in this review a number of general points emerge. Ultrasound techniques represent a large and increasing benefit to mankind, particularly in the broad fields of medicine and industry. In medicine, ultrasound has made available a diagnostic technique that is used in virtually every major medical centre in the world and has important potential for widespread use in local centres. At power levels sufficiently high to modify, damage or destroy human tissues, ultrasound is employed in many therapeutic and surgical applications. In industry, similarly, it has become the basis for an indispensable set of techniques for both investigation and processing.

Although it is difficult to build up a systematic and complete account of all of the possible occurrences and mechanisms that might lead to undesirable consequences of ultrasound exposure, it is possible to make a number of positive statements. Ultrasound has been used in diagnostic medicine for more than 25 years, and in therapeutic applications even longer. Many millions of patients have been exposed to ultrasound during this period, and no verifiable indication has arisen of any adverse effect of the use either of diagnostic procedures or, when properly controlled, of therapy. Nevertheless, because of the general impossibility of fully predicting all the conceivable consequences of human exposure to any given agent, and because of the very large numbers of individuals being exposed, particularly children and fetuses, it is essential to maintain constant vigilance for any evidence of adverse effect.

In this situation, therefore, when risk of injury is at most hypothetical and benefit of some kind is generally expected (though not necessarily always proven) it seems helpful to try to define conditions and working practices that will be "appropriate" to the application being considered. Relevant criteria will now be presented for a variety of different practical situations.

Surgery and cancer therapy

The essential objective in these applications is selective tissue destruction, often in relation to major or life-threatening disease. This is a situation that is closely analogous to the practice of radiotherapy and should thus be subject to similar criteria in determining appropriateness of particular working practices. Normally such treatments will be carried out under the supervision of a qualified physician, with the advice of an appropriate physicist, and it will be important that they and any other staff directly involved in a treatment are thoroughly knowledgeable and experienced in the field, to a degree consistent with generally accepted levels of medical practice.

Applications in physical medicine

These procedures are aimed at beneficial modification of tissue, including acceleration of repair processes, in conditions that are usually not life-threatening. It therefore becomes important to control treatment conditions within bounds that will reliably exclude tissue destruction, pain or other ill effects.

Equipment for this purpose generally operates in the frequency range 0.75–3.0 MHz and here it is generally agreed to be appropriate to confine intensity (temporal average, spatially averaged over the area of the transducer face) in a treatment to not more than $30 \, kW/m^2$. However, since some undesirable bioeffects have been reported to occur at intensities below this level, it will be appropriate to apply further limitations to treatment conditions in some particular circumstances. The pregnant uterus should

275

generally not be irradiated (e.g. as a consequence of therapeutic treatment of low back pain), nor should the epipheses of growing bone or the eye. Exposure conditions that cause pain or discomfort should be avoided, and special care should be taken in the planning of treatments that may include areas having a reduced sensitivity to temperature or pain *(58)*. More generally, treatment techniques using a stationary as opposed to a moving transducer are generally best avoided; a stationary technique, if prolonged, is more likely to give rise both to periosteal pain and also, possibly, to standing waves and the blood stasis phenomenon described by Dyson et al. *(145)*.

Again, an essential factor in attaining appropriate use of ultrasound in this application will be the training and experience of the therapist. It will also be very important that the equipment used be regularly calibrated by reference to agreed national and international standards, and be capable of accurately providing any prescribed output level. Methods for measuring and specifying acoustic output levels in this context have been described, and international agreement in the field is being negotiated through the International Electrotechnical Commission *(23)*.

Medical diagnosis and investigation
The patterns of use of ultrasound for diagnostic and investigative purposes are extremely varied, in terms both of physical exposure of the patient and of the medical problems under investigation. In many areas of medical practice, including cardiology, vascular disease, abdominal disease, obstetrics and gynaecology, paediatrics and ophthalmology, ultrasound constitutes an investigative tool of major importance, and investigations are generally carried out in expectation of substantial benefit to the patient.

In assessing the benefit of a diagnostic procedure, several levels can be discerned. First, there may be improvement in diagnostic accuracy. Second, this improved diagnosis can have a useful influence on the management of the patient. Third, these improvements in diagnosis and management can have a measurable influence on outcome.

In this context, examinations on specific medical indications are generally accepted as being entirely appropriate. However, in the field of obstetrics, a specific question arises concerning policy in relation to systematic screening of an entire pregnant population. Epidemiological evidence is insufficient both in quantity and in quality to indicate whether or not routine obstetric ultrasound examinations, in the absence of specific medical indications, are likely to be of benefit to either mother or fetus, and good evidence seems likely to be difficult to obtain. Thus, although such routine examinations are carried out quite widely in some countries, the technical quality of the examinations may be very variable and their justification derives (in common with many other aspects of the practice of medicine) from the experienced judgement of individuals rather than on statistically-based evidence. It has been recognized that psychosocial, economic and legal/ethical issues must influence such judgements, and different practices have therefore evolved in different countries. Thus an appropriate general policy is to encourage the use of individual clinical judgement in this matter

276

in the absence of a national consensus on the question of whether or not to carry out systematic screening. It is noteworthy in this connection that, for example, a consensus panel convened by the National Institutes of Health in the United States concluded that "data on clinical efficacy and safety do not allow a recommendation for routine screening at this time" *(4)*, whereas the corresponding position of the European Federation of Societies for Ultrasound in Medicine and Biology is that "routine clinical scanning of every woman during pregnancy is not contra-indicated by the evidence currently available from biological investigations and its performance should be left to clinical judgement" *(174)*. These statements are not logically incompatible and they have been followed by a number of similar statements by government officials and agencies and societies of health professionals.

The range of application of ultrasound investigative techniques is continually being extended, and such advances rely on opportunities to try out and confirm efficacy in hitherto unproven situations. Provided that considerations on minimization of exposure are observed, as outlined below, and that the trial is carried out as part of an organized research programme intended for open publication, the procedure can be regarded as appropriate.

It has been common practice to use human subjects as test objects for demonstrating (e.g. commercially) or testing (e.g. routinely) the performance of diagnostic equipment. While there may still be some situations where this practice is justifiable, in general it is inappropriate. Endorsement is here particularly given to the official policy of the World Federation for Ultrasound in Medicine and Biology strongly to discourage the employment of live subjects specifically for commercial demonstration of diagnostic equipment. Furthermore, the quality and availability of tissue-equivalent test phantoms are now reaching a stage where such phantoms can and should always be used for routine test and calibration purposes, rather than employing human subjects.

It has been pointed out earlier in this review that there is a very wide range of levels of ultrasound exposure to patients that arises from the various available diagnostic equipment (Table 2). There is no direct evidence to indicate that any exposure within this range is potentially harmful, nor is there clear evidence of a systematic relationship between exposure level and diagnostic quality. In principle, however, it will be desirable to minimize ultrasound exposure to a patient to a degree consistent with the achievement of the desired diagnostic information, and this will be an important goal at which to aim in future practice. This principle — that exposures should be "as low as reasonably achievable" (acronym ALARA) — is one that has become well accepted in the practice of ionizing radiation protection. In this context the most appropriate measures of exposure appear to be those listed above under "Measurement of sound fields", p. 254, and it will be important that the relevant measurements are made using internationally agreed methods (or other, generally agreed methods where these are not available) and by reference to generally agreed standards. Manufacturers should be encouraged to report these measurements and make them generally available.

In using a specific item of equipment to examine a patient, the ALARA principle can be partially implemented by procedures that minimize the examination time. Additionally, if the equipment is provided with means for adjusting the acoustic output, it may be possible further to minimize exposure of the patient. This would be done by reducing the output to the lowest level at which the equipment is diagnostically effective for a specific application.

This section has attempted to set out guidelines for the appropriate use of ultrasound in medical investigation. These derive from an assessment of available data and also take into account conclusions reached, in recent years, by a number of national expert bodies, such as the American Institute of Ultrasound in Medicine *(147)*, that have considered aspects of the problem.

Airborne ultrasound

Investigations of the possible damage due to airborne ultrasound have generally shown that any actual auditory effects that are produced are attributable to associated acoustic energy in the audible frequency range. Such energy may either be a normal accompaniment of the ultrasound energy or, at very high ultrasound intensities, may be generated by nonlinear processes in the ear itself *(170, 171)*.

This situation, including both the physical measurements that have been made and the correlation that can be shown with auditory effects, has been well described by Acton *(165, 168, 175, 176)*, who has developed a resulting set of exposure criteria that can be endorsed as guidelines for appropriate use. Acton's approach is to extend the range of applicability of "ISO noise rating curve number 85" (the International Organization for Standardization curve, which has been put forward as a hearing damage risk criterion for audible frequencies) to the limit of the audible frequency range and, above this, to introduce a level that takes account of the above-mentioned possibility of nonlinear generation of audible subharmonics within the ear itself. Acton's criterion has the following form:

> The permitted level is 75 dB in the octave band centred on 16 kHz, or in one-third octave bands centred on frequencies up to and including 20 kHz, or in narrow bands centred on frequencies up to 22.5 kHz; the permitted level is 110 dB in octave bands centred on frequencies of 31.5 kHz and above, or in one-third octave bands centred on frequencies of 25 kHz and above, or in narrower bands centred on 22.5 kHz and above. Decibel levels as used here are referred to a reference level of 2×10^{-5} Pa.

These levels are considered by some authorities to be conservative, particularly below 16 kHz and above 31.5 kHz, and somewhat different sets of criteria have been adopted, for example, by the American Conference of Governmental Industrial Hygienists and by some national authorities *(165)*. There is also some evidence that band widths narrower than an octave may be more appropriate for specification and measurement in this context.

Appropriate measurements can be made with a sound-level meter with an overall "linear" frequency response up to at least the fundamental

278

frequency of the ultrasound equipment. The "A-weighting" filter should not be used for this purpose, as this causes relative insensitivity to the higher frequencies. It should be noted that the above frequency range extends beyond that normally specified for sound-level meters *(177)* and care must be taken to ensure that the frequency response and tolerance of the overall equipment, including the microphone and any associated recording equipment, are adequate for these measurements. The microphone is normally placed in the position that will be occupied by the operator's ear.

PROTECTION MEASURES

An important factor that distinguishes ultrasound from ionizing radiation is that an interface between air and a liquid or solid medium presents a major barrier to the transmission of ultrasound. It is largely for this reason that the arrangement of effective protection against ultrasound is generally a fairly straightforward matter. In practical terms, protection measures can be considered separately for two rather different cases: airborne, low-frequency ultrasound, and solid- or liquid-borne ultrasound.

For airborne ultrasound, the critical organ will generally be the ear; protection measures should take this into account and should, in general, be modelled on the procedures established for the audible frequency range. The general objective will be to ensure (by means of appropriate physical measurement procedures, and the installation, where necessary, of sound-absorbing and sound-containing baffles around powerful sound sources) that ambient sound levels do not exceed the above limits. However, where this is, for some reason, not possible, appropriate and adequate protection will need to be used.

Specific protection against solid- and liquid-borne sound may seldom be necessary. Particular situations where caution should be exercised, however, are those where (as in ultrasound cleaning tanks) high power levels of ultrasound are present in liquid media offering potentially good acoustic coupling to the human body. Care should therefore be taken that, during operation, hands are not immersed in such tanks, which should be labelled accordingly.

With regard to medical exposures, guidelines on protection measures follow from the above consideration of "Criteria for appropriate use" and can be summarized as follows.

— Be aware of exposure levels[a] and use minimum exposure of a patient consistent with effective achievement of the desired clinical benefit.

— Do not deliberately expose staff or other personnel in an unnecessary manner or degree.

---

[a] For the quantification of exposure in this context it will be appropriate to employ the parameters listed on p. 257.

— Ensure that all procedures are carried out only by (or, for students, under the supervision of) well and appropriately trained staff.

When using properly designed equipment, and following these guidelines, any exposure of operating personnel should be negligible and it will not be necessary to adopt further specific measures for their protection.

CONCLUSIONS AND RECOMMENDATIONS[a]

Conclusions

Ultrasound is a form of mechanical energy. Many of its uses entail exposure of human beings, either incidentally or, as in the case of medical applications, as an essential part of a procedure. There is a wide spectrum of uses to which ultrasound techniques may be put, mainly in the broad fields of medicine and industry.

Existing literature giving evidence for the induction of biological change demonstrates the importance of valid and relevant dosimetry. This is a poorly developed aspect of the ultrasound field, and the subject of some discussion. The most commonly used dosimetric parameter to date has been intensity, i.e. the flux of energy per unit area along the direction of propagation. An ultrasound field is most easily characterized with a pressure-sensitive hydrophone giving point values of acoustic pressure, and it is anticipated that future use will be for fields to be documented in terms of acoustic pressure.

In view of the wide range of uses of ultrasound, it has been thought best to separate the applications. In medicine, ultrasound at low power levels has become a very widespread and indispensable diagnostic tool, whereas at power levels sufficiently high to modify, damage or destroy human tissues it has many therapeutic and surgical applications. Similarly, in industry, ultrasound is used both for investigation and for processing.

It is difficult to build up a systematic and complete account of all the possible occurrences and mechanisms that might lead to undesirable consequences of ultrasound exposure, but it is possible to make a number of positive statements. Although ultrasound has been used in medicine for more than 25 years, and many millions of patients have been exposed, no verifiable indication has arisen of any adverse effect of the use either of diagnostic procedures or, when properly controlled, of therapy. It is important, however, to be continually vigilant for any evidence of adverse effects.

In the situation, therefore, where risk of injury is at most hypothetical and benefit is generally expected, it seems useful to put forward the following conditions and working practices that are appropriate to particular applications. For surgery and cancer therapy, where the aim is normally

---

[a] These conclusions and recommendations are those made by the WHO Working Group on Health Implications of the Increased Use of NIR Technologies and Devices, Erice, Sicily, September 1985.

selective tissue destruction in relation to major or life-threatening disease, treatments should be carried out under the close supervision of an expert. In physiotherapy, where tissue modification is required, in conditions that are usually not life-threatening, treatment regimes should be chosen to exclude tissue destruction, pain or other ill effects. Particular care should be taken to avoid irradiating the pregnant uterus, the epipheses of growing bone or the eye. In the case of diagnostic ultrasound examinations, it is a good precaution to use exposures that are "as low as reasonably achievable" while still giving good images or other information. The decision as to whether or not to perform a diagnostic examination should be left to informed clinical judgement.

Recommendations
As long as the biological data base on which decisions about both benefit and possible adverse effects remains poor, it is only possible to recommend appropriate working practices; it is not possible to set exposure level limits. For medical exposures, the recommended guidelines are as follows.

1. Be aware of exposure levels and use minimum exposure of a patient consistent with effective achievement of the desired clinical benefit. In this context, minimization of exposure should be done with respect to the following quantities: (*a*) for diagnostic pulse–echo exposures, the spatial peak pressure amplitude, in pascals (where this is not possible, the spatial peak, temporal peak intensity in W/m^2 may be a satisfactory alternative) and (*b*) for surgical, therapeutic and Doppler diagnostic exposures, the total acoustic (temporally averaged) power, in watts.

2. Do not deliberately expose staff or other personnel in an unnecessary manner or degree.

3. Ensure that all procedures are carried out only by (or, for students, under the supervision of) well and appropriately trained staff.

4. For physiotherapy applications, it is thought appropriate to confine intensity (temporal average, spatially averaged over the area of the transducer face) in a treatment to not more than $30\,kW/m^2$.

REFERENCES

1. **French, L.A. et al.** Attempts to determine harmful effects of pulsed ultrasonic vibrations. *Cancer,* **4**: 342 (1951).
2. **Hill, C.R.** The possibility of hazard in medical and industrial applications of ultrasound. *British journal of radiology,* **41**: 561–569 (1968).
3. *Ultrasound.* Geneva, World Health Organization, 1982 (Environmental Health Criteria 22).
4. *Diagnostic ultrasound imaging in pregnancy.* Washington, US Government Printing Office, 1984 (NIH Publication 84-667).

5. *Biological effects of ultrasound: mechanisms and clinical implications.* Bethesda, MD, National Council on Radiation Protection and Measurements, 1983 (NCRP Report No. 74).

6. *Report of the RCOG Working Party on Routine Ultrasound Examination in Pregnancy.* London, Royal College of Obstetricians and Gynaecologists, 1984.

7. **Williams, A.R.** *Ultrasound: biological effects and potential hazards.* New York, Academic Press, 1983.

8. **Bergmann, L.** *Der Ultraschall* [Ultrasound]. Stuttgart, Hirzel Verlag, 1954.

9. **Gooberman, G.L.** *Ultrasonics: theory and application.* London, English Universities Press, 1968.

10. **ter Haar, G. & Hopewell, J.W.** Ultrasonic heating of mammalian tissues *in vivo. British journal of cancer,* **45**(Suppl. 5): 65–67 (1982).

11. **Carstensen, E.L. et al.** Demonstration of nonlinear acoustical effects at biomedical frequencies and intensities. *Ultrasound in medicine and biology,* **6**: 359 (1980).

12. **Duck, F.A. et al.** The output of pulse–echo ultrasound equipment; a survey of powers, pressures and intensities. *British journal of radiology,* **58**: 989–1001 (1985).

13. **Borodzinski, K. et al.** Absolute measurements of output intensity in ultrasonic c.w. and pulse Doppler instruments for blood profile determination and the effect of beam width on blood velocity estimation. *In:* Kazner, E. et al., ed. *Proceedings of the 2nd European Congress on Ultrasonics in Medicine, Munich, 12–16 May 1975.* Amsterdam, Excerpta Medica, 1975 (Excerpta Medica International Congress Series No. 363).

14. **Borodzinski, K. et al.** Quantitative transcutaneous measurements of blood flow in carotid artery by means of pulse and continuous wave Doppler methods. *Ultrasound in medicine and biology,* **2**: 189–193 (1976).

15. **Carson, P.L. et al.** Ultrasonic power and intensities produced by diagnostic ultrasound equipment. *Ultrasound in medicine and biology,* **3**: 341–350 (1978).

16. **Filipczynski, L.** Ultrasonic medical diagnostic methods. *In:* Stephens, R.W.B., ed. *Acoustics 1974.* London, Chapman and Hall, 1974, pp. 71–85.

17. **Hill, C.R.** Acoustic intensity measurements on ultrasonic diagnostic devices. *In:* Bock, J. & Ossoinig, K., ed. *Ultrasonographia medica.* Vienna, Vienna Academy of Medicine, 1970, pp. 21–27.

18. **Rooney, J.A.** Determination of acoustic power outputs in the microwatt–milliwatt range. *Ultrasound in medicine and biology,* **1**: 13–16 (1973).

19. **Stewart, H.F.** Output levels from commercial diagnostic ultrasound equipment. *Journal of ultrasound medicine,* **2**(Suppl.): 39 (1983).

20. **Stewart, H.F. et al.** *Survey of use and performance of ultrasonic therapy equipment in Pinellas County, Florida.* Rockville, MD, Bureau of Radiological Health, 1973 (DHEW Publication No. (FDA) 73-8039).

21. *Acoustical data for diagnostic ultrasound equipment: 1985.* Bethesda, MD, American Institute of Ultrasound in Medicine, 1985.
22. American Institute of Ultrasound in Medicine. Safety standard for diagnostic ultrasound equipment. *Journal of ultrasound medicine,* **2**: S1–S50 (1983).
23. *Characteristics and calibration of hydrophones for operation in the frequency range 0.5 to 15 MHz.* Geneva, International Electrotechnical Commission, 1985 (IEC Publication 866).
24. **Shotton, R.C. et al.** A PVDF membrane hydrophone for operation in the range 0.5 to 15 MHz. *Ultrasonics,* **18**: 123–126 (1980).
25. **De Reggi, A.S. et al.** Piezoelectric polymer probe for ultrasonic application. *Journal, Acoustical Society of America,* **69**: 853–859 (1981).
26. **Lewin, P.A.** Miniature piezoelectric polymer ultrasonic hydrophone probes. *Ultrasonics,* **19**: 213–216 (1981).
27. **Lewin, P.A. & Chivers, R.C.** Two miniature ceramic ultrasonic probes. *Journal of physics E: scientific instruments,* **14**: 1420–1424 (1981).
28. **Bacon, D.R.** Characteristics of a PVDF membrane hydrophone for use in the range 1–100 MHz. *IEEE transactions on sonics and ultrasonics,* SU-29: 18–25 (1982).
29. **Preston, R.C. et al.** PVDF membrane hydrophone performance properties and their relevance to the measurement of acoustic output of medical ultrasonic equipment. *Journal of physics E: scientific instruments,* **16**: 786–796 (1983).
30. **Platte, M.** A polyvinylidene fluoride needle hydrophone for ultrasonic applications. *Ultrasonics,* **23**: 113–118 (1985).
31. **Brendel, K. & Ludwig,** Calibration of ultrasonic standard probe transducers. *Acoustica,* **36**: 203–212 (1976).
32. **Herman, B.A. & Harris, G.R.** Calibration of miniature ultrasonic receivers using a planar scanning technique. *Journal, Acoustical Society of America,* **72**: 1357–1363 (1982).
33. **Rooney, J.A. & Nyborg, W.L.** Acoustic radiation pressure in a travelling plane wave. *American journal of physics,* **40**: 1825–1830 (1972).
34. **Shotton, R.C.** A tethered float radiometer for measuring the output power from ultrasonic therapy equipment. *Ultrasound in medicine and biology,* **6**: 131–133 (1980).
35. **Farmery, M.J. & Whittingham, T.A.** A portable radiation force balance for use with diagnostic ultrasonic equipment. *Ultrasound in medicine and biology,* **3**: 373 (1978).
36. **Livett, A.J. & Preston, R.C.** A comparison of the AIUM/NEMA, IEC and FDA (1980) definitions of various acoustic output parameters for ultrasonic transducers. *Ultrasound in medicine and biology,* **11**: 793–802 (1985).
37. **Bang, J.** The intensity of ultrasound in the uterus during examination for diagnostic purposes. *Acta pathologica et microbiologica scandinavica, A,* **80**: 341–344 (1972).
38. **Neppiras, E.A.** Acoustic cavitation. *Physics reports,* **61**: 159 (1980).
39. **Neppiras, E.A.** Acoustic cavitation threshold and cyclic processes. *Ultrasonics,* **18**: 201–209 (1980).

40. **Apfel, R.E.** Acoustic cavitation. *In:* Edmonds, P.D., ed. *Methods of experimental physics — ultrasonics.* New York, Academic Press, 1981, Vol. 19, pp. 356–411.

41. **Apfel, R.E.** Acoustic cavitation: a possible consequence of biomedical uses of ultrasound. *British journal of cancer,* **45**(Suppl. V): 140–146 (1982).

42. **Hill, C.R.** Ultrasonic exposure thresholds for changes in cells and tissues. *Journal, Acoustical Society of America,* **52**: 666–672 (1972).

43. **Hill, C.R.** Detection of cavitation. *In:* Reid, J.M. & Sikov, M.R., ed. *Interaction of ultrasound and biological tissues.* Washington, DC, Department of Health, Education, and Welfare, 1972, pp. 199–200 (Publication No. (FDA) 73-8008).

44. **Apfel, R.E.** Acoustic cavitation prediction. *Journal, Acoustical Society of America,* **69**: 1624 (1981).

45. **Flynn, H.G.** Generation of transient cavities in liquids by microsecond pulses of ultrasound. *Journal, Acoustical Society of America,* **72**: 1926–1932 (1982).

46. **ter Haar, G.R. & Daniels, S.** Evidence for ultrasonically induced cavitation *in vivo. Physics in medicine and biology,* **26**: 1145 (1981).

47. **ter Haar, G.R. et al.** Ultrasonically induced cavitation *in vivo. British journal of cancer,* **45**(Suppl. 5): 151 (1982).

48. **Gross, D.R. et al.** A search for ultrasonic cavitation within the canine cardiovascular system. *Ultrasound in medicine and biology,* **11**: 85–97 (1985).

49. **Jacke, S.E.** Ultrasonics in industry today. *In: Ultrasonics international 1979.* Guildford, IPC Science and Technology Press, 1979.

50. **Shoh, A.** Industrial application of ultrasound — a review. I. High power ultrasound. *IEEE transactions on sonics and ultrasonics,* SU-22: 60–71 (1975).

51. **Wood, R.W. & Loomis, A.L.** The physical and biological effects of high frequency sound waves of great intensity. *Philosophical magazine,* **4**: 417–436 (1927).

52. **Lynnworth, L.C.** Industrial applications of ultrasound — a review. II. Measurements, tests and process control using low intensity ultrasound. *IEEE transactions on sonics and ultrasonics,* SU-22: 71–101 (1973).

53. WHO Technical Report Series, No. 723, 1985 (*Future use of new imaging technologies in developing countries:* report of a WHO Scientific Group).

54. **Lehmann, J.F.** Diathermy. *In:* Krusen, F.H. et al., ed. *Handbook of physical medicine and rehabilitation.* Philadelphia, PA, W.B. Saunders, 1971, pp. 273–345.

55. **Lehmann, J.F. et al.** Therapeutic heat and cold. *In:* Umst, M.R., ed. *Clinical orthopedics and related research.* Philadelphia, PA, J.J. Lippincot, 1974.

56. **Lehmann, J.F. & Lateur, B.J.** Therapeutic heat. *In:* Licht, S., ed. *Therapeutic heat and cold,* 3rd ed. Baltimore, Williams & Wilkins, 1982.

57. **Binder, A. et al.** Is therapeutic ultrasound effective in treating soft tissue lesions? *British medical journal,* **290**: 512–514 (1985).
58. **Dyson, M.** Therapeutic applications of ultrasound. *In:* Nyborg, W.L. & Ziskin, M.C., ed. *Biological effects of ultrasound.* New York, Churchill Livingstone, 1985, pp. 121–133 (Clinics in Diagnostic Ultrasound, Vol. 16).
59. **McDiarmid, T. et al.** Ultrasound and the treatment of pressure sores. *Physiotherapy (Lond.),* **71**: 66–70 (1985).
60. **Patrick, M.K.** Ultrasound in physiotherapy. *Ultrasonics,* **4**: 10–14 (1966).
61. **Dyson, M. et al.** Stimulation of healing of varicose ulcers by ultrasound. *Ultrasonics,* **14**: 232–236 (1976).
62. *Canada-wide survey of non-ionizing radiation emitting devices. Part II. Ultrasound devices.* Ottawa, Environmental Health Directorate, Health Protection Branch, 1980 (Report 80-EHD-53).
63. **Hahn, G.** *Hyperthermia and cancer.* New York, Plenum, 1982.
64. **Corry, P.M. et al.** Human cancer treatment with ultrasound. *IEEE transactions on sonics and ultrasonics,* **SU-31**: 444–456 (1984).
65. **Kremkau, F.W.** Cancer therapy with ultrasound: a historical review. *Journal of clinical ultrasound,* **7**: 287–300 (1979).
66. **Wells, P.N.T.** Surgical applications of ultrasound. *In:* Nyborg, W.L. & Ziskin, M.C., ed. *Biological effects of ultrasound.* New York, Churchill Livingstone, 1985, pp. 157–167 (Clinics in Diagnostic Ultrasound, Vol. 16).
67. **Goliamina, I.P.** Ultrasonic surgery. *In:* Stephens, R.W.B., ed. *Acoustics 1974.* London, Chapman and Hall, 1975, pp. 63–69.
68. **Bullen, M.A. et al.** A physical survey of the ultrasonic treatment of Ménière's disease. *Ultrasonics,* **1**: 2–8 (1963).
69. **Coleman, D.J. et al.** Therapeutic ultrasound in the treatment of glaucoma. I. Experimental model. *Ophthalmology,* **92**: 339–346 (1985).
70. **Coleman, D.J. et al.** Therapeutic ultrasound in the treatment of glaucoma. II. Clinical applications. *Ophthalmology,* **92**: 347–353 (1985).
71. **Kelman, C.D.** Phaco-emulsification and aspiration. A report of 500 consecutive cases. *American journal of ophthalmology,* **75**: 764–768 (1973).
72. **Fahiq, S. & Wallace, D.M.** Ultrasonic lithotripter for urethral and bladder stones. *British journal of urology,* **50**: 255 (1978).
73. **Saunders, J.E. & Coleman, A.J.** "The lithotripter" — a non-invasive method for the disintegration of renal stones by extra-corporeally generated shock waves. *IPSM Report,* No. 47 (1986).
74. **Ewen, S.J. & Glickstein, C.** *Ultrasonic therapy in periodontics.* Springfield, IL, Charles Thomas, 1968.
75. Counsel on dental research: ultrasonics in dentistry. *Journal of the American Dental Association,* **50**: 573–576 (1955).
76. **Lees, S.** Proposed studies in potential toxicity of ultrasonics in dentistry. *In:* Reid, J.M. & Sikov, M.R., ed. *Interaction of ultrasound and biological tissues.* Washington, DC, Department of Health, Education, and Welfare, 1971, pp. 135–138 (Publication No. (FDA) 73-8000).

77. **El'piner, I.P.** *Ultrasound: physical, chemical and biological effects.* New York, Consultants Bureau, 1964.

78. **Peacocke, A.R. & Pritchard, N.J.** Some biophysical aspects of ultrasound. *Progress in biophysics,* **18**: 185–208 (1968).

79. **Clarke, P.R. & Hill, C.R.** Physical and chemical aspects of ultrasonic disruption of cells. *Journal, Acoustical Society of America,* **47**: 649–653 (1970).

80. **Combes, R.D.** Absence of mutation following ultrasonic treatment of *Bacillus subtilis* cells and transforming deoxyribonucleic acid. *British journal of radiology,* **48**: 306–311 (1975).

81. **Graham, E. et al.** Cavitational bio-effects at 1.5 MHz. *Ultrasonics,* **18**: 224–228 (1980).

82. **Morton, K.I. et al.** Subharmonic emission as an indicator of ultrasonically induced biological damage. *Ultrasound in medicine and biology,* **9**: 629–633 (1983).

83. **Coakley, W.T. et al.** Quantitative relationships between ultrasonic cavitation and effects upon amoebae at 1 MHz. *Journal, Acoustical Society of America,* **50**: 1546–1553 (1971).

84. **Clarke, P.R. & Hill, C.R.** Biological action of ultrasound in relation to the cell cycle. *Experimental cell research,* **38**: 443–444 (1969).

85. **Rott, H.-D. & Soldner, R.** The effect of ultrasound on human chromosomes *in vitro. Humangenetik,* **20**: 103–112 (1973).

86. **Fu, Y.K. et al.** Ultrasound lethality to synchronous and asynchronous Chinese hamster V-79 cells. *Ultrasound in medicine and biology,* **6**: 39 (1980).

87. **Thacker, J.** The possibility of genetic hazard from ultrasonic radiation. *Current topics in radiation research quarterly,* **8**: 235–258 (1973).

88. **Thacker, J.** Investigations into genetic and inherited changes produced by ultrasound. *In:* Nyborg, W.L. & Ziskin, M.C., ed. *Biological effects of ultrasound.* New York, Churchill Livingstone, 1985, pp. 67–76 (Clinics in Diagnostic Ultrasound, Vol. 16).

89. **Thacker, J.** An assessment of ultrasonic radiation hazard using yeast genetic systems. *British journal of radiology,* **47**: 130–138 (1974).

90. **Buckton, K.E. & Baker, N.V.** An investigation into possible chromosome damaging effects of ultrasound on human blood cells. *British journal of radiology,* **45**: 340–342 (1972).

91. **Coakley, W.T. et al.** Examination of lymphocytes for chromosome aberrations after ultrasonic irradiation. *British journal of radiology,* **45**: 328–332 (1972).

92. **Hill, C.R. et al.** A search for chromosome damage following exposure of Chinese hamster cells to high intensity, pulsed ultrasound. *British journal of radiology,* **45**: 333–334 (1972).

93. **Macintosh, I.J.C. et al.** Ultrasound and *in vitro* chromosome aberrations. *British journal of radiology,* **48**: 230–232 (1975).

94. **Watts, P.L. et al.** Ultrasound and chromosome damage. *British journal of radiology,* **45**: 335–339 (1972).

95. **Rott, H.-D.** Zur Frage der Schädigungsmöglichkeit durch diagnostischen Ultraschall. *Ultraschall,* **2**: 56–64 (1981).

96. **Kunze-Muhl, E.** Chromosome damage in human lymphocytes after different combinations of X-ray and ultrasonic treatment. *In:* Kazner, E. et al., ed. *Ultrasonics in medicine.* Amsterdam, Excerpta Medica, 1975, pp. 3–9.

97. **Gebhart, E.** Sister chromatid exchange (SCE) and structural chromosome aberration in mutagenicity testing. *Human genetics,* **58**: 235–254 (1981).

98. **Liebeskind, D. et al.** Sister chromatid exchanges in human lymphocytes after exposure to diagnostic ultrasound. *Science,* **205**: 1273–1275 (1979).

99. **Morris, S.M. et al.** Effect of ultrasound on human leucocytes. Sister chromatid exchange analysis. *Ultrasound in medicine and biology,* **4**: 253 (1978).

100. **Au, W.W. et al.** Sister chromatid exchanges in mouse embryos after exposure to ultrasound *in utero. Mutation research,* **103**: 315 (1982).

101. **Barrass, N. et al.** The effect of ultrasound and hyperthermia on sister chromatid exchange and division kinetics of BHK21 C13/A3 cells. *British journal of cancer,* **45**(Suppl. V): 187 (1982).

102. **Barnett, S.B. et al.** An investigation into the mutagenic potential of pulsed ultrasound. *British journal of radiology,* **55**: 501 (1982).

103. **Miller, D.L.** The botanical effects of ultrasound: a review. *Environmental and experimental botany,* **23**: 1–27 (1983).

104. **Wegner, R.D. et al.** Has diagnostic ultrasound mutagenic effects? *Human genetics,* **56**: 95 (1980).

105. **Miller, M.W.** Does ultrasound induce sister chromatid exchanges? *Ultrasound in medicine and biology,* **11**: 561–570 (1985).

106. **Goss, S.A.** Sister chromatid exchange and ultrasound. Report of the Bio-effects Committee of the American Institute of Ultrasound in Medicine. *Journal of ultrasound in medicine,* **3**: 463–470 (1984).

107. **Repacholi, M.H. et al.** Interaction of low intensity ultrasound and ionizing radiation with the tumour cell surface. *Physics in medicine and biology,* **16**: 221–227 (1971).

108. **Taylor, K.J.W. & Newman, D.L.** Electrophoretic mobility of Ehrlich suspensions exposed to ultrasound of varying parameters. *Physics in medicine and biology,* **17**: 270–276 (1972).

109. **Joshi, G.P. et al.** Mode of action of ultrasound on the surface charge of mammalian cells *in vitro. Ultrasound in medicine and biology,* **1**: 45–48 (1973).

110. **Chapman, I.V.** The effect of ultrasound on the potassium content of rat thymocytes *in vitro. British journal of radiology,* **47**: 411–415 (1974).

111. **Hrazdira, I.** Changes in cell ultrastructure under direct and indirect action of ultrasound. *In:* Bock, J. & Ossoinig, K., ed. *Ultrasonographia medica.* Vienna, Vienna Academy of Medicine, 1970, pp. 457–463.

112. **Dvorak, M. & Hrazdira, I.** Changes in the ultrastructure of bone marrow cells in rats following exposure to ultrasound. *Zeitschrift für mikroskopische-anatomische Forschung,* **4**: 451–460 (1966).

113. **Taylor, K.J.W. & Pond, J.B.** Primary sites of ultrasonic damage on cell systems. *In:* Reid, J.M. & Sikov, M.R., ed. *Interaction of ultrasound and*

biological tissues. Washington, DC, Department of Health, Education, and Welfare, 1972 (Publication No. (FDA) 73-8008).

114. **Watmough, D.J. et al.** The biophysical effects of therapeutic ultrasound on HeLa cells. *Ultrasound in medicine and biology,* **3**: 205–219 (1977).

115. **ter Haar, G.R. et al.** Ultrastructural changes in the mouse uterus brought about by ultrasonic irradiation at therapeutic intensities in standing wave fields. *Ultrasound in medicine and biology,* **5**: 167–179 (1979).

116. **Dyson, M. et al.** The production of blood cell stasis and endothelial damage in the blood vessels of chick embryos treated with ultrasound in a stationary wave field. *Ultrasound in medicine and biology,* **1**: 133–148 (1974).

117. **Clarke, P.R. et al.** Synergism between ultrasound and X-rays in tumour therapy. *British journal of radiology,* **43**: 97–99 (1970).

118. **ter Haar, G.R. et al.** Ultrasonic irradiation of mammalian cells *in vitro* at hyperthermic temperatures. *British journal of radiology,* **53**: 784–789 (1980).

119. **Li, G.C. et al.** Cellular inactivation by ultrasound. *Nature,* **267**: 163–165 (1977).

120. **Dunn, F. & Hawley, S.A.** Ultra high-frequency acoustic waves in liquids and their interaction with biological structures. *In:* Kelly, E., ed. *Ultrasonic energy.* Urbana, University of Illinois Press, 1965, pp. 66–76.

121. **Harvey, W. et al.** The *in vitro* stimulation of protein synthesis in human fibroblasts by therapeutic levels of ultrasound. *In:* Kazner, E. et al., ed. *Ultrasonics in medicine.* Amsterdam, Excerpta Medica, 1975, pp. 10–21.

122. **Dyson, M. & Brookes, M.** Stimulation of bone repair by ultrasound. *In:* Lerski, R.A. & Morley, P., ed. *Ultrasound '82. Proceedings of the 3rd Meeting, World Federation of Ultrasound in Medicine and Biology.* Oxford, Pergamon Press, 1983, pp. 61–66.

123. **Duarte, L.R.** The stimulation of bone growth by ultrasound. *Archives of orthopedic and trauma surgery,* **101**: 153–159 (1983).

124. **Hogan, R.D. et al.** The effect of ultrasound on microvascular hemodynamics in skeletal muscle: effects during ischemia. *Microvascular research,* **23**: 370–379 (1982).

125. **Stratmeyer, M.E. et al.** *In:* Hazzard, D.G. & Litz, M.L., ed. *Symposium on biological effects and characterization of ultrasound sources.* Rockville, MD, Bureau of Radiological Health, 1977, pp. 140–145 (DHEW Publication (FDA) 78-8048).

126. **Stolzenberg, S.J. et al.** Toxicity of ultrasound in mice: neonatal studies. *Radiation and environmental biophysics,* **18**: 37–44 (1980).

127. **Lyon, M.F. & Simpson, G.M.** An investigation into the possible genetic hazards of ultrasound. *British journal of radiology,* **47**: 712–722 (1974).

128. **Stolzenberg, S.J. et al.** Effects of ultrasound on the mouse exposed at different stages of gestation: acute studies. *Radiation and environmental biophysics,* **17**: 245–270 (1980).

129. **Stratmeyer, M.E. et al.** Effects of *in utero* ultrasound exposure on the growth and development of mice. *Ultrasound in medicine and biology,* **8**: 185 (1982).

130. **Sikov, M.R. & Collins, H.D.** Embryotoxic potential of C.W. ultrasound in the pre-implantation rat embryo. *Journal of ultrasound medicine,* **1**: 141 (1982).

131. **Akamatsu, N.** Ultrasound irradiation effects on pre-implantation embryos. *Acta obstetrica et gynaecologica japonica,* **33**: 969–978 (1981).

132. **Garrison, B.M. et al.** The influence of ovarian sonication on fetal development in the rat. *Journal of clinical ultrasound,* **1**: 316–319 (1973).

133. **McClain, R.M. et al.** Teratological study of rats exposed to ultrasound. *American journal of obstetrics and gynecology,* **114**: 39–42 (1972).

134. **Mannor, S.M. et al.** The safety of ultrasound in fetal monitoring. *American journal of obstetrics and gynecology,* **113**: 653–661 (1972).

135. **O'Brien, W.D.** Dose dependent effect of ultrasound on fetal weight in mice. *Journal of ultrasound medicine,* **2**: 1–8 (1983).

136. **Shoji, R. et al.** Influence of low intensity ultrasonic irradiation on prenatal development of two inbred mouse strains. *Teratology,* **12**: 227–231 (1975).

137. **Sikov, M.R. et al.** Intensity–response relationships following exposure of the 9 day rat embryo to 0.8 MHz C.W. ultrasound. *Journal of ultrasound medicine,* Suppl. 2, p. 41 (1983).

138. **Lele, P.P.** Ultrasonic teratology in mouse and man. *In:* Kazner, E. et al., ed. *Ultrasonics in medicine.* Amsterdam, Excerpta Medica, 1975, pp. 22–27.

139. **Abdulla, U. et al.** Effect of diagnostic ultrasound on maternal and fetal chromosomes. *Lancet,* **2**: 829–831 (1971).

140. **Ikeuchi, T. et al.** Ultrasound and embryonic chromosomes. *British medical journal,* **1**: 112 (1973).

141. **Levi, S. et al.** *In vivo* effect of ultrasound at human therapeutic doses on marrow cell chromosomes of golden hamster. *Humangenetik,* **25**: 133–141 (1974).

142. **Stella, M. et al.** Induction of sister chromatid exchanges in human lymphocytes exposed *in vitro* and *in vivo* to therapeutic ultrasound. *Mutation research,* **138**: 75–85 (1984).

143. **Zheng, H.Z. et al.** *In vivo* exposure to diagnostic ultrasound and *in vitro* assay of sister chromatid exchange in cultured amniotic fluid cells. *Obstetrics and gynecology,* **9**: 491 (1981).

144. **Edwards, M.J.** Congenital defects in guinea pigs: fetal resorptions, abortions and malformations following induced hyperthermia during early gestation. *Teratology,* **2**: 313 (1969).

145. **Dyson, M. et al.** Flow of red blood cells stopped by ultrasound. *Nature,* **232**: 572–573 (1971).

146. **Dyson, M. et al.** Stimulation of tissue regeneration by pulsed plane wave ultrasound. *IEEE transactions on sonics and ultrasonics,* **SU-17**: 133–140 (1970).

147. *Safety considerations for diagnostic ultrasound: report of the Bioeffects Committee.* Bethesda, MD, American Institute of Ultrasound in Medicine, 1984 (AIUM Publication No. 316).

148. **Selman, G.G. & Counce, S.J.** Abnormal embryonic development in *Drosophila melanogaster* induced by ultrasonic treatment. *Nature,* **172**: 503–505 (1953).

149. **Child, S.Z. et al.** Effects of ultrasound on *Drosophila*. III. Exposure of larvae to low temporal average intensity, pulsed irradiation. *Ultrasound in medicine and biology,* **7**: 167 (1981).

150. **Child, S.Z. & Carstensen E.L.** Effects of ultrasound on *Drosophila*. IV. Pulsed exposures of eggs. *Ultrasound in medicine and biology,* **8**: 311 (1982).

151. **Lehmann, J.F. et al.** The effects of ultrasound on chromosomes, nuclei and other structures of the cells in plant tissues. *Archives of physical medicine and rehabilitation,* **35**: 141–148 (1954).

152. **Selman, G.G.** The effect of ultrasonics upon mitosis. *Experimental cell research,* **3**: 656–674 (1952).

153. **Bleaney, B.I. & Oliver, R.** The effect of irradiation of *Vicia faba* roots with 1.5 MHz ultrasound. *British journal of radiology,* **45**: 358–361 (1972).

154. **Gregory, W.D. et al.** Non thermal effects of 2 MHz ultrasound on the growth and cytology of *Vicia faba* roots. *British journal of radiology,* **47**: 122–129 (1974).

155. **Miller, M.W. et al.** Chromosomal anomalies cannot account for growth rate reduction in ultrasonicated *Vicia faba* root meristems. *Radiation botany,* **15**: 431–437 (1975).

156. **Miller, M.W. et al.** Absence of an effect of diagnostic ultrasound on sister chromatid induction in human lymphocytes *in vitro. Mutation research,* **120**: 261 (1983).

157. **Gershoy, A. et al.** Intercellular gas: its role in sonated plant tissue. *In:* White, D. & Barnes, R., ed. *Ultrasound in medicine.* New York, Plenum, 1976, Vol. 2, pp. 501–511.

158. **Bakketeig, L.S. et al.** Randomised controlled trial of ultrasonographic screening in pregnancy. *Lancet,* **2**: 207–210 (1984).

159. **Eik-Nes, S.H. et al.** Ultrasound screening in pregnancy: a randomised controlled trial. *Lancet,* **1**: 1347 (1984).

160. **Hellman, L.M. et al.** Safety of diagnostic ultrasound in obstetrics. *Lancet,* **1**: 1133–1135 (1970).

161. **Koranyi, G. et al.** Follow-up examination of children exposed to ultrasound *in utero. Acta paediatrica academiae scientiarum hungaricae,* **13**: 231–238 (1972).

162. **Stark, C.R. et al.** Short and long term risks after exposure to diagnostic ultrasound *in utero. Obstetrics and gynecology,* **63**: 194–200 (1984).

163. **Kinnier-Wilson, L.M. & Waterhouse, J.A.H.** Obstetric ultrasound and childhood malignancies. *Lancet,* **2**: 997–998 (1984).

164. **Cartwright, R.A. et al.** Ultrasound examinations in pregnancy and childhood cancer. *Lancet,* **2**: 999–1000 (1984).

165. **Acton, W.I.** Exposure to industrial ultrasound: hazards, appraisal and control. *Journal of the Society of Industrial Medicine,* **33**: 107–113 (1983).

166. **Acton, W.I. & Carson, M.B.** Auditory and subjective effects of airborne noise from industrial ultrasonic sources. *British journal of industrial medicine,* **24**: 297–304 (1967).

167. **Skillern, C.P.** Human response to measured sound pressure levels from ultrasonic devices. *American Industrial Hygiene Association journal,* **26**: 132 (1965).

168. **Acton, W.I.** The effects of industrial airborne ultrasound on humans. *Ultrasonics,* **12**: 124–128 (1974).

169. **Grigor'eva, V.M.** Effect of ultrasonic vibrations on personnel working with ultrasonic equipment. *Soviet physics — acoustics,* **11**: 426 (1966).

170. **Knight, J.J.** Effects of airborne ultrasound on man. *Ultrasonics,* **6**: 39–42 (1968).

171. **Parrack, M.O.** Effects of airborne ultrasound on humans. *International audiology,* **5**: 294 (1966).

172. **Dallos, P.J. & Linnel, C.O.** Sub-harmonic components in cochlear microphonic potentials. *Journal, Acoustical Society of America,* **40**: 4 (1966).

173. **Dallos, P.J. & Linnel, C.O.** Even-order sub-harmonics in the peripheral auditory system. *Journal, Acoustical Society of America,* **40**: 561 (1966).

174. Statement on the safety evaluation of ultrasound, on behalf of the European Federation of Societies for Ultrasound in Medicine and Biology. *European medical ultrasonics,* **6**(3): 5 (1984).

175. **Acton, W.I.** A criterion for the prediction of auditory and subjective effects due to airborne noise from ultrasonic sources. *Annals of occupational hygiene,* **11**: 227–234 (1968).

176. **Acton, W.I.** Exposure criteria for industrial ultrasound. *Ultrasonics,* **14**: 42 (1976).

177. *Sound level meters.* Geneva, International Electrotechnical Commission, 1979 (Publication 651).

291

7

Regulation and enforcement procedures

F. Kossel

CONTENTS

	Page
Introduction	293
International agreements	294
Technological development and protection measures	295
Data for legislation	296
Standards	297
Exposure standards	298
Emission standards	298
Regulation and enforcement procedures	298
Education and training	299
Safe exposure limits	299
Causal model	300
Phenomenological model	303
Conservative assumptions	304
Compliance and enforcement	305
Approval of application	306
Approval of equipment	307
Remedial protection measures	307
Inspection and maintenance	308
Review of regulations	308
Exemptions	308
Information programmes	308
References	309

INTRODUCTION

Nonionizing radiation is capable of causing effects in biological systems under certain circumstances. Everybody is exposed to various forms of NIR in everyday life, but this alone would not justify the introduction of

293

administrative measures to control the production, distribution and application of NIR sources. Nevertheless, because some of the effects may lead to changes that are potentially hazardous for biological systems, and because of the rapidly expanding use of NIR for telecommunications, as well as for scientific, medical, industrial, commercial and domestic purposes, and the multiplicity of radiation-emitting devices and installations, certain regulations are necessary *(1)*.

A potential hazard to health can be produced either as a result of the exposure of the human body to NIR, or by interaction with technical devices, which are themselves affected by NIR and then give rise to health hazards (e.g. interference with electromedical devices, unintentional triggering of electrically activated detonators, and ignition of flammable materials). This is especially true if the individuals concerned are not aware of the potential hazard or are unable to identify it.

In previous chapters, where the various types of radiation are discussed in detail, guidance is provided on the application of regulatory controls. However, in many cases if such guidance, which is aimed at ensuring the effective protection of both public and individual health, is to be put into practice it will have to be given the force of law. Three successive phases may be identified in the orderly development of regulation and enforcement procedures. First, biological changes and possible hazards to health are identified and studied. Next, based on such studies, standards are derived and proposed. These are generally of two types, namely exposure standards (concerned with individuals who may be exposed either occupationally or in the course of everyday life) and emission standards (concerned with devices and installations which emit NIR, whether intentionally, incidentally or because they have developed a fault). The third phase is the translation of exposure and emission standards into appropriate procedures.

The normal way for a government to introduce legislation in a new technical field is to apply approved and accepted legislative principles to it. In many countries, the basic legislation is supplemented by ordinances and rules designed to meet specific needs. This provides adequate flexibility to permit adaptation to future developments in science and technology.

It must be accepted that the national legal practice of each individual country must take precedence. Thus, general guidance on the application of technical recommendations may often be inappropriate as a basis for making regulations to cover a specific case. For this reason, the intention here is only to give information on the methodological aspects of regulation and enforcement procedures, and to discuss their advantages and disadvantages, irrespective of their compatibility with existing legal traditions.

INTERNATIONAL AGREEMENTS

When legal provisions laying down technical standards, e.g. exposure limits, are introduced, the applications of the particular branch of science or technology concerned are restricted to those existing at the time of enactment. Whether or not those standards can be adapted to meet subsequent developments will depend on the legislative procedure. The consequences may therefore be, firstly, to inhibit the application of further technical developments and, secondly, to establish trade barriers between countries. This may occur even if their safety philosophy and protection policies are essentially the same. A more formal but typical example is provided by national requirements for the marking of certain products or components by means of symbols or colours. The failure of certain devices to comply with the requirements of a national standard can sometimes be due merely to the fact that the labelling requirements of other national standards are different. Uniformity in identification is by all means an essential safety requirement. However, the replacement of, for example, one of the conductors in a finished complete circuit, because it is necessary to change the colour of the covering or insulation to satisfy export requirements, is not only uneconomical but also gives rise to a greater risk of errors. This can easily be avoided by international agreements.

National standards will depend on the state of knowledge at the time of adoption. If the technology is rapidly developing, such national standards are likely to differ if adopted at different times. Unless internationally agreed standards for NIR are adopted in the near future, the piecemeal adoption of national regulations may lead to a situation of this kind.

The preparation of internationally accepted scientific recommendations establishing an international safety level is probably the best way of preventing such a situation from developing. A good example of such recommendations in the field of ionizing radiation are those of the International Commission on Radiological Protection (ICRP) (2).

As early as 1974 it was suggested that an international commission on protection against NIR be formed along the lines of the ICRP (3), but for various reasons such a commission has not yet been launched. However, a valuable initiative was taken by the General Assembly of the International Radiation Protection Association (IRPA) in 1977 in establishing an IRPA committee on NIR. This committee is cooperating with WHO in the preparation of a number of NIR environmental health criteria documents.

TECHNOLOGICAL DEVELOPMENT AND PROTECTION MEASURES

Any radiation protection measures, whether enforced by law or not, that are effectively implemented, limit the possibility of scientifically observing the effects of radiation exposure on the human body. Indeed, the mere suspicion of a hazard is sufficient for protection measures to be taken in order to prevent possible effects. Because of such protection measures, practical experience of effects on man is reduced to the observation of exceptional cases. A particular protection measure will be judged effective if, following its introduction, cases of radiation damage are rare. As a consequence, the necessary experience can be obtained only over long periods of time or from animal experiments correlated with observations on man. Such results will require careful interpretation.

295

Legislation should be introduced to require the reporting to a central registry of radiation accidents or incidents involving abnormal exposures. Evaluation of such reports can provide valuable insight into potential problem areas requiring regulatory control. It is necessary, however, to avoid drawing inappropriate conclusions with regard to cause and effect and radiation risks. Such conclusions are valid only if they are supported by systematic experimental and epidemiological research.

Moreover, legislation should avoid the possibility of hazards resulting from the use of a particular type of radiation and from the frequency and duration of this application. The necessary data should be obtained by whatever means may be most effective and appropriate. This may involve a survey of those cases where the introduction of protective measures has proved necessary.

Many countries have general regulations covering the safe application of new technologies. Based on such regulations, protection measures could be introduced at an early stage. Unfortunately, such cases are the exception since the control of all technical developments, before the existence of any potential hazard has been confirmed, is very expensive and time consuming and often not in harmony with existing political principles. Yet, for most of the applications of NIR, technical developments are so far advanced that the necessary protection measures are already embodied (or will be in the near future) in special regulations for the type of radiation concerned. In this connection, reference may be made to the legislation on protection against NIR which already exists in a number of countries.

DATA FOR LEGISLATION

As mentioned previously, there is a risk that rapid technical development will result in a gap between the present state of technology and existing regulations, since the translation of advanced knowledge into adequate legal controls always requires a considerable period of time. Introduction of regulatory controls when technical development has reached an advanced stage is a poor alternative. From experience with air and water pollution control in large industrial centres, it has been learned, for example, that it becomes rather difficult to introduce more stringent protection measures without hurting already existing industries.

Only in the last few years has it been realized that the development of new technologies should go hand in hand with systematic research on potential health hazards and the introduction of adequate protection measures. One step in the right direction is substantial governmental support to appropriate fundamental investigations to meet the most urgent needs. This support should complement the resources already allocated by industry and public funding for improving the design of technical equipment and installations which are or include NIR sources. Great efforts will be necessary if a dangerous increase in the gap between the state of technological development and the available protection measures is to be avoided.

296

Although scientific publications on the causes of NIR effects or on the development of protection measures are based on systematic research, no fundamental general strategy exists for the generation of adequate scientific information necessary for the drafting of legal provisions on radiation protection.

The general objective is to use NIR for the benefit of mankind while avoiding potentially harmful effects on health and the environment. Even the first step towards this objective, however, leads to a quantitative problem. Not every biological effect caused by NIR is harmful. The question therefore arises of limits for radiation exposure below which no harmful effects will be produced in the exposed individual or group. Knowledge of such limits is thus essential in order to ensure, by means of legislation, that they are not exceeded. A review of the relevant literature for each type of radiation does not always reveal agreed numerical values for these limits.

The establishment of exposure limits and acceptable regulatory procedures for the safe use of NIR poses a number of problems. Not only are there deficiencies in the scientific literature, as already noted, but the limits of validity and accuracy of the scientific data presented must also be recognized. Furthermore, it may be necessary to take action even though a causal relationship has not been established. Consequently, and despite such deficiencies, an effort must be made to assess the relative hazards from NIR with regard both to the severity of the harm resulting from a particular operation and to its risk (i.e. the probability that such harm will occur). That probability must also be related to the population exposed, which may be either an occupational group or the general public. The information thus obtained will provide the scientific basis for the establishment of legislation and/or appropriate regulatory procedures. Where serious hazards are involved the legislator may require an assessment with respect to their severity and risk.

STANDARDS

The philosophy and principles underlying the development of exposure and emission standards for ionizing radiation protection have been thoroughly treated by ICRP. Essentially, the aim has been to develop exposure guidelines, i.e. basic and derived limits for human exposure to ionizing radiation.

Similar approaches may be appropriate in the development of standards for NIR protection. However, because present knowledge of the fundamental mechanisms of NIR interaction and the associated biological effects is still less developed, and because it is not known in all cases whether stochastic as well as non-stochastic effects will occur, existing protection measures must be regarded as provisional. In addition, it is still not certain whether the mechanism for the transmission and expression of hereditary traits is affected. In cases of uncertainty, protection measures should err in the direction of providing greater safety. In this case, the philosophy adopted by ICRP (4), namely the ALARA (as low as readily achievable) principle, might provide an interim and conservative basis for a standard until more quantitative data on biological effects become available.

The setting of emission or exposure standards implies the ability to measure the particular type of NIR concerned. To obtain valid and compatible measurements, it is necessary to ensure that measuring equipment is properly calibrated and recalibrated by means of specified techniques, and that the measurement procedures followed are in accordance with those prescribed.

Exposure standards

Exposure standards may include basic limits and derived limits, as set out in ICRP Publication 26 *(2)*. When the purpose of establishing limits is public health protection, it would be appropriate to refer to them as limits of permissible exposure rather than safety levels. Although the concept of permissible exposure is designed to limit risk, it may not ensure absolute safety because of the lack of definite information on the biological effects of chronic or repeated low-level exposure to NIR. Protection limits may be set both for those who are occupationally exposed and for the general population.

Emission standards

Emission standards are protection standards designed to limit the risk from products or devices that may emit NIR hazardous to human health. They will also limit leakage of radiation, where this is incidental to the primary functions of the product or device. Such standards may require the incorporation of safety features designed to prevent or minimize human exposure to radiation (e.g. screens and interlocking devices).

Regulation and enforcement procedures

These represent the translation of exposure and emission standards into the control system of a country. They will vary according to the magnitude and the probability of the hazard to health, and the details will depend on the legislative practices of the country. Some such common procedures, arranged in descending order of stringency, include the following.

Licensing of installations or devices. This type of procedure requires that facilities and institutions using NIR devices be licensed and comply with the appropriate standards, including exposure and emission standards, in order to retain their licensed status. This may include the designation of controlled areas around installations. Access to such areas may be restricted and, in certain circumstances, residence in them may be prohibited.

Statutory regulations. These are mandatory rules having the force of law which require compliance by the user; it is the user's duty to know the law and to obey it. The regulations are generally based on accepted exposure and emission standards. The user is required to provide equipment and to make contingency plans for dealing with accidents and emergencies.

Registration. It is sometimes the case that the introduction of a hazardous technology or process may have to be notified to and registered by the enforcing authority, which may then grant permission for its use.

Notification. This is a less stringent requirement and does not include the granting of permission for use.

Voluntary procedures. These may be applied to both exposure and emission standards; the former may be embodied in codes of practice, while the latter may be developed by industry by consensus, with self-monitoring. Should a risk to human health be identified as a result of the violation of the voluntary standards, or should a clear danger to human health be associated with radiation emitted by a product not covered by a standard, action can be taken under a "defective product procedure". Such a defective product would be identified as one emitting radiation at intensities of such magnitude as to constitute a hazard to human health. The responsibility for demonstrating that the product was not defective would lie with the manufacturer, who would also be responsible for taking the necessary corrective action.

Guidelines and recommendations. These represent a voluntary approach to radiation protection. Guidelines and recommendations may cover equipment, procedures, facilities and the conduct of physical and medical examinations.

Quality assurance programmes. Such programmes are designed to encourage practices among manufacturers and users of radiation-emitting products and devices that will effectively reduce exposure and/or any risk that may be associated with the use of the product or device.

Certification. Certification of professionals and technicians can serve to ensure that only individuals with appropriate training are permitted to operate or service equipment that may be potentially hazardous.

Education and training
Education and training programmes serve to teach professionals, technicians and members of the general public about the risks that may be associated with exposure to NIR. Through education, users at all levels may better appreciate the risks to which they may be exposed, and therefore be amenable to taking those steps that may be effective in limiting risk, such as reduction of exposure for diagnostic examinations, avoidance of routine exposure, and use of equipment in conformity with safety recommendations.

SAFE EXPOSURE LIMITS

Many devices emitting NIR are already being produced in large quantities by modern industrial methods. The question thus arises once again as to which devices can be regarded as safe, even if this decision has to be taken in the absence of a complete knowledge of the biological hazards. A few relevant concepts are presented in what follows, with the object of stimulating the development of appropriate strategies from which existing legal practices can be derived for special cases.

For the purposes of these more fundamental considerations, it is assumed that the relation between cause and effect has been investigated sufficiently to enable a hazard limit to be defined. The difficulties involved in defining such a limit are discussed below to demonstrate the resulting consequences.

Causal model
The irradiation of a biological system, e.g. a human body, is shown diagrammatically in Fig. 1A. Increasing irradiation is shown by the direction of the arrow on the abscissa, on which the level of irradiation which causes injuries to living tissue can be plotted, and is considered to be the hazard limit. It is not an absolute limit but represents the most probable value above which irradiation may cause biological damage. Because of variations in biological response, there must always be some degree of uncertainty as to the actual value of the hazard limit.

A safe exposure limit, i.e. one about which there is no uncertainty — and this is a requirement that has often to be satisfied before such a limit can be incorporated in legislation — must by definition be lower than the lowest value for which any uncertainty does exist. Because scientific knowledge is incomplete, it will also be necessary to include a safety factor. The difference between the safe exposure limit and the level of irradiation at which damage is most likely to occur indirectly indicates the state of scientific knowledge.

The diagram can be completed by plotting on the ordinate the severity of the biological effect produced by irradiation (Fig. 1B). This is done on the assumption that a true causal relationship exists between irradiation and effect.

In a first approach, it is assumed that low exposures do not cause a measurable effect. With increasing irradiation, however, the effect also increases, at first slowly and then more rapidly, up to a level above which the effect cannot be increased by more intense irradiation. This level is reached when all life functions of the biological system have been destroyed. Thus, a sigmoid curve is obtained under these circumstances, the assumed hazard limit being located at a certain point on the curve (Fig. 1B). Irradiation above this level is considered to constitute a health hazard.

If one disregards the important question of whether this simplification of the relationship is admissible for each type of NIR, the fixing of a hazard limit and thus of an exposure limit is in any case significant only if what is meant by a harmful effect has been clearly defined. This problem cannot be solved without discussing which physiological or biological parameter is to be used as a measure of the effect and which physical parameter as a measure of irradiation. It must be remembered that Fig. 1 is only a model but, with all its uncertainties, it does demonstrate the fundamental problems involved in the assessment of exposure limits, independently of the true relationship between response and irradiation.

The definition of the effect constituting the hazard is of the utmost importance; it is because of the lack of a clear and uniform definition that many of the data given in the literature are not comparable. For example, the WHO Constitution itself gives a definition of health which may not

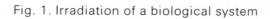
Fig. 1. Irradiation of a biological system

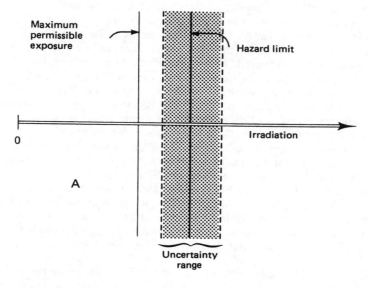

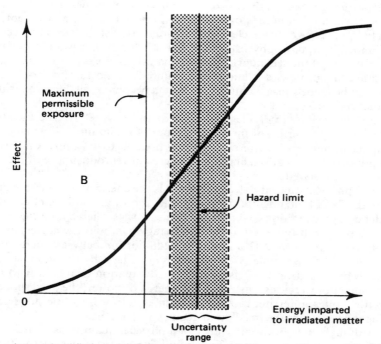

Note. A = a simplified model; B = a model with allowance for the severity of the biological effect.

always be appropriate to quantitative scientific research. According to this definition, health is a state of complete physical, mental and social wellbeing and not merely the absence of disease or infirmity. In the light of this definition, different values for a hazard limit can be obtained depending upon whether it is intended merely that disease is to be avoided or whether wellbeing has to be ensured. There is no doubt, however, that both kinds of definition are justifiable. This can be seen, for example, in the case of certain microwave standards, where the level for uncontrolled exposure is so low that no adverse effect on wellbeing is to be expected, whereas under well controlled working conditions, higher exposure levels are permitted for a certain period of time. Thus the replacement of conventional sources of heat by microwave devices in industrial processes, e.g. in the drying of foodstuffs, may improve working conditions considerably, even if the wellbeing of the workers is marred by a certain residual discomfort.

In the model shown in Fig. 1B, a value is postulated for the ordinate which can be determined objectively and which is causally related to the irradiation. It is also possible to quantify the effect in terms of subjective reactions, but a statistical survey is then necessary in order to eliminate the effect of other factors that can produce similar effects (e.g. headache, fatigue).

As far as irradiation is concerned, it would be desirable for a physical quantity to be selected which is correlated as closely as possible with the effect observed. Some other secondary physical quantities may sometimes be more suitable from the point of view of normal routine assessment, but will be acceptable only if a clear-cut relationship exists between the secondary and the primary physical quantities. If this relationship is not fully understood but the secondary physical quantity is one that is indispensable for practical purposes, a relationship can initially be postulated. Corrections can then be made subsequently as valid data become available as a result of further research.

For the abscissa in Fig. 1B, a physical quantity is preferred because both the physical properties of the various types of radiation and the measurement techniques are well known. It is thus possible to draw up the necessary quantitative specification, and a high degree of reproducibility of the data can thus be ensured.

The physical quantity often preferred is the energy imparted to matter, since this is of fundamental significance in almost all kinds of effect. Only under certain conditions, when a mechanism exists whereby the energy can be dissipated out of the irradiated body region, would the corresponding quantity be the power. This implies an equilibrium between energy input and output. An example is the production of heat by the interaction of microwave radiation with tissue and the dissipation of this heat to non-irradiated body regions. Instead of the energy or power, the corresponding specific values, referring to mass, volume or surface, will be preferred if independence from the irradiation geometry is important.

Although an appropriate choice of a physical quantity as a measure of irradiation would seem an obvious requirement, many reports on the relationship between cause and effect have been published in which the data

302

have been expressed in terms of other quantities more easily measured under experimental conditions. If the authors then neglect to specify the relationship between the data obtained and any precisely defined fundamental quantity, or if the paper provides insufficient information to permit the necessary relationship to be derived from the experimental conditions, the value of the research, for purposes of comparison or the assessment of hazard limits, is reduced. Even if important observations are made regarding the general problem of evaluating potential health hazards, failure to indicate the reference quantity for the irradiation of an object may limit the significance of otherwise excellent investigations.

An effect which changes significantly in magnitude when the intensity of the irradiation is changed would be suitable for use in experiments and can thus easily be observed. If possible, there should be a clear-cut relationship between effect and irradiation. In the selection of a parameter which is a measure of the hazard to the irradiated biological system, the effect to be observed must be related to its life functions. Finally, if the investigations are not carried out on man, extrapolation of the results to man must be feasible. This requirement may seriously limit the effects which can be considered, but must be satisfied if a valid model is to be obtained.

As already pointed out, the biological response mechanism, postulated in the model shown in Fig. 1B, starts with a threshold level for the irradiation below which no effect is observable. Above a certain level of irradiation, further increase does not produce an increased effect. A sigmoid curve for the relationship between effect and energy imparted to matter would thus be expected. Whatever the mechanism underlying the model, the threshold level might be very low, and possibly even so low as not to be observable. The curve would then pass through the origin.

The model will not correctly represent the situation if the wrong reference quantity has been chosen, e.g. the temperature of the body is a closely temperature-regulated physiological system, as in human beings. A threshold level would then be simulated, although there would definitely be other effects, such as stimulation of the entire metabolism, without a rise in body temperature. A rise in temperature would already be a danger sign and the rise could continue even if the individual were already dead. However, if a cell population was involved, the model would be satisfactory if the number of dead cells were counted. Below the threshold level all the cells would survive the irradiation. The death rate would then rise until all the cells had been killed. A further intensification of the irradiation could not increase the effect.

If it is assumed that a clear-cut cause–effect relationship exists, these examples demonstrate that the relationship between the dose–effect curve and the ordinate and abscissa is determined by the choice of parameters for observation.

Phenomenological model
Because of the multiplicity of observable parameters of radiation effects, a more pragmatic approach is, in general, unavoidable. Procedures will therefore initially be based on observable phenomena. For each parameter it will

be necessary to investigate which phenomenon occurs when the irradiation level is increased. In this phase, no attempt need be made in the investigation to explain the observed phenomenon or to determine the relationship between the changes in different parameters and the radiation level.

These methods, although they are well accepted and scientifically valid, should be used with caution in the determination of hazard limits or exposure limits. It is necessary to ensure that, in the course of the investigation, all the parameters relevant to the hazard concerned have been examined. Among the numerous phenomena there may be effects that are of no importance from the point of view of defining a hazard limit. This is the case if health or life functions are not adversely affected. It would also be inappropriate to use a particular effect for the determination of hazard limits if a different effect occurs at lower irradiation levels.

The objective of defining radiation hazard limits can be achieved only if a well thought out strategy exists. If parameters are examined at random, an important parameter may well be overlooked.

The phenomenological approach, i.e. the observation of effects occurring during or after radiation exposures but without the construction of a hypothesis as to the response mechanism, is often the only possibility in view of the extreme complexity of the physiological reactions involved. However, there is always the danger than an apparent correlation between cause and effect may be found that does not really exist. This will lead to serious errors if conclusions are subsequently drawn from such a correlation.[a]

Thus changes in the skin, such as the repeated development of erythema following time-correlated successive occupational exposure, establishes a cause–effect relationship, although similar skin changes may occur, for example, after exposure to the sun. The causal relationship often cannot be established with certainty if the parameter selected is an ill effect which becomes apparent after a fairly long latency period and if this ill effect may be caused by other factors. In this case, only long-term, large-scale surveys can help in detecting quantitative relationships which are statistically valid and significant. The scientifically important statement that a particular type of radiation in sufficient quantity can cause ill effects is not sufficient to define quantitatively the hazard limit.

The above discussion, though brief, shows that the objective of a research strategy for obtaining scientifically based data for determining the permissible exposure and the exposures to be prohibited, should be to investigate effects unequivocally caused by the radiation concerned. Sound decisions can then be made as to the requirements that may be imposed by law.

Conservative assumptions
Since the scientifically based dose–effect data for most of the NIR are still fragmentary, and therefore not yet suitable for regulation purposes, other

---

[a] A good example is the statistically significant correlation between the decline in the birth rate in certain countries and the decrease in the number of storks. This does not, however, prove that babies are brought by storks, as children used to be told.

methods for limiting radiation hazards must be found. It is sometimes helpful to define a maximum permissible exposure which is lower than the lowest of all expected threshold levels for radiation hazards (Fig. 1B). Where the relationship between irradiation and effect is unknown, the most favourable case is assumed, and a safety factor is used (in the model, this is the distance between the hazard limit and the maximum permissible exposure). If, subsequently, as a result of systematic investigations, a more refined approach becomes feasible, the exposure limit can be modified.

However, the foregoing method may sometimes be inappropriate. If the basic data are uncertain, this type of assessment can lead to exposure limits which are too low for practical application, and a different approach must be adopted. If special protective measures are prescribed for sensitive organs or parts of the body, higher exposure limits may be acceptable.

Where specific reactions develop in organs or parts of the body after irradiation, the exposure limits are often determined by the effects on such "critical organs". An example of such an organ, in relation to the visible spectral region, is the eye, and especially the retina, as far as laser beams are concerned. If protection of the critical organ is possible, consideration can be given to relaxing the exposure limit. In the foregoing example, the eye can be protected from irradiation by enclosing the beam or by providing protective glasses.

It is sometimes sufficient to avoid exposure only during certain periods. For instance, if a particular radiation danger exists for the fetus, occupational exposure of pregnant women should be discouraged or, if necessary, access to certain areas should be restricted for women of reproductive capability. Many examples of restrictions of this kind are found in occupational hygiene. All of them show that restrictions have to be adapted to each specific situation and that their observance must be controlled. Thus, the critical analysis of existing scientific data leads to the requirement that legally supported regulatory and enforcement procedures must be introduced, though this is certainly not the only reason for doing so.

COMPLIANCE AND ENFORCEMENT

The success of any proposed regulation and enforcement procedure will depend finally on whether it can be applied in practice. There are cases where well established procedures break down simply because the institutions responsible for their enforcement are overloaded with work, the resources inadequate or the necessary experts not available. The administrative and economic limits are quickly reached when the applications concerning devices, installations or radiation sources are so numerous as to exceed the ability to control them. Moreover, resources are not being effectively used when they are allocated to the surveillance of applications involving a small degree of risk, while more hazardous applications remain uncontrolled.

In order to achieve the aims of protection, it is advisable that the enforcement procedures for the proposed regulations be critically reviewed in advance. Enforcement is subject not only to legal provisions but also to technological and administrative measures. It is also important to understand that legislation is more likely to be effective if it has the support of public opinion. This may be achieved by informing and educating the public, or at least those directly affected by the legislation. The feasibility of legislation is generally predetermined by the political and administrative structure and the technical and professional organizations available to support enforcement of the proposed provisions. Because of the correlation between enforceable legal provisions and technical requirements which are in accordance with the state of science and technology, certain other generally applicable regulatory methods may be considered.

Approval of application

Approval of application implies that neither the device nor the installation but rather the user is subject to regulatory procedures. The requirements for authorization should be carefull defined. The provisions for authorization should include the requirement for an official examination of the installation by a professionally competent person.

Where authorization is required, prior approval of the installation containing the NIR equipment should also be obtained. The complete procedure is time-consuming and can be recommended only if there is a significant risk to the environment, to the public, or to persons exposed to radiation in the course of their work. This procedure is not only effective from the point of view of health and safety but can be of advantage to the applicant, because he can calculate the cost of the necessary protective measures before he begins operation. Thus, the investment required can be assessed at the planning stage.

It is also possible to apply this procedure in stages. Hence, the plans for the installation and the necessary devices can be approved separately before the installation is constructed, and operation will then be authorized after the protection measures have been finally accepted as adequate. Thus, the considerable financial risk involved in the construction of large installations, such as failure to obtain authorization after construction has been completed, can be reduced. A typical example is high-power radar air traffic control equipment located near a city, which may affect residential areas in the vicinity, and possibly also life-saving electronic equipment in nearby hospitals.

Moreover, it is appropriate to introduce regulatory controls covering the use or operation of NIR equipment, even if the extent of such use is insufficient to warrant a system of prior approval. However, the requirements to be satisfied by the user or operator should be clearly set out, and should specify the objectives to be achieved. Prior notification of use may also be required.

Regulatory control with or without prior notification allows the user to commence operation without waiting for any action on the part of the regulatory authority. Provided the user meets the requirements laid down, he is secure in the knowledge that he will not subsequently be required to introduce additional protection measures or to suspend operations. However, the user lacking in expertise will often find it necessary to seek assurances

306

from the regulatory authority or from some other advisory body that he has interpreted the requirements correctly and is applying them appropriately.

Approval of equipment

The approval procedure can be simplified if it is possible to fit safety devices to the radiation source, which either completely prevent any exposure or reduce it to a prescribed level. If protection can be ensured by construction and design, the equipment itself will satisfy the regulatory requirements. The technical performance necessary, safety requirements, information and labelling also need to be specified. The performance will, in many cases, be specified in terms of an exposure limit, measured under certain prescribed conditions. Such exposure limits differ fundamentally from permissible exposure limits for individuals, in so far as they are derived limits fixed in such a way as to ensure that the permissible exposure limits for individuals are not exceeded under any conditions of use. It is therefore advisable to give such an exposure limit a different name, such as emission limit. In specifying values for emission limits, account should be taken of such aspects of protection as the possibility of operational breakdown as a consequence of defects which cannot be totally excluded by technical means.

If acceptable emission limits cannot be achieved by suitable design measures, additional protection measures will then be necessary, such as the provision of additional shielding, restriction of access and/or the introduction of administrative controls.

The mass production of devices calls for other methods in order to simplify the regulatory procedure. One possibility would be to test a prototype of the device and require the manufacturer to ensure that each device produced was in conformity with that prototype. It must be pointed out that this procedure can be limited to the production process only if the equipment was designed with adequate protective devices, and can be used according to the specifications without any additional protective devices. It is also necessary to ensure that exposure cannot exceed the maximum permissible level for members of the general public. Inspection is thus reduced to the surveillance of a small number of manufacturers instead of a large number of applicants. The reliability of the manufacturers and the technical safety of the equipment will determine whether each individual product will need to be inspected for conformity with the prototype, or whether the testing of random samples will be sufficient.

Remedial protection measures

All the procedures previously described are preventive in character. They are intended to reduce the extent of potential hazards to a certain known and acceptable level or to avoid them completely. Another procedure that will reduce the need for administrative controls would prescribe, in technical standards, either the design and properties of the device or the maximum permissible radiation emission. Legislative measures would follow only if there was reason to believe that these standards had not been complied with during the production process. This procedure is particularly effective when devices are produced in large quantities by different manufacturers. If it

307

becomes generally known that the products of any particular manufacturer are unsafe, his economic position will be jeopardized. As a result, this system is self-regulating.

Inspection and maintenance

One of the main problems in radiation protection is to ensure that equipment remains safe after it has been in use for a long period. If the equipment or installation is subject to regulation, the regulatory administration or the officially authorized institutions must be empowered to confirm, either periodically or by means of random checks, that the requirements are still being complied with. If responsibility has been assigned to the manufacturer, however, further provisions will be necessary to ensure that the safety devices remain effective after delivery of the equipment to the user. Periodic checking and maintenance visits must take place, if necessary. It is also important to ensure that repairs are carried out by qualified personnel, so that protection measures are not impaired as a result of such repairs.

Review of regulations

The rapid technological developments in the field of NIR, together with the increased understanding of its bioeffects, are likely to lead to changes in exposure and emission standards. It may, therefore, be prudent for the regulations based on those standards to remain in force only for a limited period of time so as to provide an opportunity for review in the light of new knowledge.

Changes in the standards and in the regulations based on them may make certain types of equipment unacceptable to the regulatory authority. Therefore, subject always to the overriding considerations of health and safety, consideration may also be given to the gradual introduction of revised regulations. This would enable a reasonable period of time to be allowed for alterations and adjustments to be made to existing equipment.

Exemptions

An effective method of rationalizing the regulation and enforcement procedures is the introduction of exemptions for particular applications, namely, those which are considered to be harmless. Any such exemption limit can be withdrawn if necessary, e.g. for devices which are widely used or which can be used for purposes other than those for which they have initially been designed.

Information programmes

No system of controls depending on regulatory and administrative procedures can be made completely safe if a user is determined to ignore or to circumvent the safety measures designed to protect against potentially hazardous NIR. It is therefore essential that information and education programmes be introduced to complement and supplement other aspects of the control programme. A comprehensive and effective programme of information and/or education for users at all levels should alert them to the potential hazards of exposure to NIR, heighten their awareness of the need to avoid unnecessary exposure, and promote respect for any instructions, recommendations and guidelines that may be issued for the purpose of promoting safe practices.

Situations may indeed exist where recommendations and guidelines, effectively developed and publicized, may constitute the entire programme necessary to ensure the protection of public health and safety. The potential adequacy of such an approach to protection should be explored in any case where it may be applicable.

REFERENCES

1. *Health effects of ionizing and non-ionizing radiation.* Copenhagen, WHO Regional Office for Europe, 1972 (document EURO 4701).
2. **International Commission on Radiological Protection.** *Recommendations of the International Commission on Radiological Protection.* Oxford, Pergamon Press, 1977 (ICRP Publication 26).
3. **Suess, M.J.** International cooperation on non-ionizing radiation protection. *In:* Michaelson, S.M. et al., ed. *Fundamental and applied aspects of nonionizing radiation.* New York, Plenum, 1975, pp. 447–457.
4. **International Commission on Radiological Protection.** *Implications of Commission recommendations that doses be kept as low as readily achievable.* Oxford, Pergamon Press, 1973 (ICRP Publication 22).

GLOSSARY[a]

As in most relatively new fields, the terminology used in studies of the health effects of NIR is in many respects confused. With the exception of physical quantities — for most of which standardized names have existed for many years — there is little international standardization of NIR terminology. In some cases, different authors use a given term with different meanings, while in others they use different terms with the same meaning. Even in the few areas for which internationally agreed, standardized terminology exists, many authors do not follow the recommendations.

An attempt is now being made to achieve agreement on terminology for the more important health-related aspects of NIR. Pending the outcome of that effort, the short glossary that follows is included in this publication in order to draw attention to those terms that have already been standardized internationally. Most of these terms are the names of physical quantities, and in such cases the recommended symbol(s) and units of measurement are included. The expression "reserve symbol" refers to a symbol that, while not recommended, may be used (to avoid confusion) in any context in which the preferred symbol is used for some other purpose. The sources of the recommendations are listed at the end of the glossary.

---

absorptance, spectral See *absorption factor, spectral*

absorption coefficient, linear The part of the linear attenuation coefficient that is due to absorption. (ISO, *4*) Symbol, *a*; Unit, reciprocal metre, m^{-1} See *attenuation coefficient, linear*

absorption factor A weighted average of the spectral absorption factor. Symbol, α; Unit, 1 (dimensionless, a ratio)

absorption factor, spectral Ratio of the spectral concentration of radiant or luminous flux absorbed to that of the incident radiation. (ISO, *4*) Symbol, $\alpha(\lambda)$; Unit, 1 (dimensionless); Synonym, *spectral absorptance*

attenuation coefficient, linear The relative decrease in spectral concentration of the radiant or luminous flux of a collimated beam of electromagnetic radiation during traversal of an infinitesimal layer of a medium, divided by the length traversed. (ISO, *4*) Symbol, μ; Unit, reciprocal metre, m^{-1}; Synonym, *linear extinction coefficient*

concentration of radiant energy density, spectral See *radiant energy density, spectral*

emissivity Ratio of *radiant exitance* of a thermal radiator to that of a full radiator (black body) at the same temperature. (ISO, *4*) Symbol, ε; Unit, 1 (dimensionless)

energy surface density See *radiant exposure*

---

[a] Compiled for the first edition by Mr D.A. Lowe, Chief, Technical Terminology Service, World Health Organization, Geneva, Switzerland.

311

exposure See *radiant exposure*

extinction coefficient, linear See *attenuation coefficient, linear*

impedance, acoustic At a surface, the complex representation of sound pressure divided by the complex representation of volume flow rate. (ISO, *5*) Symbol, Z_a; Unit, pascal second per metre cubed, Pa·s/m³

impedance, specific acoustic At a surface, the complex representation of sound pressure divided by the complex representation of particle velocity. (ISO, *5*) Symbol, Z_s; Unit, pascal second per metre, Pa·s/m

impedance of a medium, characteristic At a point in a medium and for a plane progressive wave, the complex representation of sound pressure divided by the complex representation of particle velocity. (ISO, *5*) Symbol, Z_c; Unit, pascal second per metre, Pa·s/m

irradiance At a point of a surface, the radiant energy flux incident on an element of the surface, divided by the area of that element. (ISO, *4*) Symbol, E; Reserve symbol, E_e; Unit, watt per square metre, W/m²; Deprecated synonym, *power surface density*

phantom A volume of material behaving in essentially the same manner as tissue, with respect to absorption and scattering of the radiation in question. (IEC, *2*)

power surface density See *irradiance*

radiant emittance See *radiant exitance*

radiant energy Energy emitted, transferred, or received as radiation. (ISO, *4*) Symbol, Q or W; Reserve symbols, U, Q_e; Unit, joule, J

radiant energy density Radiant energy in an element of volume, divided by that element. (ISO, *4*) Symbol, w; Reserve symbol, u; Unit, joule per cubic metre, J/m³

radiant energy density, spectral The radiant energy density in an infinitesimal wavelength interval, divided by the range of that interval. (ISO, *4*) Symbol, w_λ; Unit, joule per metre to the fourth power, J/m⁴; Synonym, *spectral concentration of radiant energy density*

radiant energy fluence rate At a given point in space, the radiant energy flux incident on a small sphere, divided by the cross-sectional area of that sphere. (ISO, *4*) Symbol, φ or ψ; Unit, watt per square metre, W/m²; Synonym not recommended, *radiant flux density*

radiant energy flux Power emitted, transferred, or received as radiation. (ISO, *4*) Symbol, P or Φ; Reserve symbol, Φ_e; Unit, watt, W; Synonym, *radiant power*; Synonym not recommended, *radiant flux*

radiant exitance At a point of a surface, the radiant energy flux leaving an element of the surface, divided by the area of that element. (ISO, *4*) Symbol, M; Reserve symbol, M_e; Unit, watt per square metre, W/m²

radiant exposure Radiant energy incident on a surface divided by the area of the surface. Symbol, H; Unit, joule per square metre, J/m²; Synonym, *exposure* (in part); Deprecated synonym, *energy surface density* (*Note.* The term "exposure" has many different meanings, depending on context.)

radiant flux See *radiant energy flux*

radiant flux density See *radiant energy fluence rate*

radiant intensity In a given direction from a source, the *radiant energy flux* leaving the source, or an element of the source, in an element of solid angle containing the given direction, divided by that element of solid angle. (ISO, *4*) Symbol, *I*; Reserve symbol, I_e; Unit, watt per steradian, W/sr

radiant power See *radiant energy flux*

reflectance See *reflection factor*

reflectance, spectral See *reflection factor, spectral*

reflection factor A weighted average of the spectral reflection factor. Symbol, ρ; Unit, 1 (dimensionless, a ratio); Synonym, *reflectance*; Deprecated synonym, *reflectivity*

reflection factor, spectral Ratio of the spectral concentration of radiant or luminous flux reflected to that of the incident radiation. (ISO, *4*) Symbol, $\rho(\lambda)$; Unit, 1 (dimensionless); Synonym, *spectral reflectance*

reflectivity See *reflection factor*

threshold (*Note.* This term should not be used without qualification.)

threshold, stimulus Minimum value of a sensory stimulus needed to give rise to a sensation. (ISO, *6*) Synonym, *detection threshold*

threshold limit value A concentration (in air) of a material, or a level of noise or radiation, to which most workers can be exposed daily without adverse effect. Threshold limit values, established by the American Conference of Governmental Industrial Hygienists, are time-weighted values for a 7- or 8-hour workday and 40-hour workweek. In most cases exposures exceeding the limit are permissible provided there are equivalent compensatory exposures below the limit during the workday (or in some cases the week). For a few materials the limit is given as a maximum permissible concentration. (After *1*)

transmission factor A weighted average of the spectral transmission factor. Symbol, τ; Unit, 1 (dimensionless, a ratio); Synonym, *transmittance*

transmission factor, spectral Ratio of the spectral concentration of radiant or luminous flux transmitted to that of the incident radiation. (ISO, *4*) Symbol, $\tau(\lambda)$; Unit, 1 (dimensionless); Synonym, *spectral transmittance*

transmittance See *transmission factor*

transmittance, spectral See *transmission factor, spectral*

Sources

1. AMERICAN CONFERENCE OF GOVERNMENTAL INDUSTRIAL HYGIENISTS *TLVs: threshold limit values for chemical substances in workroom air.* Cincinnati, Ohio (published annually).
2. INTERNATIONAL ELECTROTECHNICAL COMMISSION *International electrotechnical vocabulary,* 2nd ed. *Group 65: radiology and radiological physics.* Geneva, 1964.
3. INTERNATIONAL ELECTROTECHNICAL COMMISSION *International electrotechnical vocabulary,* chapter 391. *Detection and measurement of ionizing radiation by electric means.* Geneva, 1975.
4. INTERNATIONAL ORGANIZATION FOR STANDARDIZATION *Quantities and units of light and related electromagnetic radiations.* Geneva, 1980 (international standard ISO 31/6).

5. INTERNATIONAL ORGANIZATION FOR STANDARDIZATION *Quantities and units of acoustics*. Geneva, 1978 (international standard ISO 31/VII).
6. INTERNATIONAL ORGANIZATION FOR STANDARDIZATION *Sensory analysis — vocabulary — part 3*. Geneva, 1979 (international standard ISO 5492/3).

Special acknowledgements: first edition

The World Health Organization is grateful to the specialists listed below who collaborated in the preparation of this book.[a] Their comments on various chapters were taken into consideration during the revision of those chapters and the finalization of the manuscript.

Dr W.I. Acton, Senior Consultant, Wolfson Unit, Institute of Sound and Vibration Research, The University, Southampton, United Kingdom

Dr V.J. Akimenko, Chief, Laboratory of Biological and Hygienic Investigations, A.N. Marzeev Institute of General and Community Hygiene, Kiev, USSR

Dr E. Albert, Professor of Anatomy and Director of Histology, George Washington University Medical Center, Washington, DC, USA

Dr R.M. Albrecht, Division of Biological Effects, Bureau of Radiological Health, US Public Health Service, Rockville, MD, USA

Dr F.A. Andersen, Chief, Standards Support Staff, Division of Biological Effects, Bureau of Radiological Health, US Public Health Service, Rockville, MD, USA

Dr Maria Anguelova, Senior Researcher, Section of Physical Factors, Department of Occupational Hygiene, Institute of Hygiene and Occupational Health, Sofia, Bulgaria

Dr J. Bang, Ultrasound Department, St Joseph Hospital, Copenhagen, Denmark

Mr F.S. Barnes, Professor and Chairman, Department of Electrical Engineering, University of Colorado, Boulder, CO, USA

Mr H.I. Bassen, Division of Electronic Products, Bureau of Radiological Health, US Public Health Service, Rockville, MD, USA

Mr M.W. van Batenburg, Director, Physics Laboratory, The National Defence Research Organization TNO, The Hague, Netherlands

---

[a] The affiliation given for each specialist is that which obtained during the period of his or her collaboration.

Mr P.F. Beaver, Superintending Inspector, Nuclear Branch 5, Health and Safety Executive, London, United Kingdom

Dr D. Beischer, Professor and Chief, Biomedical Division, Naval Aerospace Medical Research Laboratory, Pensacola, FL, USA

Mrs Deirdre A. Benwell, Non-ionizing Radiation Section, Radiation Protection Bureau, Environmental Health Centre, Health and Welfare Canada, Ottawa, ON, Canada

Dr I. Berenblum, Emeritus Professor, Former Head, Department of Experimental Biology, The Weizmann Institute of Science, Rehovot, Israel

Dr M.C. Bessis, Director, Institute of Cellular Pathology and Experimental Cancerology, Bicêtre Hospital, Le Kremlin-Bicêtre, France

Dr K. Bischoff, Head, Radiometry, Federal Institute for Engineering Physics and Technology, Braunschweig, Federal Republic of Germany

Dr G. Bittner, Director, Department 2, Federal Institute for Physics and Technology, Braunschweig, Federal Republic of Germany

Dr N. Bom, Faculty of Medicine, Erasmus University, Rotterdam, Netherlands

Mr R.G. Borland, Head, Biophysics, RAF Institute of Aviation Medicine, Farnborough, United Kingdom

Dr K. Brendel, Department 5, Federal Institute for Physics and Technology, Braunschweig, Federal Republic of Germany

Dr D.H. Brennan, Department of Visual Science, Institute of Ophthalmology, University of London, United Kingdom

Dr M.S. Bruma, National Centre of Scientific Research, Bellevue Laboratories, Bellevue, France

Dr J. Cabanes, Medical Committee, Electricité de France — Gaz de France, Paris, France

Mr E.L. Carstensen, Professor, Department of Electrical Engineering, College of Engineering and Applied Science, The University of Rochester, NY, USA

Dr Maria Choutchkova, Institute of Hygiene and Occupational Diseases, Sofia, Bulgaria

Mr C.L. Christman, Bureau of Radiological Health, US Public Health Service, Rockville, MD, USA

Dr S.F. Cleary, Professor of Biophysics, Department of Biophysics, Medical College of Virginia, Virginia Commonwealth University, Richmond, VA, USA

Dr C.J. Clemedson, Professor and Retired Surgeon-General, The Swedish Armed Forces, Stocksund, Sweden

Dr W.T. Coakley, Lecturer, Department of Microbiology, University College, Cardiff, United Kingdom

Dr J.W. Copeman, Managing Physician, Thomas J. Watson Research Center, IBM Corporation, Yorktown Heights, NY, USA

Mr E.A. Cox, Principal Inspector, Health and Safety, Nuclear Installations Inspectorate, Health and Safety Executive, London, United Kingdom

Mr O.H. Critchley, Principal Inspector, Health and Safety, Nuclear Installations Inspectorate, Health and Safety Executive, London, United Kingdom

Dr A.P. Cullen, Associate Professor, College of Optometry, University of Houston, TX, USA

Dr W.H. Cyr, Bureau of Radiological Health, US Public Health Service, Rockville, MD, USA

Dr Wanda Czerska, Senior Research Worker, Quantum Electronics Institute, Military Technical Academy, Warsaw, Poland

Dr P.A. Czerski, Professor and Head, Department of Genetics, National Research Institute of Mother and Child, Warsaw, Poland

Dr C.J. van Daatselaar, Head, Nuclear Department, General Directorate of Labour, Voorburg, Netherlands

Dr F. Devik, Chief Medical Officer, Medical Section, State Institute of Radiation Hygiene, Oslo, Norway

Dr F. Dukes-Dobos, Physical Agents Effects Branch, National Institute of Occupational Safety and Health, US Public Health Service, Cincinnati, OH, USA

Dr F. Dunn, Professor, Bioacoustics Research Laboratory, Department of Electrical Engineering, University of Illinois, Urbana, IL, USA

Dr G.C. Dutt, Head, Laser and Optical Radiations Unit, Radiation Protection Bureau, Environmental Department of Health Centre, Health and Welfare Canada, Ottawa, ON, Canada

Dr Mary Dyson, Anatomy Department, Guy's Hospital Medical School, London, United Kingdom

Dr P.D. Edmonds, Senior Research Physicist, Electronics and Bioengineering Laboratory, Stanford Research Institute, Menlo Park, CA, USA

Dr J. Elder, Experimental Research Center, US Environmental Protection Agency, Research Triangle Park, NC, USA

Mr. R.J. Ellis, Physical Agents Effects Branch, National Institute of Occupational Safety and Health, US Public Health Service, Cincinnati, OH, USA

Dr E.W. Emery, Medical Physics Department, Royal Postgraduate Medical School, Hammersmith Hospital, London, United Kingdom

Dr E.A. Emmett, Associate Professor, Department of Environmental Health, Division of Clinical Studies, College of Medicine, University of Cincinnati Medical Center, Cincinnati, OH, USA

Mr H. Eriskat, Chief, Division of Public Health and Radiological Protection, Directorate of Health and Security, Commission of the European Communities, Luxembourg

Dr M. Faber, Professor and Director, The Finsen Laboratory, The Finsen Institute, Copenhagen, Denmark

Mr J.R. Fancher, Commonwealth Edison, Chicago, IL, USA

Dr L. Filipczynski, Professor, Institute of Fundamental Technological Research, Polish Academy of Sciences, Warsaw, Poland

Dr H.M. Frost, Bureau of Radiological Health, US Public Health Service, Rockville, MD, USA

Dr H. Gaebelein, Leader, Working Group on Climate of Working Places, Central Institute of Occupational Medicine, Berlin, German Democratic Republic

317

Dr W.D. Galloway, Bureau of Radiological Health, US Public Health Service, Rockville, MD, USA

Mr R. Genève, Deputy Director, Electronics Laboratory for Applied Physics, Philips Research Centre, Limeil, France

Mr A. Glansholm, National Board of Occupational Safety and Health, Stockholm, Sweden

Dr Z.R. Glaser, Director, Special Occupational Hazards, Priorities and Research Analysis Branch Review Program, National Institute of Occupational Safety and Health, US Public Health Service, Rockville, MD, USA

Dr L. Goldman, Director, Laser Laboratory, and Professor and Chairman, Department of Dermatology, College of Medicine, University of Cincinnati Medical Center, Cincinnati, OH, USA

Dr Zinaida V. Gordon, Professor and Head, Laboratory of Electromagnetic Fields, Institute of Industrial Hygiene and Occupational Diseases, USSR Academy of Medical Sciences, Moscow, USSR

Dr E.H. Grant, Professor, Physics Department, Queen Elizabeth College, University of London, United Kingdom

Dr L.I. Grossweiner, Chairman and Professor, Department of Physics, Lewis College of Sciences and Letters, Illinois Institute of Technology, Chicago, IL, USA

Mr H.M. Grove, Chairman, Committee on Man and Radiation, Institute of Electrical and Electronics Engineers Inc., Washington, DC, USA

Dr A.W. Guy, Professor, Department of Physical Medicine and Rehabilitation, School of Medicine, University of Washington, Seattle, WA, USA

Dr Ludmila Gvozdenka, Senior Research Scientist, Laboratory of Industrial Microclimate, Institute of Industrial Hygiene and Occupational Diseases, Kiev, USSR

Dr Gail ter Haar, Department of Physics, Institute of Cancer Research, Royal Marsden Hospital, Sutton, Surrey, United Kingdom

Dr D. Harder, Professor, Institute for Medical Physics and Biophysics, University of Göttingen, Federal Republic of Germany

Mr F. Harlen, Principal Scientific Officer, National Radiological Protection Board, Harwell, United Kingdom

Dr H.G. Häublein, Director, Central Institute of Occupational Medicine, Berlin, German Democratic Republic

Dr R. Hauf, Professor and Scientific Director, Research Institute of Electropathology, Freiburg, Federal Republic of Germany

Dr W. Hauser, Department of General Technical-Scientific Services, Federal Institute for Engineering Physics and Technology, Braunschweig, Federal Republic of Germany

Dr A. Henschel, Division of Technical Services, Center for Disease Control, National Institute of Occupational Safety and Health, US Public Health Service, Cincinnati, OH, USA

Dr C.R. Hill, Department of Physics, Institute of Cancer Research, Royal Marsden Hospital, Sutton, Surrey, United Kingdom

Dr F. Hillenkamp, Department of Coherent Optics, Society for Radiation and Environmental Research, Neuherberg, Federal Republic of Germany

Dr K. Hishimoto, The First Surgical Department, University of Tokyo, Japan

Mr H.S. Ho, Bureau of Radiological Health, US Public Health Service, Rockville, MD, USA

Mr N.G. Holmer, Malmö General Hospital, Sweden

Mr B. Holmgren, Chief, Division of Electrotechnical Interference Problems, Swedish State Power Board, Vällingby, Sweden

Dr R.C. Honey, Group Manager — Optics, Electronics and Bioengineering Laboratory, Stanford Research Institute, Menlo Park, CA, USA

Mr D. Hoogeveen, Saskatchewan Power Corporation, Regina, SK, Canada

Mr S.M. Horvath, Director and Professor, Institute of Environmental Stress, University of California, Santa Barbara, CA, USA

Dr J. Hrazdira, Assistant Professor, Department of Biophysics, Faculty of Medicine, Purkyne University, Brno, Czechoslovakia

Dr D.E. Hughes, Professor of Microbiology, Department of Microbiology, University College, Cardiff, United Kingdom

Dr H. Iida, National Institute of Radiological Sciences, Anagawa, Chibashi, Japan

Mr M. Izrael, Institute of Hygiene and Occupational Health, Sofia, Bulgaria

Dr H. Jammet, Director, Health Protection Service, Institute of Protection and Nuclear Safety, Nuclear Studies Centre — Atomic Energy Commission (CENCEA), Fontenay-aux-Roses, France; representing the International Radiation Protection Association (IRPA)

Mr D.E. Janes, Chief, Electromagnetic Radiation Analysis Branch, Environmental Analysis Division, US Environmental Protection Agency, Silver Springs, MD, USA

Dr T.V. Kalada, Chief, Department of Hygiene, Institute of Industrial Hygiene and Occupational Health, Leningrad, USSR

Dr F. Kaloyanova, Professor and Director, Medical Academy, Institute of Hygiene and Occupational Health, Sofia, Bulgaria

Dr I. Kaplan, Clinical Professor of Plastic Surgery, Department of Plastic Maxillo-Facial Surgery, Tel-Aviv University Medical School and Beilinson Hospital, Tel-Aviv, Israel

Mr G. Karches, Physical Agents Effects Branch, National Institute of Safety and Health, US Public Health Service, Cincinnati, OH, USA

Dr Kebbel, Medical Technology Sector, Siemens AG, Erlangen, Federal Republic of Germany

Dr T. Kecik, Associate Professor and Head, Clinic of Ophthalmology, Warsaw Academy of Medicine, Poland

Mr V.E. Kinsey, Institute of Biological Sciences, Oakland University, Rochester, MI, USA

Dr E. Kivisäkk, Senior Radiation Protection Officer, National Institute of Radiation Protection, Stockholm, Sweden

Dr B. Kleman, Research Director and Professor of Physics, National Institute of Defence, Stockholm, Sweden

Dr B. Knave, Professor and Director, Department of Occupational Hygiene, National Board of Occupational Safety and Health, Stockholm, Sweden

Dr K. Koren, Director, State Institute of Radiation Hygiene, Osterås, Norway

Dr H. Kornberg, Program Manager, Environmental Assessment Department, Electric Power Research Institute, Palo Alto, CA, USA

Dr F. Kossel, Director and Professor, Division of Medical Radiation Technology and Radiation Protection, Institute of Radiation Hygiene, Federal Health Office, Neuherberg, Federal Republic of Germany

Dr D. Krastel, Scientific Worker, Central Institute of Occupational Medicine, Berlin, German Democratic Republic

Dr K. Krell, Bureau of Radiological Health, US Public Health Service, Rockville, MD, USA

Mr N. Kroo, Central Research Institute for Physics, Hungarian Academy of Sciences, Budapest, Hungary

Dr J. Kupfer, Leader, Occupational Hygiene Standardization, Central Institute of Occupational Medicine, Berlin, German Democratic Republic

Mr C. Lancée, Faculty Echocardiographic Group, Erasmus University, Rotterdam, Netherlands

Mr R. Landry, Bureau of Radiological Health, US Public Health Service, Rockville, MD, USA

Dr L. Lang, Scientific Deputy Head, Physics Institute, Budapest Technical University, Hungary

Dr R. Latarjet, Section of Biology, Radium Institute, Curie Foundation, Paris, France

Dr W.M. Leach, Chief, Experimental Studies Branch, Division of Biological Effects, Bureau of Radiological Health, US Public Health Service, Rockville, MD, USA

Dr H.K. Lee, Radiation Devices Section, Radiation Protection Bureau, Health and Welfare Canada, Ottawa, ON, Canada

Mr W. Lee, Bureau of Radiological Health, US Public Health Service, Rockville, MD, USA

Dr W.R. Lee, Professor and Director, Department of Occupational Health, University of Manchester, United Kingdom

Dr S. Leeman, Royal Postgraduate Medical School, Hammersmith Hospital, London, United Kingdom

Dr J.F. Lehmann, Professor and Chairman, Department of Rehabilitation Medicine, School of Medicine, University of Washington, Seattle, WA, USA

Dr J.C. van der Leun, Institute of Dermatology, State University of Utrecht, Netherlands

Dr K. Liden, Radiophysics Institute, Lund University, Sweden

Dr K. Lindström, Department of Industrial Medicine, Malmö General Hospital, Sweden

Mr M.S. Little, Bureau of Radiological Health, US Public Health Service, Rockville, MD, USA

Dr J. de Lorge, Biomedical Division, Naval Aerospace Medical Research Laboratory, Pensacola, FL, USA

Mr D.A. Lowe, Chief, Technical Terminology Service, World Health Organization, Geneva, Switzerland

320

Dr C.D. Lytle, Bureau of Radiological Health, US Public Health Service, Rockville, MD, USA

Dr I.A. Magnus, Professor, Department of Photobiology, Institute of Dermatology, Postgraduate Medical Federation, University of London, United Kingdom

Dr J.F. Malone, Lecturer in Medical Physics, Physics Department, College of Technology, Dublin, Ireland

Dr K. Marha, Laboratory of Electromagnetic Fields, Institute of Hygiene and Epidemiology, Prague, Czechoslovakia

Dr J. Marshall, Senior Lecturer, Department of Visual Science, Institute of Ophthalmology, University of London, United Kingdom

Dr D.I. McRee, Research Physicist, National Institute of Environmental Health Sciences, Research Triangle Park, NC, USA

Dr J.A. Medeiros, Centre for Interdisciplinary Studies in Chemical Physics, University of Western Ontario, London, ON, Canada

Dr R.G. Medici, Brain Research Institute, The Center for the Health Sciences, University of California, Los Angeles, CA, USA

Dr W.H. Mehn, Medical Director, Commonwealth Edison, Chicago, IL, USA

Dr E. Mešter, Professor and Director, Second Surgical Clinic, Semmelweis University, Budapest, Hungary

Dr D. Methling, Chief, Light Laboratory, Scientific-Technical Centre of Industrial Hygiene, Ministry of Construction, Berlin, German Democratic Republic

Dr S.M. Michaelson, Professor, Department of Radiation Biology and Biophysics, School of Medicine and Dentistry, University of Rochester, NY, USA

Dr L. Miro, Professor, Laboratory of Biophysics, Medical Faculty, Nîmes, France

Dr K. Mohan, Bureau of Radiological Health, US Public Health Service, Rockville, MD, USA

Mr J.C. Monahan, Research Psychologist, Division of Biological Effects, Bureau of Radiological Health, US Public Health Service, Rockville, MD, USA

Mr C.E. Moss, Physical Agents Effects Branch, National Institute of Occupational Safety and Health, US Public Health Service, Cincinnati, OH, USA

Dr A.M. Muc, Consultant, Non-ionizing Radiation, Special Studies and Services Branch, Radiation Protection Service, Ontario Ministry of Labour, Toronto, ON, Canada

Mr W.E. Murray, Physical Agents Effects Branch, National Institute of Occupational Safety and Health, US Public Health Service, Cincinnati, OH, USA

Dr Z.A. Naprstek, Chief, Cardiostimulation Division, Institute of Clinical and Experimental Surgery (IKEM), Prague, Czechoslovakia

Dr K.V. Nikonova, Senior Scientific Worker, Laboratory of Electromagnetic Waves of Radiofrequencies, Scientific Research Institute of Labour, Hygiene and Occupational Diseases, Moscow, USSR

Dr B. Nižetić, Regional Officer for Public Health Ophthalmology, World Health Organization Regional Office for Europe, Copenhagen, Denmark

Mr. W.T. Norris, Research Division, Central Electricity Research Laboratories, Central Electricity Generating Board, Leatherhead, Surrey, United Kingdom

Dr W.L. Nyborg, Professor, Department of Physics, University of Vermont, Burlington, VT, USA

Dr Å. Öberg, Professor of Biomedical Engineering, Department of Medical Engineering, Linköping University, Sweden

Dr A. Oksala, Professor of Ophthalmology, Department of Ophthalmology, University Central Hospital, Turku, Finland

Dr R. Oliver, Professor, Department of Medical Physics, Royal Postgraduate Medical School, Hammersmith Hospital, London, United Kingdom

Dr J.M. Osepchuk, Consulting Scientist, Research Division, Electron Beam Devices, Raytheon Company, Waltham, MA, USA

Dr W.H. Parr, Chief, Physical Agents Effects Branch, National Institute of Occupational Safety and Health, US Public Health Service, Cincinnati, OH, USA

Dr M.A. Pathak, Principal Associate, Department of Dermatology, Harvard Medical School, Boston, MA, USA

Dr Jana Pazderova, Research Worker, Clinic of Occupational Diseases, Prague, Czechoslovakia

Dr Perdriel, General Inspector and Director, School of Application, Army and Air Health Service, and Research Centre of Aeronautical Medicine, Ministry of Defence, Paris Armées, France

Mr R.W. Peterson, Bureau of Radiological Health, US Public Health Service, Rockville, MD, USA

Dr C.H. Powell, Assistant to the Director, Special Programs, National Institute of Occupational Safety and Health, US Public Health Service, Rockville, MD, USA

Dr A. Priou, Office for Studies of Radiotechnology, Toulouse Studies and Research Centre (CERT), Toulouse, France

Dr J. Prokopenko, A.N. Sysin Institute of General and Communal Hygiene, USSR Academy of Medical Sciences, Moscow, USSR

Dr Z. Puzewicz, Professor, Institute of Quantum Electronics, Military Technical Academy, Warsaw, Poland

Dr M.F. Quinn, Lecturer, Physics Department, College of Technology, Dublin, Ireland

Dr J.D. Ramsey, Professor of Industrial Engineering, Texas Technical University, Lubbock, TX, USA

Mr H.J.L. Rechen, Bureau of Radiological Health, US Public Health Service, Rockville, MD, USA

Dr R. Reiter, Institute of Atmospheric Environmental Research, Garmisch-Partenkirchen, Federal Republic of Germany

Dr M.H. Repacholi, Head, Non-ionizing Radiation Section, Radiation Protection Bureau, Environmental Health Centre, Health and Welfare Canada, Ottawa, ON, Canada

Dr S.C. Rexford-Welch, Procurement Executive, Atomic Weapons Research Establishment, Ministry of Defence, Aldermaston, United Kingdom

Mr W.V. Richings, Dawe Instruments Ltd, London, United Kingdom

Dr C.R. Ricketts, Medical Research Council, Industrial Injuries and Burns Unit, Birmingham Accident Hospital, United Kingdom

Mr R.J. Rockwell, Associate Professor of Laser Sciences, Laser Laboratory, Department of Dermatology, College of Medicine, University of Cincinnati Medical Center, Cincinnati, OH, USA

Dr J. Rooney, Physics Faculty, University of Maine, Orono, ME, USA

Dr S.W. Rosenthal, Associate Professor, Electrical Engineering and MRI, Polytechnic Institute of New York, Farmingdale, NY, USA

Dr H.D. Rott, Institute of Human Genetics and Anthropology, University of Erlangen-Nürnberg, Erlangen, Federal Republic of Germany

Dr D.E. Rounds, Pasadena Foundation for Medical Research, Pasadena, CA, USA

Mr W. Ruth, Department of Occupational Health, National Board of Occupational Safety and Health, Stockholm, Sweden

Dr G.M. Samaras, Neuro-Oncology Research Laboratories, Department of Radiation Therapy, School of Medicine, University of Maryland, Baltimore, MD, USA

Dr C.L. Sanders, Division of Physics, National Research Council, Ottawa, ON, Canada

Dr B.M. Savin, Professor and Head, Laboratory of Electromagnetic Waves of Radiofrequencies, Scientific Research Institute of Labour, Hygiene and Occupational Diseases, Moscow, USSR

Dr P.C. Scheidt, Bureau of Radiological Health, US Public Health Service, Rockville, MD, USA

Dr S.O. Schiff, Chairman, Department of Biology and Graduate Program in Genetics, George Washington University, Washington, DC, USA

Dr K.H. Schneider, Professor and Director, Research Association for High Voltage and High Current Technology, Mannheim, Federal Republic of Germany

Dr H.P. Schwan, Professor, Department of Bioengineering, Moore School of Electrical Engineering, University of Pennsylvania, Philadelphia, PA, USA

Dr S.A. Sebo, Professor, Dreese Laboratory, Department of Electrical Engineering, Ohio State University, Columbus, OH, USA

Dr B. Servantie, Section of Cellular Biology, Naval Centre for Applied Biophysical Studies and Research, Sainte-Anne, Toulon, France

Mr A.E. Sherr, Coordinator, Toxic Chemicals Registration, American Cyanamid Company, Bound Brook, NJ, USA

Dr M.L. Shore, Director, Division of Biological Effects, Bureau of Radiological Health, US Public Health Service, Rockville, MD, USA

Dr Charlotte Silverman, Deputy Director, Division of Biological Effects, Bureau of Radiological Health, US Public Health Service, Rockville, MD, USA

Dr Yvette Skreb, Chief, Laboratory of Cellular Biology, Institute of Medical Research and Occupational Health, Zagreb, Yugoslavia

Dr N.A. Slark, Senior Principal Scientific Officer, Department of Health and Social Security, London, United Kingdom

Mr D.H. Sliney, Physicist and Chief, Laser Hazards Branch, Laser-Microwave Division, US Army Environmental Hygiene Agency, Aberdeen Proving Ground, MD, USA

Dr F. Stenbäck, Associate Professor, Department of Pathology, University of Oulu, Finland

Dr H.F. Stewart, Chief, Acoustics Branch, Bureau of Radiological Health, US Public Health Service, Rockville, MD, USA

Dr J.A.J. Stolwijk, Professor of Epidemiology, Yale University Medical School, and Associate Director, John B. Pierce Foundation Laboratory, New Haven, CT, USA

Dr M.E. Stratmeyer, Bureau of Radiological Health, US Public Health Service, Rockville, MD, USA

Dr M.A. Stuchly, Physicist, Non-ionizing Radiation Section, Radiation Protection Bureau, Environmental Health Centre, Health and Welfare Canada, Ottawa, ON, Canada

Dr M.L. Swicord, Division of Electronic Products, Bureau of Radiological Health, US Public Health Service, Rockville, MD, USA

Dr K. Szymczykiewicz, Professor and Director, Institute of Occupational Medicine in the Mining and Metallurgical Industry, Sosnowiec, Poland

Dr N.C. Telles, Bureau of Radiological Health, US Public Health Service, Rockville, MD, USA

Dr B.M. Tengroth, Professor and Chairman, Department of Ophthalmology, Karolinska Institute and Hospital, Stockholm, Sweden

Dr J. Thacker, Scientist, Radiobiology Unit, Medical Research Council, Harwell, United Kingdom

Mr B. Thalèn, Technical Product Manager, International Catering Equipment Division, Philips AB, Norrköping, Sweden

Dr J.R.E. Thuerauf, Institute of Occupational and Social Medicine, University of Erlangen-Nürnberg, Erlangen, Federal Republic of Germany

Dr F. Urbach, Professor and Director, The Center for Photobiology, Temple University School of Medicine, Philadelphia, PA, USA

Dr D. Utmischi, Research Worker, Institute for High Voltage and Installation Technology, Technical University, Munich, Federal Republic of Germany

Mr J.C. Villforth, Director, Bureau of Radiological Health, US Public Health Service, Rockville, MD, USA

Dr J.J. Vos, Head, Vision Department, Institute for Perception TNO, Soesterberg, Netherlands

Dr J.C. Wang, Bureau of Radiological Health, US Public Health Service, Rockville, MD, USA

Dr C.G. Warren, School of Medicine, University of Washington, Seattle, WA, USA

Dr F. Weill, Professor, Department of Radiology, Hospital Centre, University of Besançon, France

Dr R. Wever, Professor, Max-Planck Institute of Behavioural Physiology, Andechs, Federal Republic of Germany

Mr G.H. Whipple, Department of Environmental and Industrial Health, School of Public Health, University of Michigan, Ann Arbor, MI, USA

Dr A. Wiskemann, Professor, University Skin Clinic, Hamburg, Federal Republic of Germany

Dr M.L. Wolbarsht, Director, Ophthalmic Research, Department of Ophthalmology, Duke University Medical Center, Durham, NC, USA

Dr J.K. Zieniuk, Institute of Fundamental Technological Research, Polish Academy of Sciences, Warsaw, Poland

Special acknowledgements: second edition

The World Health Organization is grateful to the specialists listed below who collaborated in the preparation of this book.[a] Their comments on various chapters were taken into consideration during the writing, revision or updating of those chapters and the finalization of the manuscript.

Dr W.R. Adey, Associate Chief of Staff for Research, Research Service 151, Pettis Veterans' Administration Hospital, Loma Linda, CA, USA

Dr L.E. Anderson, Program Manager, Bioelectromagnetics, Biology and Chemistry Department, Battelle Pacific Northwest Laboratory, Richland, WA, USA

Dr J. Bang, Ultrasound Department, State Hospital, Copenhagen, Denmark

Dr F.S. Barnes, Professor, Department of Electrical Engineering, University of Colorado, Boulder, CO, USA

Dr J.H. Bernhardt, Professor and Director, Institute for Radiation Hygiene, Neuherberg, Federal Republic of Germany

Dr S. Bly, Physicist, Non-ionizing Radiation Section, Research and Standards Division, Bureau of Radiation and Medical Devices, Environmental Health Directorate, Health and Welfare Canada, Ottawa, ON, Canada

Dr J.A. Bonnell, Medical Advisor, Central Electricity Generating Board, London, United Kingdom

Dr K. Brendel, Head of Group, Acoustics Division, Physical Acoustics, Federal Institute of Metrology, Braunschweig, Federal Republic of Germany

Dr J. Cabanes, General Administration, Committee for Medical Studies, Electricité de France — Gaz de France, Paris, France

Dr E.L. Carstensen, Professor, Department of Electrical Engineering, University of Rochester, NY, USA

---

[a] The affiliation given for each specialist is that which obtained during the period of his or her collaboration.

Dr S. Charschan, Western Electric Engineering Research Laboratory, Princeton, NJ, USA

Dr C.K. Chou, Head, Biomedical Engineering, Department of Radiation Research, City of Hope National Medical Center, Duarte, CA, USA

Dr S. Cleary, Professor, Department of Biophysics, Virginia Commonwealth University, Richmond, VA, USA

Dr P.A. Czerski, Research Scientist, Molecular Biology Branch, Division of Life Sciences, Office of Science and Technology, Center for Devices and Radiological Health, Food and Drug Administration, Rockville, MD, USA

Dr T.G. Davis, Smith Kline Beckman Corporation, Philadelphia, PA, USA

Dr J.D.Y. Deslauriers, Biophysicist, Laser and Electro-Optics Unit, Non-ionizing Radiation Section, Research and Standards Division, Bureau of Radiation and Medical Devices, Environmental Health Directorate, Health and Welfare Canada, Ottawa, ON, Canada

Dr F. Dunn, Chairman and Professor, Bioacoustics Research Laboratory, Department of Electrical Engineering, University of Illinois, Urbana, IL, USA

Dr C.H. Durney, Professor, Department of Electrical Engineering, University of Utah, Salt Lake City, UT, USA

Dr Mary Dyson, Anatomy Department, Guy's Hospital Medical School, London, United Kingdom

Dr P.D. Edmonds, Senior Research Physicist, Bioengineering Research Laboratory, Stanford Research Institute International, Menlo Park, CA, USA

Dr J.A. Elder, Chief, Cellular Biophysics Branch, and Acting Director, Experimental Biology Division, US Environmental Protection Agency, Research Triangle Park, NC, USA

Dr A. Fleischer, Department of Radiology and Radiological Sciences, Vanderbilt University, Nashville, TN, USA

Mr J.K. Franks, Laser-Microwave Division, US Army Environmental Hygiene Agency, Aberdeen Proving Ground, MD, USA

Dr O.M.P. Gandhi, Professor, Department of Electrical Engineering, University of Utah, Salt Lake City, UT, USA

Dr E.H. Grant, Professor, Physics Department, King's College, University of London, United Kingdom

Dr A.W. Guy, Professor and Director, Bioelectromagnetics Research Laboratory, Center for Bioengineering, School of Medicine, University of Washington, Seattle, WA, USA

Dr Gail ter Haar, Joint Department of Physics, Institute of Cancer Research, Royal Cancer Hospital, in association with Royal Marsden Hospital, Sutton, Surrey, United Kingdom

Mr F. Harlen, Principal Scientific Officer, Physics Department, National Radiological Protection Board, Chilton, Didcot, United Kingdom (deceased)

Dr D.J. Hart, Director, Smith Kline Beckman Corporation, Philadelphia, PA, USA

Dr R. Hauf, Professor and Scientific Director, Research Institute for Electropathology, Freiburg, Federal Republic of Germany

Dr C.R. Hill, Professor and Head, Joint Department of Physics, Institute of Cancer Research, Royal Cancer Hospital, in association with Royal Marsden Hospital, Sutton, Surrey, United Kingdom

Mr M. Ide, Professor, Department of Electronics and Communication Engineering, Musashi Institute of Technology, Tokyo, Japan

Dr K. Joyner, Australian Radiation Laboratory, Yallambie, Australia

Dr D.R. Justesen, Director, Laboratory for Experimental Neuropsychology, Veterans' Administration Hospital, Kansas City, MO, USA

Dr W.T. Kaune, Battelle Pacific Northwest Laboratory, Richland, WA, USA

Dr H.A. Kornberg, Program Manager, Environmental Assessment Department, Electric Power Research Institute, Palo Alto, CA, USA

Dr F. Kossel, Director and Professor, Division of Medical Radiation Technology and Radiation Protection, Institute of Radiation Hygiene, Federal Health Office, Neuherberg, Federal Republic of Germany

Dr H. Kreibich, Director, Central Institute of Occupational Medicine, Berlin, German Democratic Republic

Dr F.W. Kremkau, Professor and Director, Center for Medical Ultrasound, Bowman-Gray School of Medicine, Winston-Salem, NC, USA

Dr J. Kupfer, Chief, Department of Ergonomics, Central Institute of Occupational Medicine, Berlin, German Democratic Republic

Mr R.J. Landry, Electro-Optics Branch, Division of Electronic Products, Center for Devices and Radiological Health, Food and Drug Administration, Rockville, MD, USA

Dr J. Lary, Research Biologist, Microwave Radiation Department, National Institute of Occupational Safety and Health, Cincinnati, OH, USA

Dr J.C. van der Leun, Professor, Institute of Dermatology, State University of Utrecht, Netherlands

Dr F.L. Lizzi, Research Director, Biomedical Engineering Laboratories, Riverside Research Institute, New York, NY, USA

Dr J.C. Male, Central Electricity Research Laboratory, Surrey, United Kingdom

Dr A.F. McKinlay, Head, Physics Department, National Radiation Protection Board, Chilton, Didcot, United Kingdom

Mrs Patricia McKinney, Yorkshire Regional Cancer Organization, Cookridge Hospital, Leeds, United Kingdom

Dr S.M. Michaelson, Professor, Department of Radiation Biology and Biophysics, School of Medicine and Dentistry, University of Rochester, NY, USA

Dr K.H. Mild, First Research Engineer, National Board of Occupational Health and Safety, Umeå, Sweden

Dr M.W. Miller, Associate Professor, Radiation Biology and Biophysics Department, University of Rochester, NY, USA

Mr C.E. Moss, Physical Agents Effects Branch, National Institute of Occupational Safety and Health, US Public Health Service, Cincinnati, OH, USA

329

Dr A.M. Muc, Supervisor, Special Studies and Services Branch, Occupational Health and Safety Division, Ontario Ministry of Labour, Toronto, ON, Canada

Dr Z.A. Naprstek, Chief, Cardiostimulation Division, Institute of Clinical and Experimental Surgery, Prague, Czechoslovakia

Dr P. Nicolini, Manager, Enel — Electrical Research Center, Cologno Monzese, Italy

Mr W.T. Norris, Research Division, Central Electricity Research Laboratories, Central Electricity Generating Board, Leatherhead, Surrey, United Kingdom

Dr W.L. Nyborg, Professor, Department of Physics, University of Vermont, Burlington, VT, USA

Dr W.D. O'Brien, Jr., Associate Professor, Bioacoustics Research Laboratory, Department of Electrical Engineering, University of Illinois, Urbana, IL, USA

Dr Mary Ellen O'Connor, Associate Professor, Psychology Department, University of Tulsa, OK, USA

Dr A. Oksala, Professor, Department of Ophthalmology, University Central Hospital, Turku, Finland

Dr J.M. Osepchuk, Research Division, Electron Beam Devices, Raytheon Company, Lexington, MA, USA

Dr Dianne B. Pettiti, Division of Family and Community Medicine, University of California School of Medicine, San Francisco, CA, USA

Dr R.D. Phillips, Director, Experimental Biology Division, US Environmental Protection Agency, Research Triangle Park, NC, USA

Dr R.C. Preston, Radiation and Acoustics Division, National Physical Laboratory, Teddington, Middlesex, United Kingdom

Mr W.V. Richings, Dawe Instruments Limited, London, United Kingdom

Dr J.A. Rooney, Jet Propulsion Laboratory, California Institute of Technology, Pasadena, CA, USA

Dr H.D. Rott, Institute of Human Genetics and Anthropology, University of Erlangen-Nürnberg, Erlangen, Federal Republic of Germany

Dr T.C. Rozzell, Program Manager, Bioelectromagnetics, Office of Naval Research, Arlington, VA, USA

Dr D.E. Rounds, Director, Cell Biology and Laser Laboratories, Huntington Medical Research Institute, Pasadena, CA, USA

Dr M.I. Rudnev, A.N. Marzeev Research Institute of General and Communal Hygiene, Kiev, USSR

Dr B.M. Servantie, Director for Academic Affairs, School of Military Medicine, Bordeaux, France

Dr A.R. Sheppard, Research Physicist, Neurobiological Research, Research Service 151, Pettis Veterans' Administration Hospital, Loma Linda, CA, USA

Dr M.L. Shore, Associate Director for International Affairs, Center for Devices and Radiological Health, Food and Drug Administration, Rockville, MD, USA

Dr Charlotte Silverman, Associate Director for Human Studies, Division of Biological Effects, Office of Science and Technology, Center for Devices and Radiological Health, Food and Drug Administration, Rockville, MD, USA

Mr D.H. Sliney, Physicist and Chief, Laser Hazards Branch, Laser-Microwave Division, US Army Environmental Hygiene Agency, Aberdeen Proving Ground, MD, USA

Dr G. Stingl, Vienna Academy of Postgraduate Medical Education, Austria

Dr M.E. Stratmeyer, Center for Devices and Radiological Health, Food and Drug Administration, Rockville, MD, USA

Dr Maria A. Stuchly, Research Scientist, Non-ionizing Radiation Section, Research and Standards Division, Bureau of Radiation and Medical Devices, Environmental Health Directorate, Health and Welfare Canada, Ottawa, ON, Canada

Dr E. Sutter, Light and Radiation, Federal Institute of Metrology, Braunschweig, Federal Republic of Germany

Dr. K.J.W. Taylor, Professor, Department of Diagnostic Radiology, Yale University, New Haven, CT, USA

Dr T.S. Tenforde, Physiology Group Leader, Lawrence Berkeley Laboratory, Berkeley, CA, USA

Dr B.M. Tengroth, Professor and Chairman, Department of Ophthalmology, Karolinska Institute and Hospital, Stockholm, Sweden

Dr F. Urbach, Professor and Acting Chairman, The Center for Photobiology, School of Medicine, Temple University, Philadelphia, PA, USA

Dr K. Vandenberghe, Department of Obstetrics and Gynaecology, University Hospital, University of Leuven, Gasphalsberg, Belgium

Dr M. Wagner, Regional Officer for Maternal and Child Health, World Health Organization Regional Office for Europe, Copenhagen, Denmark

Dr R.A. Weale, Director, Department of Visual Science, Institute of Ophthalmology, University of London, United Kingdom

Dr A.R. Williams, Department of Medical Biophysics, University of Manchester, United Kingdom

Dr M.L. Wolbarsht, Professor of Ophthalmology and Biomedical Engineering, Department of Psychology, Duke University, Durham, NC, USA

Dr J.K. Zieniuk, Institute of Fundamental Technological Research, Polish Academy of Sciences, Warsaw, Poland

Dr M.C. Ziskin, Professor, Radiology and Medical Physics, School of Medicine, Temple University, Philadelphia, PA, USA

Dr J. Zuclich, Technology Incorporated, San Antonio, TX, USA

Lists of working groups

First edition

Working Group on Health Effects of Lasers, Dublin, 21–24 October 1974

Mr R.G. Borland
Dr J.W. Copeman
Dr Wanda Czerska
Dr M. Faber
Dr L. Goldman (*Chairman*)
Mr. F. Harlen
Dr F. Hillenkamp
Dr H. Jammet
Dr I. Kaplan
Dr B. Kleman
Dr J. Kupfer
Dr J.F. Malone
Dr J. Marshall
Dr J.A. Medeiros
Dr E. Mešter
Dr S.M. Michaelson
Dr L. Miro
Dr Z.A. Naprstek (*Vice-Chairman*)
Dr B. Nižetić
Dr W.H. Parr
Dr C.H. Powell
Dr M.F. Quinn (*Rapporteur*)
Mr R.J. Rockwell
Mr D.H. Sliney
Dr M.J. Suess (*Secretary*)
Dr B.M. Tengroth
Dr J.J. Vos
Dr M.I. Wolbarsht

Working Group on Health Effects of Exposure to Ultrasound Radiation, London, 18–21 October 1976

Dr W.I. Acton
Dr J. Bang
Dr K. Brendel
Dr W.T. Coakley
Dr F. Dunn (*Chairman*)
Dr L. Filipczynski
Mr R. Genève
Dr Gail ter Haar
Dr D. Harder
Dr C.R. Hill
Dr J. Hrazdira (*Vice-Chairman*)
Dr D.E. Hughes
Mr. C. Lancée
Dr W.L. Nyborg
Dr M.H. Repacholi (*Rapporteur*)
Mr W.V. Richings
Dr H.D. Rott
Dr N.A. Slark
Dr H.F. Stewart
Dr M.J. Suess (*Secretary*)
Dr J. Thacker

Working Group on Health Effects of Exposure to Ultraviolet and Infrared Radiation, Sofia, 21–24 February 1978

Dr F.A. Andersen
Dr Maria Anguelova
Dr K. Bischoff
Dr Maria Choutchkova
(*Vice-Chairman*)
Mr E.A. Cox (*Rapporteur*)
Dr G.C. Dutt
Dr M. Faber
Mr A. Glansholm
Dr Ludmila Gvozdenka
Mr F. Harlen

Mr M. Izrael
Dr I.A. Magnus
Mr C.E. Moss
Dr W.H. Parr
Dr J. Prokopenko
Mr D.H. Sliney
Dr J.A.J. Stolwijk
Dr M.J. Suess (*Secretary*)
Dr K. Szymczykiewicz
Dr B.M. Tengroth (*Chairman*)
Dr J.R.E. Thuerauf

Working Group on Health Effects of Exposure to Electric and Magnetic Fields at Power Frequencies, and Regulation and Enforcement Procedures, Freiburg, 22–26 May 1978

Dr V.J. Akimenko
Dr J. Cabanes
Mr E.L. Carstensen
Dr S.F. Cleary
Mr O.H. Critchley
Dr P.A. Czerski
Mr H. Eriskat
Mr F. Harlen
Dr R. Hauf
Mr B. Holmgren
Dr E. Kivisäkk (*Vice-Chairman*)
Dr B. Knave
Dr K. Koren

Dr H. Kornberg
Dr F. Kossel
Dr J. Kupfer
Dr W.R. Lee
Dr R.G. Medici
Dr S.M. Michaelson
Dr M.H. Repacholi (*Rapporteur*)
Dr K.H. Schneider (*Vice-Chairman*)
Dr S.A. Sebo
Dr M.L. Shore (*Chairman*)
Dr M.J. Suess (*Secretary*)
Dr D. Utmischi
Dr R. Wever

Review Group on Health Effects of Exposure to Microwave Radiation, Washington, DC, 30 October – 3 November 1978

Dr E. Albert
Dr R.M. Albrecht
Mr H.I. Bassen
Dr S.F. Cleary
Dr P.A. Czerski
Dr J. Elder
Dr Z.R. Glaser
Dr A.W. Guy
Mr F. Harlen (*Rapporteur*)
Mr D.E. Janes
Dr W.M. Leach
Dr D.I. McRee

Dr S.M. Michaelson
Mr J.C. Monahan
Dr J.M. Osepchuk
Dr M.H. Repacholi (*Chairman*)
Dr G.M. Samaras
Dr S.O. Schiff
Dr B. Servantie
Dr M.L. Shore
Dr Charlotte Silverman
Mr D.H. Sliney
Dr M.L. Swicord

Second edition

Working Group on Health Implications of the Increased Use of Nonionizing Radiation Technologies and Devices, Ann Arbor, MI, 13–18 October 1985 (for Chapters 1–5)

Dr L.E. Anderson
Dr F.S. Barnes[a]
Dr J.H. Bernhardt[b]
Dr P.A. Czerski
Dr J.D.Y. Deslauriers (*General rapporteur and subgroup rapporteur for Chapters 1–3*)
Dr J.A. Elder
Mr F. Harlen (*Subgroup rapporteur for Chapters 4–5*)
Dr R. Hauf[b]
Dr J. Kupfer
Dr J.C. van der Leun
Dr S. Michaelson[b]
Dr K.H. Mild

Dr A.M. Muc
Dr T.C. Rozzell[a]
Dr M.I. Rudnev[b]
Dr B.M. Servantie
Dr A.R. Sheppard
Dr M.L. Shore (*Chairman*)
Dr Charlotte Silverman
Mr D.H. Sliney (*Subgroup leader for Chapters 1–3*)
Dr Maria A. Stuchly (*Subgroup leader for Chapters 4–5*)
Dr M.J. Suess (*Secretary*)
Dr B.M. Tengroth (*Vice-Chairman*)
Dr F. Urbach[a]
Dr M.L. Wolbarsht[a]

Working Group on Health Implications of the Increased Use of Nonionizing Radiation Technologies and Devices, Erice, 16–17 September 1985 (for Chapter 6)

Dr J. Bang
Mrs Deirdre A. Benwell-Morison (*Co-Secretary and subgroup rapporteur*)
Dr Gail ter Haar
Dr C.R. Hill
Mrs Patricia McKinney

Dr W.L. Nyborg (*Subgroup leader*)
Dr M.E. Stratmeyer
Dr K. Vandenberghe
Dr A.R. Williams
Dr M.C. Ziskin

---

[a] Part-time participation.

[b] Unable to attend.

INDEX

Acoustic fields, 256, 257, 262

Acoustic impedance, 248, 249 (table)

Acoustic transients, 54, 143

Actinic keratosis, 32

ALARA principle, 277, 278, 297

Albinism, 34

Aluminium screening, IR control, 108

Alternating current, 180

American Conference of Governmental Industrial Hygienists (ACGIH), airborne ultrasound levels, 278
   IR threshold limit values, 106
   laser installation guides, 74
   UV threshold values, 36, 37, 40

American Institute for Ultrasound in Medicine (AIUM), 254, 257, 278

American National Standards Institute (ANSI), personnel exposure standard, laser users, 74, 79, 105

American National Standards Institute (ANSI), protection standards, 63, 123, 124, 151, 153, 154

p-Aminobenzoic acid, 38

Applications, laser(s), 52–53

Applications, ultrasound, 260–264, 275–276

Atmospheric electric field, 178

Auditory effects, "RF hearing", 136, 138, 143–144, 157

Behaviour, animal studies, 208

Bioelectromagnetics, 118

Biological effects, ELF fields, 207–213, 221
   infrared radiation, 89–105
   laser(s), 53–57
   radiofrequency radiation, 120, 134, 138–151, 156
   ultrasound, 256, 265–274

Biological system, irradiation of, 300–303, 301 (fig.)

Biophysical analysis, electric field coupling, 189–195

Biophysical analysis, magnetic field coupling, 203–205

Black light, 14
   lamps, 35
   UV emitters, 15

Blepharitis, 103

Blue-spectrum laser radiation, eye damage due to, 59

Brain cells, sensitivity to extracellular electrical environment, 194, 206–207

Broad-band dosemeter, 36

Broadcasting installations, 151

Brunescent cataract, 30

Bureau of Radiological Health, meeting (1978), 10

Cancer, 146–147, 151, 215, 219, 223, 224
   diagnostic ultrasound, 273, 280
   hyperthermia therapy, 124, 263
   of the eye, 34
   of the skin, 32–34

Carbon arc, 15

Carbon arc, retinal lesions due to, 50

Cardiac pacemakers, 153, 220

Carotene, 38

Cataract, bilateral cortical, 98 (fig.), 99 (fig.)
   brunescent, 30
   furnace workers', 93
   glassblowers', 50, 93, 94, 96 (table)
   IR-produced, 77, 93, 94
   photochemical, 55 (table), 80
   RF-produced, 140–141, 150, 156, 157
   surgery, 263–264
   UV-produced, 30, 32, 41

Cells, heat killing of, 134, 269

Cellular studies, electric fields, 205–207

Chlorofluorocarbon emissions, action on ozone layer, 21

Chromosomal aberrations, ELF fields and, 219

Chromosomal aberrations, ultra-
  sound and, 8
Clothing, protection from UV, 38
Communication systems, 122, 123, 182
  see also Transmitters
Congenital anomalies, 147, 219, 273
Conjunctivitis, IR-produced, 98
Conjunctivitis, UV-produced, 29
Cornea, burn, 55 (table), 56
  IR effects on, 90–92
  tumours of, 34
  UV damage, 29–30, 62, 63 (fig.)
Corneal epithelium, percentage of
  energy incident on, and other
  ocular media, 25 (table)
Currents, electric fields, 176–181,
  184–202
Currents, magnetic fields, 181–183,
  203–205

Detectors, electromagnetic fields, 125
Diathermy equipment, 124, 128,
  145, 153
Discharges, electric current,
  195–202, 219
DNA, excision repair (dark repair)
  of defects, 26
  infrared and, 105
  protein cross-links, 26
  strand breaks, 25
  ultrasound and, 266, 267, 268
  UV absorption and, 25
Doppler effect, diagnostic
  uses, 257, 258, 262, 281
Dosemeter, personal, for UV, 36
Down's syndrome, 147
Dry eye, 98, 110

Eddy currents, 203
Education and training, electric and
  magnetic fields, 223
  exposure to NIR, 299
  infrared sources, 111
  laser(s), 80
  radiofrequency radiation, 158

Electrical discharges, personnel
  protection, 195–202, 219
Electric and magnetic fields, 118–121
  animal studies, 207–215
  at extremely low frequencies,
  175–243
  biological effects, 182
  cardiac pacemaker interference, 220
  conclusions, 220–222
  human studies, 215–220
  laboratory experiments, 205–207
  magnetic flux density, 183 (fig.),
  205, 217, 220
  recommendations, 222–224
  see also Radiofrequency radiation
Electric field, behaviour, effects on, 208
  cancer, 219
  currents, 176–181, 184–202
  discharges, 195–202, 219
  exposure of workers to ELF
  fields, 217–220
  measurements, 177–178
  protective measures, 195, 198
  shielding, 195, 222
  shock hazards, 194
  short-circuit currents, 185 (table),
  198, 202 (table)
  threshold current densities, 189
  threshold values, 198
  voltage and waveforms, 196–198
Electro-explosive devices, 153
Electromagnetic fields, detectors, 125
  medical exposure, 124
  surveys, 126–128
Electromagnetic interference, 153
Electromagnetic waves, 118,
  119 (fig.), 120–121
Electronic equipment, 153, 306
  see also Cardiac pacemakers
Energy absorption, radiofrequency, 134
  UV radiation, 23–26
  see also Specific absorption
  rate (SAR)
Epidemiological studies, 223, 272–273
Erythema (sunburn), 27, 36, 38,
  103, 304
Erythema (sunburn), action
  spectrum, 35, 37

338

Erythropoietic protoporphyria,
β-carotene for, 38

European Federation of Societies
for Ultrasound in Medicine
and Biology, 277

Experimental models, electric
fields, 185

Experimental models, RF radiation, 142

Exposure limits, standards, 36, 105,
294, 295, 307
laser(s), 57–66
radiofrequency radiation, 138, 154,
155 (tables), 156, 157

Eye, adverse effects of optical
radiation, 55–56
composite absorption of optical
radiation by deeper
tissues, 91 (fig.)
cross-section, 30 (fig.)
energy absorption of deeper
tissues, black body radiation at
different temperatures, 92 (fig.)
energy absorption of deeper
tissues, spectral distribution
of sunlight, 92 (fig.)
laser injury, 78
late non-stochastic effects
of UV, 32
late stochastic effects of UV, 32–34
lens defects, RF exposure, 138,
140–141, 157
melanoma, 34
minimum exposure factors, 59
penetration by UV, 23
protection against IR exposure,
108, 109 (fig.)
spectral transmittance of ocular
media, 90, 91 (fig.)
threshold limit value (TLV)
to IR, 106
UV effects in man, 29–31, 40–41
see also Cornea and Retina

Faraday's law, 204

Fibrosarcoma of cornea, 34

Flash tubes, 15

Fluorescent tubes, 15, 29, 35

Fluorocarbons, 21

Furnace workers' cataract, 93

Gas tungsten arc welding, emission
spectrum, 17 (fig.)

Glassblowers' cataract, 50, 93,
94, 96 (table)

Glassworkers, 103

Glaucoma, surgery, 264

Goggles for IR protection, 108–109

Haemangioendothelioma of cornea, 34

Hague, The (1971) working group, 9

Halogen-containing fluorocarbons, 21

Health, WHO definition, 300–301

Heating devices, 105, 123, 184

HeLa cells, repair in, 26

Herpes, treatment of, exposure
to UV, 35

High-voltage overhead lines, 8
see also Transmission lines

High-voltage workers, 195, 218

Human exposure, effects of RF
radiation, 138–148, 158
laboratory experiments, electric
and magnetic fields, 215–217
ultrasound, 247, 254
see also Occupational exposure

Hydrogen and deuterium lamps, 15

Hydrophones, 256

Hyperbilirubinaemia, phototherapy of
infants with, 35

Immune system, effects of electric
fields, 214–215

Induced currents in body, 186–195

Information programmes, 41, 308

Infrared radiation, 2 (table), 8, 85–115
aluminium screening, 108
apnoea in infants exposed to radiant
warmers, 105
aqueous humour, effects on, 92
biological effects, 89–105

339

Infrared radiation (contd)
  blepharitis, 103
  carcinogenesis and, 105
  cataracts due to, 93–98
  characteristics, 86
  conjunctivitis, decreased
    lachrymation, 98
  control measures, 107–109
  cornea, effects on, 90–92
  experimental threshold
    exposures, 94
  eye threshold limits, 106, 110
  eyelid, effects on, 90, 103
  genetic effects, 105
  hazard evaluation, 111
  high temperature heater, emission
    curve, 99, 100 (fig.)
  immunological reactions, 104
  industrial exposures, 94–98
  instrumentation, 88–89
  intestinal injury and adhesions,
    post-operative, 105
  iris, effects on, 93
  kidneys, vascular congestion, 104
  laser radiation exposure, 56, 76
  lens, effects on, 93
  nerve conduction blocked, 104
  occupational exposure, 87–88
  ocular hazards, 89–98, 106
  penetration into Caucasian/Negroid
    skin, 102, 103 (fig.)
  photon detector, 88
  problems and recommendations,
    109–111
  production, 94
  retina, effects on, 94
  skin hazards, 98–104, 110, 139
  sources, 86–88
  spectroradiometers, 89
  sperm count reduced, 105
  spleen, vascular
    congestion, 104
  standards for exposure,
    105–107, 110
  testicle damage, 104
  thermal detector, 88
  upper respiratory disease in
    foundry workers, 104
International agreements, 294–295

International Commission on
    Radiological Protection
    (ICRP), 295, 297
International Electrotechnical
    Commission (IEC), 40, 74, 80
International Labour Office
    (ILO), 152
International Organization for
    Standardization (ISO), 106, 278
International Radiation Protection
    Association (IRPA), 40, 80,
    152, 154, 157, 295
International Symposium on
    Biological Effects and Health
    Hazards of Microwave Radiation
    (Warsaw, 1973), 10
Iris, melanoma of, 34
Irradiation, maximum permissible
    exposure, 301 (fig.)
  of biological system, 300–303
  physical quantity, 302

Laboratory, cellular studies, 205–207
  electric field exposure, 178, 216, 217
  testing of cardiac pacemakers, 220
  ultrasound processing
    equipment, 258
Laser(s), 2–3 (table), 8, 49–83
  acoustic transients, 54
  action, 50–51
  applications, 52–53
  biological effects, 53–57
  control measures, 73–75, 80
  denaturation of protein, 53–54
  elastic transient pressure wave, 54
  emission duration, 51
  environmental hazard, 52
  exposure, long-term (chronic),
    77, 79
  exposure, low-level, 76
  exposure, minimum factors of, 59
  exposure, ultra-short, 77
  exposure limits, 57–66
  exposure limits, direct ocular
    exposures, 64 (table)
  exposure limits, for viewing diffuse
    reflection, 65 (table)
  exposure limits, minimum limiting
    angle of extended source,
    65 (table)

Laser(s) (contd)
  exposure limits, skin
    exposure, 65 (table)
  eye damage, 55 (table), 56, 78
  eye protection, 59, 305
  hair growth and, 57
  hazards, 66–73
  hazards, classification, 67–70
  hazards, detailed analysis, 70–73
  hazards, environmental factors,
    71–72
  hazards, evaluation, 79
  hazards, multiple wave-
    length/multiple source, 70
  hazards, output parameters, 68
  hazards, personnel exposed,
    72–73, 80
  indoor operations, 71
  medical assessment, 75–76
  mode-locked, 76
  outdoor operations over extended
    distances, 71–72
  pathophysiology, 54–57
  phagocytosis index and, 57
  photochemical reactions, 54
  photosensitizing chemicals and, 57
  recommendations for further
    investigations, 76–79
  retinal injury, 55–56, 59,
    75–76, 80
  retinal injury, threshold for
    extended sources, 60 (fig.)
  retinal injury, threshold for
    minimal image condition,
    59 (fig.), 59–60
  safety code "all-purpose", 59
  secondary effects, 51
  skin, adverse effects on, 56–57, 75,
    79, 80
  thermal coagulation
    necrosis by, 56
  thermal effects, 56
  thermo-acoustic transient pressure
    wave, 54
  types of pulsed, 51
  use in medicine, 52
  wound healing and, 57
  see also Optical radiation
Laser Institute of America, laser
  installations guide, 74
Legislation, data for, 294–297, 306

Lens of eye, damage, 30, 78,
  138, 140–141
Lenticular cataract, 8
Lorentz force, 181, 204

Magnetic field, biophysical
  analysis, 203–205
  cancer and, 219–220
  currents, 181–183, 203–205
  human/animal studies, 207–220
  man-made sources, 182
  natural, 182
  pulsed, in medical care, 183, 217
  sensors, 125
  transmission lines, 182
Magnetic flux density, 183 (fig.), 205,
  217, 220
Magnetic resonance imaging
  (MRI), 124
Man, resonant frequency for, 129
Measurements, electric fields,
  177–178, 221
  infrared radiation, 88
  laser emissions, 79
  magnetic fields, 182, 221
  radiofrequency radiation, 125
  sound fields, 254–258
  ultraviolet radiation, 35–36
Melanin pigment of skin, UV
  effect, 27–28
Melanoma, 32, 33, 34, 40
Ménière's disease of the inner ear, 264
Mercury lamp, 15, 16–17 (figs)
8-Methoxypsoralen, 34
Microwave ovens, 8, 122, 128, 151, 302
Microwave radiation, see Radiofrequency radiation
Minimal erythema dose (MED), 27
Modification factor, relative
  absorption by retina
  and, 60, 61 (fig.)
Modification factor, repetitively pulsed
  lasers having pulse durations shorter
  than 10 s, 61, 62 (fig.)
Mortality studies, RF radiation,
  145–147
Mouth, UV effects in man, 29

National Council on Radiation
Protection and Measurements
(NCRP), 251, 257

Neonatal jaundice, phototherapy
of infants with, 35

Nervous system, effects on,
electric and magnetic
fields, 207–208, 209–213

Noble gas lamps, 15

Non-melanoma skin cancer, ozone
layer thickness and, 21

Occupational exposure, electric
fields, 222
infrared radiation, 87–88
laser(s), 72–76
of pregnant women, 305
RF energy, 144–147, 155 (table),
158, 305
ultrasound, 174
ultraviolet, 34–35

Optical radiation, 49–83
emission from black bodies at
various temperatures, 87 (fig.)
eye penetration, 50
nonionizing, 50
pathophysiological responses, 54–57
protection standards, 62–63
relative absorption by retina and
modification factor, 61 (fig.)
skin penetration, 50
surface action, 50
see also Laser(s)

Ozone layer, upper atmosphere, 19–21

Photo-allergy, 31

Photodiode detector, 36

Photodynamic dyes, 35

Photokeratitis, 29, 37, 55 (table),
56, 80

Photokeratitis, threshold, 63 (fig.)

Photomultiplier, 36

Photosensitization, chemical, 31

Photosensitizer, 31

Phototherapy, 35

Phototoxicity, 31

Phototube, 36

Physiotherapy, ultrasound, 262–263,
280, 281

Piezoelectric hydrophone, 256

Post-replication repair, 26

Protection measures, capacitative
discharges and contact
currents, 198
electric field coupling, 195
legislation data, 295–297
magnetic field coupling, 205
radiofrequency, 151–154
remedial, 307
technological development
and, 295–296
ultrasound, 279–280
ultraviolet, 38–39

Psoralens, naturally occurring, 31

Psoralens, psoriasis treated by, 35

Psoriasis, 35

Pulse-echo diagnostic equipment, 252,
257, 259, 262, 281

PUVA, 34, 35, 38

Pyrimidines, 25, 26

Radar(s), 122, 123, 138, 141, 143,
151, 306

Radar(s), occupational exposure
surveys, 144–147

Radiation accidents, reporting on, 296

Radiofrequency radiation, 3 (table),
8, 117–173
ambient environment, 122
athermal interaction mechanisms,
136–137
auditory effects ("RF-hearing"),
138, 143–144, 157
biological effects, 120, 134,
138–151, 156
blood-forming system, effects
on, 149, 157
burn, 137
cancer, 124
cardiovascular changes, 144
cataracts produced by, 140–141,
150, 156, 157
cellular effects, 134–135
code of practice for protection of
workers, 151

Radiofrequency radiation (contd)
  congenital anomalies, 147
  cutaneous perception, effects
    on, 138–140
  devices, 121–124
  diathermy devices, 124, 128, 145
  dosimetry, 128–133
  Down's syndrome, 147
  electric shock, 137
  electroencephalographic
    changes, 157
  electromagnetic field surveys,
    126–128
  exposure limits, 138, 154,
    155 (tables), 156
  frequencies, 129–134, 136, 138
  genetics and mutagenesis, 150
  hazardous fields, 126–128
  heating equipment, 123, 263
  high-power sources, 122
  immune system, effects on, 149, 157
  instrumentation, 125–128, 154
  interaction mechanisms, 134–137
  measurement, 125–128
  medical equipment, 124, 153
  microwave irradiation, 145–146, 302
  mortality and morbidity,
    145–147, 156
  nervous system, effects on, 149,
    157, 158
  non-human studies, 148–151
  occupational exposure, 123, 124,
    128, 144–147, 158
  protective measures, 151–154
  pulse-modulated, 143, 149, 157
  quantities and units, 118, 121
  relay systems, 122
  reproductive effects, 147, 148, 157
  research, 158
  sealers, 123, 126, 128
  shielding, 151, 153, 158
  solitons, 137
  sources, 121–124, 156
  specific absorption rate (SAR),
    128–134, 130–133 (figs), 141,
    142, 148, 149, 154, 156–158
  standards, 154, 157
  surveys, potentially hazardous
    fields, 126–128
  synergistic effects, 157
  thermal effects, 134–135, 137, 158

Radiofrequency radiation (contd)
  thermoregulatory responses, 139,
    141–143, 148
  threshold temperature, 139
  urban environment, 122
  see also Electric and magnetic fields
Regulation and enforcement
    procedures, 293–314
  approval of application, 306
  approval of equipment, 307
  common procedures, 298–299
  compliance and enforcement,
    305–309
  exemptions, 308
  information programmes, 308
  inspection and maintenance, 308
  legislation, data for, 294–297
  protection measures, 295–296
  review of regulations, 308
  safe exposure limits, 299–305
  safe exposure limits, causal
    model, 300–303
  safe exposure limits,
    phenomenological model, 303
  standards, 297–299

Resonant frequency, 129

Retina, IR effects on, 94, 95
  laser injury, 55–56, 59–62
  ophthalmoscopic examination,
    75–76
  photochemical injury, 59
  relative absorption of optical
    radiation, 60
  thermal injury, 55 (table), 56, 59
  thermo-acoustic injury, 59
  threshold mechanisms of injury, 59
  UV damage, 30

Retinoic acid, 38

Robertson-Berger apparatus, 38

Safe exposure limits, 299–309
  causal model, 300–303
  conservative assumptions, 304
  phenomenological model, 303

Short-circuit current, 185–189, 198

Skin, basal-cell carcinoma, 32, 33
  cancer, 8, 21, 32–34
  cell growth and UV, 34
  DNA defects after UV, 34

Skin (contd)
erythema (sunburn), 27, 36, 57
exposure from a laser beam, limits,
66 (table)
hazards due to IR, 98–104, 110, 139
heat effects, 103, 139, 156
IR penetration, Negroid/Caucasian
skin, 102, 103 (fig.)
laser action on, 56–57, 75, 79, 80
late non-stochastic effects
of UV, 32
late stochastic effects of UV, 32–34
malignant tumours after UV, 34
melanin pigment and UV, 27–29
minimal reaction levels, 58 (table)
pigmentation and reflectance,
99–101, 100 (fig.)
premature aging due to UV, 32
properties, 98
reflection curves, 99, 100 (fig.)
scattering of transmitted energy,
101 (fig.), 102
squamous cell carcinoma, 32–33
structure, 28 (fig.), 98
thermal coagulation necrosis, 56
transmission spectrum, 101,
102 (fig.)
UV effects in man, 27–34, 40–41
Snow blindness, 55
Solar activity, 182
Sound fields, 251–254
Sound speed, 247–248
Sources, electric field, 178–181
infrared radiation, 86–88
magnetic field, 182–183
radiofrequency radiation, 121–124
Specific absorption rate (SAR),
128–134, 130–133 (figs), 141, 142,
148, 149, 154, 156, 157
Spectacles for IR protection, 109
Spectral weighting function
for assessing retinal hazards
from broad-band optical
sources, 107 (table)
Spectroradiometers, 89
Standards, 297–299
personnel exposure, laser users, 74,
79, 105
protection, 63, 123, 124, 151–154

Submarine detection, pulse-echo
method, 261
Sun, source of UV, 14, 19–21
Sun, spectrum, outer surface,
atmosphere/sea level, 18 (fig.)
Sun-blocking substances, 38
Sunburn, 27–29, 80
see also Erythema
Suntan, 27–29, 35

Thermal absorption, 134–135
Thermoelastic expansion
mechanism, 143
Thermoregulatory responses, 138,
141–143, 148
Thunderstorms, 182
Thymidine, 25, 26
Tissue damage (denaturation), heat
produced, 103
Traffic radars, 122, 306
see also Radar(s)
Transmission lines, electric field,
178, 179 (figs), 180, 180 (table),
181, 198, 205, 218, 223
Transmission lines, magnetic field, 182
Transmitters, radio and television, 122,
124, 153, 158
Transverse electromagnetic cell, 126
Tungsten halogen lamps, 15

Ultrasound energy, 8, 246–291
absorption, 250, 251
acoustic impedance, 248, 249 (table)
airborne, effects of, 258, 278–279
applications, 260–264, 275–276
attenuation, 249–250, 254
biological effects, 256, 265–274
biomolecules, action on, 265–266
cancer therapy, 263, 275, 280
cavitation, 258, 258–260, 264,
267, 271
cell death, 267
cells, action on, 267–269
cells, chromosomal changes,
267–268
cells, genetic damage, 267–268

Ultrasound energy (contd)
  cells, structural changes, 268
  DNA, action on, 266
  Doppler and therapy band, 265, 266 (table)
  hearing loss due to, 274
  human epidemiology, 272–273
  hydrophone measurement, 256
  in dentistry, 264
  industrial/laboratory processing equipment, 258
  insects, action on, 271–272
  intensity, 254, 256, 257
  mammals, action on, 270–271
  measurement applications, 261
  measurement of sound fields, 254–258, 277
  medical diagnosis, 254, 257, 259, 261–262, 274, 276–278, 281
  medical equipment, 255 (table), 279–280
  multicellular organisms, action on, 270–272
  occupational exposures, 273–274
  physical medicine, 275–276
  physical properties, 247
  physiotherapy, 262–263, 280, 281
  plants, action on, 272
  pressure thresholds, 259
  properties of various media, 249 (table)
  protection measures, 279–280
  sound speed, 247
  standards, 258
  structure of sound fields, 251
  surgery, use in, 263–264, 280
  testing applications, 261, 277
  therapy applications, 262–263, 277
  threshold for thermal tissue damage, 251
  transmission through interfaces, 248
  underwater target location, 261

Ultraviolet radiation, 2 (table), 8, 13–48
  absorption, 23–26
  arcs emitting, 34
  areas of risk from over-exposure, 34–35
  artificial production in mines and cellars, 14

Ultraviolet radiation (contd)
  black light region (UV-A), 14
  broad-band source maximal permissible exposure, 37
  chemical photosensitization, 31
  destructive effect on skin and eye, 14
  dosimetry, 35–36
  epidemiological and experimental studies, 40–41
  erythema dose, minimal (MED), 27, 36
  exposure of body to, 34, 39
  fluorescent lamps (UV-A), 29
  germicidal region (UV-C), 15
  in graphic reproduction techniques, 34
  incandescent sources, 15
  irradiance/cancer incidence, measured by Robertson-Berger method, 38
  late non-stochastic pathological effects on skin and eye, 32
  late stochastic effects on skin and eye, 32–34
  mice, action on skin, 29
  mixed sources, 15, 16
  mouth, effects on, in man, 29
  ozone layer, 21
  pathological effects in man, 26–34, 40
  pathological effects, immuno-logical, 31
  pathological effects, non-stochastic, 26–31
  pathological effects, stochastic, 32–34
  personal dosemeters, 36
  photochemical effects, 19, 23–26, 40
  physical description, 14
  polluted air effect, 21
  production, 13–14
  production, mixed sources, 15–16
  protection from industrial sources, 39
  protection from solar UV radiation, 38
  relative levels from illumination sources, 37
  safety standards, 36–38
  skin erythemal region (UV-B), 15, 36

Ultraviolet radiation (contd)
    sterilization of food and air, 34
    sun as source, 14, 19–21, 27
    threshold limit values and relative
        spectral effectiveness by wave-
        length, 8-hour exposure,
        37 (table)
    transmission in biological tissue, 23
    vacuum, 50
    vitamin D_3, 14, 36

Video-display equipment, 182
Visible light radiation, 2 (table)
Vitamin D_3, 14, 36

Wavelengths, tumour induction
    suitable, 32–33
Welders' flash, 55
Welders, protection from IR, 108–109
Welders, protection from UV, 39

Welding, arcs, 34
    curtains for IR control, 108
    emission spectrum, 17
    masks/hoods, 39
    retinal injury from, 95
    ultrasound in, 260
Women, pregnant, occupational
    exposure, 305
Workers, code of practice on
        protection against RF
        radiation, 151
    studies of electric and magnetic
        fields exposure, 218
    see also Occupational exposure
Working Group on Health
    Effects from Lasers
    (Dublin 1974), 10
World Federation for Ultrasound in
    Medicine and Biology, 277

Xeroderma pigmentosum, 26, 34